D1129758

Mercantile States and the World Oil Cartel, 1900–1939

A volume in the series

Cornell Studies in Political Economy

Edited by Peter J. Katzenstein

A full list of titles in the series appears at the end of the book.

Mercantile States and the World Oil Cartel, 1900–1939

GREGORY P. NOWELL

CORNELL UNIVERSITY PRESS

Ithaca and London

First published 1994 by Cornell University Press.

Printed in the United States of America

♾ The paper in this book meets the minimum requirements of the American National Standard for Information Sciences—Permanence of Paper for Printed Library Materials, ANSI Z39.48-1984.

Library of Congress Cataloging-in-Publication Data

Nowell, Gregory P. (Gregory Patrick), 1954–
Mercantile states and the world oil cartel, 1900–1939 / Gregory P. Nowell.
p. cm. — (Cornell studies in political economy)
Includes bibliographical references and index.
ISBN 0-8014-2878-5
1. Petroleum industry and trade—History—20th century.
2. Petroleum industry and trade—France—History—20th century. 3. Cartels—History—20th century. 4. Cartels—France—History—20th century.
I. Title. II. Series.
HD9560.5.N67 1994
338.8′7—dc20 93-38020

Contents

Acknowledgments vii
Abbreviations ix
1. Transnational Structuring and the World Order 1
2. Transnational Structuring before World War I: The Pattern of State Monopoly Attempts and Market Interventions 45
3. The Great War and the Struggle for Commercial Advantage, 1914–1921 80
4. France and the Development of the World Oil Cartel, 1922–1939 148
5. The World Hydrocarbon Cartel, 1922–1939 223
6. Conclusion: Transnational Structuring and the World Order 280
Bibliography 302
Index 319

Acknowledgments

Charles Beard prefaced his study of the Constitution with the remark that the "letters, papers, and documents" of the United States' founders gave him "the shock of my life." With similar naivete I approached the French parliamentary documents, the main primary source available in the United States, expecting a simple confirmation of the statist thesis that France had the ability and will to control its oil companies for the national good. On the heels of the oil shock of 1979–1980, I imagined that proud, nationalist France might show us how the puissant oil industry could be tamed. I was disillusioned; but it was clear that much more had happened than merely a tangle between oil and the state with oil emerging as the victor. Determining just what had happened, and how to describe it, became my personal odyssey over the next twelve years.

Along the way I have incurred a great number of debts. At MIT my work was greatly influenced by Hayward Alker, Jr., Suzanne Berger, Thomas Ferguson, and William Griffith. A more eclectic and stimulating group of minds is difficult to imagine. In France, I enjoyed the help of French scholars including Maurice Lévy-Leboyer, André Nouschi, and Lionel Zinsou-Derlin. Roberto Nayberg very generously supplied me with a copy of his dissertation; Philippe Mullerfeuga also made some of his research available. Stéphane Desmarais, who in the 1950s ran the family oil firm that is extensively discussed in this book, was also generous with his time.

Others have contributed: Morris Adelman at MIT; Ed Morse, publisher of *Petroleum Intelligence Weekly;* Glyn Short of Imperial Chemical Industries; Daniel Yergin at Cambridge Energy Research Associates; Deborah Avant, Robert Frost, Steven Livingston, Todd Swanstrom, and Ann Hildreth, all of the State University of New York at Albany; Richard Kuisel of SUNY Stony Brook; Peter Katzenstein at Cornell University; Marjorie Beale at the University of California at Irvine; Risa

Mednick and Roger Haydon at Cornell University Press. Megan Morrison-Sawyer supplied the world oil map. Celeste Newbrough compiled the index. Douglas Lowell and Richard Helley helped with the final proofreading.

I am grateful to the French government for financial support in the form of a Bourse Chateaubriand and the luxurious free accommodations provided by the Ecole Normale Supérieure as part of its exchange with MIT. Pierre Gille at the Ecole Normale was a great help in my second year of research. American institutional support included a Fulbright Travel Fellowship and a grant from the Institute for the Study of World Politics.

I thank Cambridge friends whose moral support and occasional criticism have contributed to this book: Steven Anderson, Cathie Jo Martin, Ashutosh Varshney, Edwige Leclerq, Andrew Barnaby, Camillo Imbimbo, Terese Lyons, Gavan Duffy, Erik Devereux, Lily Ling, Roger Hurwitz, David Gibbs, Ahmed Hashim, Ann Grazewski, Mary Ann Lord, Lola Klein, Frances Powell, Helen Ray, and Eva Nagy.

I would also like to thank my colleagues in the Alternative Fuels Group at Acurex Environmental Corporation in Mountain View, California, for their stimulating insights and wealth of technical knowledge. I am particularly indebted to Carl Moyer and Michael Jackson.

About twenty years ago my mother began working with the Sierra Club for the preservation of forest lands in southern California. Her efforts, and those of the people she worked with, came to fruition in the Los Padres Condor Range and River Protection Act, introduced in 1989 and passed by Congress in June 1992. This major wilderness legislation established five new national wilderness areas and made large additions to two others. Four hundred thousand acres of wilderness were protected, from Ventura to Monterey. It has been my privilege to watch this project unfold as I threaded my way through the operations of the world oil cartel. Books are not the only projects that seem to take forever, though it is easy to lose sight of this fact. My mother and stepfather, Sally and Les Reid, have shown admirable persistence in their work for the environment and the Sierra Club; this book is dedicated to their example.

Gregory P. Nowell

Albany, New York

Abbreviations

Abbreviations, principally used in notes, are given for each archival or primary source. A few common French usages have been kept in tables:
Cie for Compagnie, or company
Sté for Société, an alternative legal name for company
Fse for française, the feminine adjective for French
FF for francs français, or French francs

A few abbreviations occasionally occur in source documents, of which the most important are:
CFP for the Compagnie Française des Pétroles
CFR for the Compagnie Française de Raffinage
IPC for the Iraq Petroleum Company
TPC for the Turkish Petroleum Company, the predecessor of the IPC

I have avoided using a host of acronyms for oil companies of yesteryear. Three exceptions are:

Socony: Standard Oil Company of New York, also Socony-Vacuum, now Mobil Oil

Epu: Europäische Petroleum Union, marketing syndicate of the Rothschilds, the Nobels, the Deutsche Bank, and Royal Dutch–Shell before World War I

Socal: Standard Oil of California, now Chevron

A few other companies are not abbreviated but are referred to frequently:

Anglo-Persian: Later Anglo-Iranian, now known as British Petroleum, a name lifted from the German-owned marketing subsidiary of Epu in World War I

Bnito: Russian initials of the Rothschilds' company, the Société de la Caspienne et de la Mer Noire, active before World War I

Burmah Oil: Still a major shareholder in British Petroleum, began producing oil in Burma in the nineteenth century

Abbreviations

Desmarais Frères: Largest French national oil firm, bought out by Compagnie Française des Pétroles in 1962

I. G. Farben: Pre–World War II German chemical combine

Pétrofina: Often known in Europe simply as "Fina"; oil company of Belgian origin

Royal Dutch–Shell: Often known simply as Shell or Shell Oil in the United States

Saint Gobain: French chemical company

Shell Trading and Transport: Shell Company before amalgamation with Royal Dutch

Standard Oil: Standard Oil of New Jersey; SONJ; in Europe, often Esso, from the abbreviation SO; today, Exxon—largest oil company in the world

Steaua Romana: Deutsche Bank company in Romania; after World War I, part of Banque de Paris et des Pays Bas investments

The following abbreviations are used in the notes for archives and primary sources:

Arch.Nat	Archives Nationales, Paris
Ass.Nat.	Archives of the Assemblée Nationale, Château de Versailles
CFP	Archives of the Compagnie Française des Pétroles, Paris
Déb.Ch.	*Débats Parlementaires de la Chambre des Députés*
Déb.Sén.	*Débats Parlementaires du Sénat*
Doc.Ch.	*Documents Parlementaires de la Chambre des Députés*
Doc.Sén.	*Documents Parlementaires du Sénat*
Fin.	Ministère des Finances, Paris
FTC1	*International Petroleum Cartel* (Staff Report to the Federal Trade Commission, 1952)
FTC2	Report of the Federal Trade Commission on Petroleum Cartels, Harry S. Truman Library, Independence, Missouri
Hoover Inst.	Hoover Institution Archives, Stanford, California
MAE	Archives of the Ministère des Affaires Etrangères, Paris
Min.Ind.	Ministère de l'Industrie
Pat.Hearings	*Hearings before the Committee on Patents,* 1942
Rock.Arch.	Rockefeller Archive Center, Tarrytown, New York

Mercantile States and the World Oil Cartel, 1900–1939

In Paris they were still haggling over the price of blood, squabbling over toy flags, the riverfrontiers on reliefmaps, the historical destiny of peoples, while behind the scenes the good contract players, the Deterdings, the Zahkaroffs, the Stinnesses sat quiet and possessed themselves of the raw materials.

—John Dos Passos, *Nineteen Nineteen*

Chapter One

Transnational Structuring and the World Order

Scattered throughout the world today, in France and elsewhere, are institutional remnants of the great world hydrocarbon cartel of 1928. More than a mere device for controlling petroleum production and prices, the world hydrocarbon cartel embedded the competition for power among nations in an equally important dynamic, the transition from coal to oil in the world economy.

The world hydrocarbon cartel was a phase in the political struggle that accompanied this transition. It had three major effects on the world order: the first at the level of national, strategically oriented regulatory policies, the second at the level of the interrelationships of the members of the oil industry, and the third at the level of the oil industry as a whole and its relation to the coal industry. French national energy policy was an artifact of the cartel, but only one of many. The world hydrocarbon cartel distributed technology to some nations and withheld it from others. "Local" institutional arrangements, such as French national policy, were characteristic of the broader scheme. The cartel was indifferent to "strong state" or "weak state" traditions; it operated with only minimal reference to "reasons of state" or "national security"; it disregarded the "core" or "peripheral" status of regions brought under its sway.

The cartel has been extensively studied in its various aspects, both by scholars and by government investigatory commissions, but its vast activities have defied a comprehensive approach. There are many studies of oil corporations and of national energy policies, but few studies of the whole. The 1952 investigation of the world oil cartel by the Federal Trade Commission is well known and often cited in the oil literature;[1] but the more important hearings, held by the Senate Committee on Patents in

[1] Cited in this book as FTC1 and FTC2; see chapter 4 and bibliography.

1942, are scarcely mentioned.[2] The multinationals have sponsored lengthy corporate histories, and they have also been the subject of muckraking jeremiads, which may provide useful information but are often simplistic. Studies of the oil industry are frequently derived from diplomatic history and speak of how states use corporations rather than how corporations use states. Other studies offer a sectoral focus on how oil companies compete with oil companies and ignore the main event of this century—how the oil industry competed against the coal industry. These approaches are insufficient. During the development of the oil industry, companies learned to use and manipulate states; the companies' competition with one another was only a part of the market struggle among producers of rival sources of hydrocarbons.

This book examines the transnational structuring of world hydrocarbon markets. Understanding the world hydrocarbon cartel is an enterprise of international dimensions: hearings before the U.S. Senate's Committee on Patents can tell us as much about what happened in France as can documents from France. Conversely, documents from France provide a window into the broader operations of the world hydrocarbon market and its political struggles. Purely national histories of French policy wrongly argue that French energy policies before World War II were designed to foster national independence. This is the mythology of French etatism and the factors that allegedly caused the French state to devise policies to "control" its energy market. In France it was not the state that controlled the energy market, but actors in the energy market that controlled the state. What is remarkable is that this situation was typical of the period: France is a useful example because what happened was ordinary. There the confluence of national and international state politics, of national and international capitalist interests, has left a clear picture of worldwide patterns.

We think of the nation-state and the national interest as the primary unit and motive force of "international politics." This view is reinforced by the newspapers and by historical writing that portrays nations and their leaders as the primary actors in the world. The operations of the world hydrocarbon cartel bring to light the behavior of nonstatal actors and leaders bent on nonstatal objectives. Corporations structured the world of states not as nationally based lobbies striving for occasional modifications to purely national laws and policies, but as agents for the

[2] Cited in this book as Pat.Hearings. See bibliography and chapter 5. Hexner 1946, Borkin 1978, and Engler 1976 are among the exceptions.

control of international economic development and the dispersion of international technological abilities.

Transnational Structuring

Transnational structuring differs from previous theoretical descriptions of the international political economy. I begin with a rapid introduction to the concept: the world oil industry provides the full historical particulars of the analogy that follows.

As an example, then, let us consider the struggle over regulating indoor smoking in the United States.[3] Various cities have passed highly restrictive ordinances. To combat the proliferation of these ordinances, the American Tobacco Institute has lobbied state legislatures to assume regulatory control over indoor smoking, hoping to get less restrictive measures enacted. The American Lung Association, a national group, has fought these efforts. Each group counts on local support: restaurant and bar owners usually back the tobacco lobby, whereas local antismoking groups have allied with the American Lung Association. A continental confrontation occurs between the association and the tobacco lobby not through the federal government, but through the multiple authorities of the state legislatures.

States such as Alabama and Virginia tend to pass very lenient smoking laws. States such as California and New York pass very stringent ones. Others fall somewhere between, or in spite of heavy lobbying from both sides, do nothing at all. There is every reason to expect that, in the confrontation between two national groups, the results will partially depend on local constituencies as they support or contest the plans of the national lobbies. Obviously the presence or absence of tobacco growers and the number of smoking voters in the various states are relevant to the outcome. And even where nothing at all is done, the political situation is changed: the local authorities have been through all that; they have seen and heard both sides, and may be tired of the issue. This affects the future costs of action for both pro- and antitobacco forces; the political environment is altered no matter what the outcome.

Imagine now that we have fifteen states that heavily regulate public tobacco consumption. Fifteen other states do not; what is more, they heavily subsidize tobacco crops. Another twenty states have mixed outcomes that are difficult to categorize. All three of the outcomes result

[3] This is an illustrative fiction with partial roots in fact. See Phillips 1992 for the real politics of tobacco versus antismoking groups.

from the same process, just as in chess the same rules of engagement can lead to a victory of black or white, or to a draw. It would be silly to say, whether we are talking about a state with few regulations or one with many, that looking purely at a given state's local politics and institutions "explains" the regulatory outcome. A state's indoor smoking policies are only one expression of the broader struggle between the American Tobacco Institute and the American Lung Association, and to portray one state's policies without reference to the larger context is to miss the fundamental process. Each state's constituencies and traditions of governance are contributing, not causal, factors.

Let us consider now this same process as an international struggle by oil companies to control oil markets. The world constitutes a single chessboard for an elaborate regulatory struggle. Multinational firms have objectives tailored to the country in question. If the firms want regulated markets, they will favor production controls in a country with low consumption and high production and market sharing in a country with low production and high consumption. Firms that think such controlled markets are not in their interest will strive to create the institutional support needed to sustain a "free" market.

Powerful multinational corporations may either combine forces or fight one another in every major market of the world. But even if they combine forces, a uniform world regulatory structure does not necessarily result. As with indoor smoking regulations in the United States, results vary enormously with local conditions. In some American states the American Lung Association might create a powerful "statist" regulatory authority that strictly controls tobacco sales, imposes high tobacco taxes, and advertises the dangers of smoking with revenues earned from tobacco taxes. These "strong states" would contrast sharply with the "weak states" that did little to regulate tobacco sales and had only nominal regulatory authority. We do not "explain" the strong and weak states by referring solely to local conditions, institutions, culture, traditions, and interest groups: we must consider the larger struggle.

In the world oil market, many varying institutional arrangements for the control of oil are the product of international struggle by rival corporations. The French oil industry's "statist" character is one instance of how the process of international struggle for market control, or *transnational structuring,* can powerfully affect the local (national) institutional outcome. Though a corporation may exert a decisive influence on the creation of a regulatory authority—or, by contrast, on the deregulation of a market (since objectives vary)—this influence does not endure for all time, or even for a long time. Institutions created by transnational structuring become part of the apparatus of state power that rival groups seek to influence in later conflicts. Nothing precludes a reversal of fortune for even the most politically influential corporation.

The state's interventionist power derives from the interest group struggle. Transnational structuring creates state powers, regulates markets, and effects changes in the number of active members in a market and in their objectives. State power is a constituent element, but it also is created by the process of structuring.[4] State power may be adapted for purposes other than those its creators intended. Nonetheless, the power of states to control markets often depends on the authority that actors in those markets have helped to bestow upon those states. This holds true for strong intervention and also for deregulation. The French oil case is clearly illustrative, for it has been through phases of both "free markets" and "strong state intervention," all because of the different outcomes of the struggle for market control.

"Unregulated" markets are as much a product of interest group struggle as regulated ones. "Transnational structuring" describes a process, not an outcome. In contrast to "sovereignty at bay" portraits of corporations as a challenge to national power, the attempt to secure regulation actually increases the state's power to intervene. Transnational structuring is the means by which the normal political and economic lobbying activities of capitalism can lead to the creation of new state institutions.

The dominant political coalition always furnishes a "public interest rationale" for its policies, whether regulation or deregulation is being promulgated. In transnational structuring, the state broadcasts the version of the public interest most suited to the dominant parties: not only is the form of the state mutable, as in the institutional power, so are its professed objectives.

Everywhere transnational structuring occurs, it looks like a local (national) political event. Multinational corporate interests usually enlist local national elites, such as bankers, to represent their case. They pursue their local objectives without needing to make any statements about the larger picture, except perhaps on rare occasions. Reporters write up the story from the local, national perspective, and historians writing about oil regulation in Romania will seldom bother to look at what happened in the United States.

[4] Huntington 1973, pp. 363–364, offers a parallel argument: many transnational corporations bidding for entry into a state's market give the state the option of choosing among bidders, increasing its power relative to the firms. This is not transnational structuring: Huntington does not describe worldwide rent seeking, and he assumes government autonomy from the transnational corporations.

Transnational structuring may lead to worldwide cartels.[5] Such a cartel is truly a world historical event; but though vast in scope, it is hidden. Roosevelt, Stalin, Hitler, and Churchill are among the great and well-known leaders of the nation-state system. Gulbenkian, Deterding, Teagle, Cadman, and others are the more obscure agents of transnational structuring in world hydrocarbons: their names are known principally to specialists in oil history.

A world cartel eventually falls apart—all cartels do. In falling apart it may also lose control of the various regulatory agencies it helped create. The loss of power may be partial and slow; one or several members of the cartel may retain influence with regulatory authorities. The political cooperation between industry and members of a given state does not have to correspond to the logic of international market competition. But whether cartel and control are preserved or disintegrate, the world has been changed. Markets are regulated where before they were not; a scrupulous examination of taxation, tariff, and subvention policies across countries would also produce strong evidence of transnational structuring at work.

Transnational structuring is continuous. An international agreement among firms, enforced by state regulatory policies, can disintegrate and lose its coherent, organized center. What is left behind is neither anarchy (among states) nor a free market (in the economy) nor even a tendency to these, but a world market that has at least some regions where economic activity is more rationalized and regulated than it was before. This may make the next cartel effort less difficult or may obviate the need for it altogether: if enough of the largest markets have developed practices that dampen multinational firms' worries about intense competition, the ascendancy of a particular group in a particular region is secondary to the overall smooth functioning of the process.

In transnational structuring we do not generalize about the state's "autonomy" from "business interests" when we find companies opposed to regulatory policies. Antigovernment coalitions follow logically from the coalitions that back the government.

What if the documents needed to prove transnational structuring are destroyed, lost, or kept from the public? Corporations see historical records as a cost to maintain and even, when they document collusive activity, as a threat to the firm. There are many reasons to minimize evidence of a firm's politicking. One is to keep other firms from learning about it; ignorant rivals will not mount opposition, lowering the cost of getting regulation. Another incentive is to minimize perceived

[5] Hexner's 1946 study of international cartels is excellent, but he says little about their political lobbying. Machlup 1952 is also illuminating.

violations of laws against influence seeking that rivals might try to have enforced. Corporations therefore do not systematically preserve their political records the way nations do.[6] When there are no records, proving transnational structuring may be difficult or impossible.

When records are missing, can we conclude that transnational structuring has not occurred? No: the erasure of a game animal's trail does not mean it did not come to the brook and drink. Transnational structuring cannot be ruled out because the obvious trail is gone. But what *is* left behind? What is left behind is a series of national policies that look as if they were created in circumstances specific to each country. Even when evidence is relatively abundant, as in the world hydrocarbon case, the state-centered biases of researchers may leave the general process undiscussed for decades; where the evidence is scarcer, it is easy to fall into a state-centric interpretation, creating a "local" history for a worldwide process.

Imagine independent scholars in Britain, Spain, and Romania writing histories of their own national languages, each with no knowledge of the existence of the Roman Empire. Their honest efforts might be accurate in describing local linguistic evolution yet completely miss the point. The example seems absurd because the Romans boasted about everything they did: they left monuments and documents aplenty. But oil companies do not boast when they get a prized privilege from the government. They erect no monuments. Their commissioned histories stress their positive contributions to each nation where they do business. Many national and business oil histories are the equivalent of European linguistic histories without knowledge of the Roman Empire.

Finding evidence of transnational structuring requires a mixture of luck, persistence, and the help of nation-states that are more assiduous about preserving documents, though even these are often purged and doctored. Records of transnational structuring are found in the business press sometimes; in corporate archives rarely; in nation-state archives sometimes. Documents relative to the French oil industry may be found in American sources and vice versa. Research must be transparent to national boundaries even though state departments, foreign affairs offices, national archives, and university specialties are organized by such categories.

In the absense of documentation, the interpretive bias of economists and historians will be on the specifics of the nation-state's striving to optimize some policy in the "anarchical society" of the world order.

[6] Corporations usually destroy rather than falsify documents, for they are not expected to have "histories" like nations. The French government's archival purging and falsification are widely known. See Goodspeed 1977, p. 137.

Even where there is evidence supporting transnational structuring, the bias of nation-oriented thinking and analysis—the dominance of realist notions—precludes the search for transnational structuring. This bias is so strong that even pluralists with a business focus, Marxists who study particular nations, Marxists who work on a world-system scale, and theorists of rent seeking have entirely missed transnational structuring.

Transnational structuring is real. This book makes the point on a grand scale, with commodities of central concern to anyone who studies world political economy—oil and coal. Transnational structuring operates around and through states. A Third World nation such as Iraq or Venezuela may be caught up in this process, but no more so than France, the United States, or Germany: dependency and core/periphery issues are not central. Hegemonic power may be nice for nation-states to have, but is not a prerequisite for transnational structuring.[7] States do not hold an international conference, sign a treaty, or create a "regime" to signal the existence of transnational structuring.

States that are being transformed by transnational structuring have an incentive to legitimize their actions by puffing themselves up with rhetoric about pursuing the national interest with all the authority and resources at their disposal. Those groups whose pursuit of their own interest leads to transnational structuring have every incentive to keep quiet while national politicians make these absurd but useful claims. Because of the smoke screen of national interest rhetoric, world capitalist markets can be rationalized under our noses while simultaneously becoming difficult to study. Indeed, the prima facie evidence will support the opposite view: market regulation will usually appear to be due to purely local factors.

THE THEORETICAL BACKGROUND OF TRANSNATIONAL STRUCTURING

There is nothing new in the contention that business interests affect state policies, or that multinational firms influence the policies of other states. Transnational structuring builds upon these well-known effects.

[7] "Small" nations with big firms can play at transnational structuring. This was the case with Holland's Royal Dutch–Shell; the 1970s scandal concerning Korean influence purchasing in the United States is also apropos. Oil-producing states have tended to participate in the markets, and no doubt the politics, of the "core": Kuwait's "Q8" chain in Europe is an example. Cf. Johnson 1972, p. 163: "The huge American, British, German, French, and Japanese capitalist structures all reach into Canada, while a few equally powerful capitalist structures originating in Canada reach into other countries."

But to the usual approaches to business and the state I shall add three qualifications that change a great deal.

First, in transnational structuring private interests reinforce state power even as they use that power for their own ends. These ends are simple to describe even though they result in complex political strategies: control of market entry, control of production, control of technology, access to resources, and so forth. Private interests can first win, then lose, control of the state regulatory apparatus they helped create.[8] Nonetheless, the political process has changed the world economic environment by leaving behind regulated markets. The cumulative effect on international trade can be powerful.

Second, transnational structuring includes defeat as well as victory for business groups that seek to use state power to regulate markets. As such, it explains divergent as well as convergent outcomes. Under the simple convergence, structural-functionalist, realist-mercantilist theories outlined below, the nonexistence of a regulated market is a nonproblem. Their underlying assumption is that the "free market" is natural and therefore does not need to be explained. If we ask why the state regulated the steel sector, the answer would be something like, "there was a compelling fiscal or strategic need." These needs are deduced from the requirements of states—for example, their competition for power, their chronic revenue hunger. Regulation, not nonregulation, is the object to be explained.

Transnational structuring is most spectacular where it materializes into identifiable rules and policies across states. The appearance of this regulation does need to be explained. But the nonappearance of regulation is equally important. The political economy of a "free market" in which a major effort to achieve regulation has been made and has failed is fundamentally different from that of a "free market" in which no such effort has ever been made. For example, in the case of the oil industry, the appearance of regulation in France and elsewhere in the late 1920s is scarcely understandable unless one understands the deregulation that preceded it. "Free" markets and regulated markets alike are proper objects of inquiry from the perspective of transnational structuring.

Third, transnational structuring requires a comprehensive view of the sector in question. Oil is analyzed in relation to coal and other competitors; large-scale commercial goods may affect the balance of payments and hence the value of the national currency, which can bring in banks as major interest groups. Seldom do all firms in a political

[8] Cf. Huntington 1973, pp. 366–367, who offers three phases for transnational organization without saying why they occur.

struggle have the same objectives. Nonetheless, consensus among key firms can be established, and in the case of the world hydrocarbon cartel it is clear that corporate vocabulary, thinking, and strategy reflected a global perspective: the nation-state was a local province or unit in a world market whose chief use was to secure the kind of regulation, or "local" (national) cartel arrangement desired. This regulation controlled not just oil markets, but also rival hydrocarbon technologies that might have let the coal industry challenge oil.

These differences aside, transnational structuring shares many traditional approaches to the effects of business on the operations of the state in capitalist society. This is an ancient analytical tradition: Hobson cites Thomas More: "The business interests of the nation as a whole are subordinated to those of certain sectional interests that usurp control of the national resources and use them for their private gain. This is no strange or monstrous charge to bring; it is the commonest disease of all forms of government. The famous words of Sir Thomas More are as true now as when he wrote them: 'Everywhere do I perceive a certain conspiracy of rich men seeking their own advantage under the name and pretext of the commonwealth.'"[9]

Such trenchancy is a refreshing foil to the dull jargon of more recent social science. The juxtaposition of Thomas More (who was born in the fifteenth century) with Hobson (who died in the twentieth) underscores that business groups have attempted to turn the state to their own advantage for a long time—as long as the capitalist state itself has existed. Corruption in some form has been with us in every age; the asceticism of Saint Francis of Assisi was a revolt against the opulent luxury of the church. Nonetheless, bribing the sovereign power used to be the modus operandi of international relations: when the queen of Sheba visited Solomon she bore extravagant gifts—"camels that bare spices, and very much gold, and precious stones"—in an undisguised effort to purchase his favor.[10]

Modern capitalism has different norms about purchasing influence than precapitalist and mercantile capitalist societies. The queen's bribe was not corruption, it was high politics: she followed accepted procedure. This was also true under the mercantile absolutist state, where the king shared with capitalists the advantage of privileged markets and monopoly rents. Today, gifts between rulers of states are strictly ceremonial, and where they are more they can cause problems. Emperor Bokassa's gifts of diamonds to French president Valéry Giscard d'Estaing brought the latter dishonor that contrasts sharply with Solo-

[9] Hobson 1902, p. 51.
[10] 1 Kings 10:2.

mon's glory. That the Hebrew king had become important enough to merit a bribe from another sovereign made the event worthy of glorification. But today the explicit sharing of wealth among privileged mercantile interests and the state has also disappeared in modern capitalism: the governing norms are "free markets." Firms cannot justify a privileged relationship with the state because it enriches them as well as the president or prime minister. Numerous excluded groups would protest. The norm has shifted to keeping the state away from favoritism, unless that favoritism can be justified as the "national interest."

National Interest in an Environment of Business Conflict and Rent Seeking

Much of modern politics is argument over what constitutes the national interest. There is no answer. The illusion of an answer responds more to an emotional need for rightness in the world than to any objective "interest," and it complicates historical analysis with normative judgments that are extrinsic to the relationship between the state and business interest groups. Some theories concur that "the state" legitimizes its decisions by standing above the private interests of society; other theories try to prove this is untrue and to delegitimize state actions—an objective shared, as the economist Richard Posner notes, "by an odd mixture of welfare state liberals, muckrakers, Marxists, and free-market economists."[11]

Those in Posner's catalog of critics and reformers of the capitalist state all share a view of the illegitimacy of the state's intervention in the economy to favor or regulate economic activity on behalf of one group. Some privileged group gets a "rent" or profit that comes out of someone else's pocket: this is what galls the liberals,[12] muckrakers,[13] and free-market economists[14] that Posner describes. For Marxists, the extraction of monopoly rents is one more feature of what they consider an inherently exploitative mode of production; they do not deny that rent seeking occurs, but this is one of a host of capitalism's problems,

[11] Posner 1974, pp. 335–336.

[12] E.g., Hobson 1902. I could add Beard and Smith 1934, Schattschneider 1935, V. O. Key's description of business interests and the polity (1967, pp. 76–102), and others, including Leone 1986 and Gibbs 1991.

[13] E.g., Tarbell 1904 and Blair 1976. Muckrakers are really another form of liberal, with a populist tinge.

[14] E.g., Posner 1974, Stigler 1975, Buchanan, Rowley, and Tollison 1987, Colander 1984, Bhagwati 1982, and many others.

which range from falling real wage rates for the proletariat to imperialism.[15]

Business interests' effects on the state have therefore been studied repeatedly, from many theoretical perspectives. Several well-developed traditions—including liberal pluralists, the rent-seeking economists, and Marxists—argue that state regulation is often developed with the intention of favoring special interests, and that revolving doors, *pantouflage*, influence purchasing, and electioneering are tools to this end. Implicitly, and often explicitly, the activity of business groups shapes the institutional character, power, staffing, and authority of the state. We also have a wide range of literature that deals with transnational corporations and their negotiations with states, whether from the perspective of regimes, "sovereignty at bay," dependency, or world systems. Yet no theory discusses how the efforts of multinational enterprises, through their political activities and goals, shape the character, power, staffing, and authority of many states simultaneously. In the absence of one world authority to lobby, the obvious recourse of a multinational firm is to lobby all the countries that matter to it.

There is no reason the extensively described national lobbying of business should not also be international. But the justifying ideology of statecraft in the world order forces private interests to garb their actions in the rhetoric of local public interest (or national security) even if they are part of a global strategy. "The true national interest may or may not be served either by statecraft or by private enterprises" Beard and Smith write, but "this much is certain, namely, whenever the question arises in its simple, direct form: What is the supreme, ultimate motive underlying a given process, policy, or act in public or private life?—the answer is given just as simply, that the primary motive is the national interest, or to use variants, 'national welfare,' 'public interest,' or the 'general welfare.'"[16]

Since *all* state actions are garbed in national public interest rationales, this rhetoric helps little in explaining why specific policies are undertaken. Starting with realist assumptions about the world order and pursuit of the national interest, Beard and Smith move away from this exogenous level of constraint when it comes to explaining the particulars of policy. Examining two opposed "national interest" policies, the authors conclude that "both reflect deep division of domestic politics and interests and are affected by the oscillations and movements of economic power within the country. Neither stands out as a transcendent commitment of the nation beyond the reach of controversy and

[15] E.g., Hilferding 1970, Lenin 1977, Wallerstein 1984, etc.
[16] Beard and Smith 1934, p. 407.

diversity of opinion." The "definition of the terms of nation and interest" therefore constitutes the political struggle for control over state policy.[17] As a result, policy reflects "the outcome of specific demands by particular persons placed high in the scale of political influence," *or* policy reflects specific bureaucratic interests of state officials, represented as the "national interest:"[18] "Thus political institutions become inextricably involved in private institutions and practices until it is impossible to tell where one begins and another ends, or whether economic forces outside the government are driving it or political forces within the government are stirring up and enlisting business support for foreign policies."[19] Even when policymakers are under sharp constraints because of realpolitik, the causes of particular economic regulatory policies must refer to the specifics of the case: who is trying to maximize what, for what purpose, at a given time.[20]

Studies of domestic policies more often question the state's integrity than do broad surveys of politics among nations. We question whether the public interest is served when we see rate setting in the utilities industry or a tariff granted to sugar producers. We criticize special privileges like this because of the widespread notion that nonregulated trade is the "norm" that on occasion is violated. A violation is "bad." Special interest groups use two strategies to defend themselves against accusations of cheating the "natural" workings of the market. First, they are secretive. The less competitors and the general public know about their lobbying, the less energy and money firms need to secure privileges. Second, they elaborate a public interest argument to justify their gains. Hence arise the familiar litanies that the loss of jobs would be disruptive, that protection is needed only for a short time while an adjustment is made, or, what is frequently the most powerful argument of all, that the industry seeking a privileged position has a special claim because of "national security."

[17] Beard and Smith 1934, p. 88.

[18] Beard and Smith 1934, pp. 113, 454.

[19] Beard and Smith 1934, p. 119. Schattschneider 1935, p. 213, expresses a similar view, arguing that "the distinction between business and politics and between pressure politics and party politics tends to vanish. Some businesses are highly political; and some business men are occupied with politics so continuously that they are substantially politicians. Business interests have agents in office as well as in the lobbies, while economic pressure groups are sometimes substantially adjuncts of the parties." For Marxists, the "impossibility" of articulating a clear distinction between private and public interests translates into the theory of "relative autonomy," where "structural" factors induce officials to cooperate with capitalist interests, although their power gives them more freedom to act "independently" than an instrumentalist or rent-seeking model might allow. See Block 184, pp. 32–46.

[20] Ashley 1984 argues that realism is a theory of praxis. Is not realism more "realistic" when it includes economic pressures?

The state system objectively produces wars that threaten the security of citizens and nations; subjective policies evolve in response to the objective condition. Realist-mercantilist doctrines of state intervention "make sense" precisely because they are ad hoc and tailored to the circumstances of the moment; where threats to national territory appear greatest, there too are the greatest opportunities for special interests to masquerade as public interests. The affective appeal of these masquerades offers no objective basis for measurement. A nation's success or failure in war may provide an ultimate test, but only superficially. A nation can win a war even though encumbered with private-interest fraud. Football teams also can win even though some of their members are injured or not playing optimally. Victory does not vindicate the performance of each player or of each policy. Seemingly, victory justifies the policies of a nation; but the justification has no relation to the real value of the individual policies. The reverse is also true: Nazi Germany's synthetic fuel policies may have benefited the war machine in spite of Hitler's ultimate defeat. Neither victory nor defeat in war tells us how a nation adopted its economic strategies and equipped itself to carry them out.

Let us consider a case where a nation decides to produce one thousand tanks and two thousand airplanes. It wins in battle. But it might have also won, perhaps with a different sequence of victorious battles, with two thousand tanks and one thousand airplanes. We may find that not the national interest but the political competition between tank and airplane manufacturers determined the selection process.[21] Perhaps rulers sought what was best; they did so in an environment saturated by the efforts of private interests to define their way as "best." Well-motivated politicians may plan for war objectively to the best of their abilities, but the usual case is different: the interested parties produce reams of documentation, expert witnesses, and sophisticated arguments; they exert influence on the politicians and the bureaucrats through campaign contributions, offers of technical help, or simply by having long-standing ties that affect the promotions, salary, and career path of the typical bureaucrat or company officer.

How then can we determine whether a state policy is the product of well-motivated officers of the national interest or of a tawdry "conspiracy of rich men" as Thomas More once lamented? If we consider countries with different national oil policies, can we say these were the discrete products of individual nations reaching similar conclusions

[21] Cf. Gourevitch 1978, p. 900: "A country can face up to the competition or it can fail. Frequently more than one way to be successful exists. . . . The explanation of choice among the possibilities therefore requires some examination of domestic politics."

about the need for energy independence? Or were they the result of transnational structuring, which at the level of any one nation will look like a local political product ballyhooed with the usual national interest rhetoric?

This is a twofold problem. First, there has been no satisfactory resolution of the problem of private versus public interest in domestic economic regulation.[22] Second, when considering international policies, it is hard to make a positive identification of transnational structuring because the contrary, dominant mode of interpretation—state-centric, realist-mercantilist theory—has no real specifications about economic policy and can be bent any way we choose. I will return to this point below.

Transnational structuring contains elements of rent-seeking theory but is broader. Simply stated, rent-seeking theory argues that state regulation of the economy is provided by the state at the instigation of private business interests. The chief benefit is a limit to market entry: this might be a licensing scheme, a tariff, or some other exclusionary policy.[23] The extra profit achieved by this regulation is a surplus beyond what would exist in "normal" competitive circumstances. The surplus profits, less the costs of the political lobbying to achieve them, constitute the sought-after "rent." An elegant formulation comes from the economist J. A. Hobson, whom the neoteric rent-seeking theorists have yet to acknowledge as one of theirs. Hobson describes "trade . . . fenced against the intrusion of outside capital and labour," so that "the marginal supply may exchange at a premium against other classes of goods produced in trades which capital and labour can enter more freely." "Fenced" is how Hobson refers to entry barriers; the "premium" is a rent, which in fact he defines as such, referring to "scarcity gains or 'rents.'"[24] In the pluralist tradition, V. O. Key observes that in a regulated business sector "newcomers have to negotiate an administrative hurdle to go into business, and often the hurdle is high . . . important elements of the industry saw the possibility of eliminating 'cutthroat'

[22] A good discussion of why this is likely to remain the case is in Przeworski 1990.

[23] In recent years rent seeking has broadened to include any government initiative one disapproves of, for "rents" appropriated by one group at the expense of another are, simply put, transfer payments, which account for almost all of government expenditure. Rent-seeking theorists now debate not just market restrictions on firms, but the whole field of taxes and state expenditures. Social security for the elderly disrupts "normal" market incentives to save for retirement; the social security system thus transfers a "rent" to the old. Keynesian economists are also seen as rent seekers who justify transferring to today's beneficiaries the revenue of tomorrow's citizens. See Buchanan, Rowley, and Tollison 1987; Lee 1987, p. 304, discusses social security; Rowley 1987, p. 153, calls Keynesians rent seekers.

[24] Hobson 1966, pp. 38–39, 41.

competition by a restriction of entry to the business and by the fixing of rates by public authority."[25]

The theory of transnational structuring differs from that of rent seeking in several respects. First, it attributes importance to failed efforts at regulation as well as to those that are successful. Second, it interprets the practice across many nation-states at once, which thus far the rent-seeking theorists have not done. Third, it makes no claim that regulatory structures are permanent or always benefit those who created them.[26] Fourth, the whole notion of "rents" in relation to "the normal rate of profit" is extremely problematic; in some cases politically active firms desire downward movement in prices. Fifth, the preoccupation of rent-seeking theory with how much money firms decide to spend on politics relative to what they might gain from that expenditure is misspecified. Rent-seeking theorists are mystified as to why politicians' gains from handing out favors to business groups do not correspond with the actual size of the profits that firms take in as a result of those favors.[27]

I have already discussed the first three points, how transnational structuring extends across many countries and may also have as a possible outcome a deregulated local market rather than a regulated one, and how regulation, once achieved, may result in the unseating of the initial rent seeker by other rent seekers. The fourth point, regarding efforts to secure rents above the normal rate of profit, is misspecified in much of the rent-seeking literature. The normative bias of the literature focuses on government regulations, but firms with high efficiency may launch an attack on a cartelized or regulated market by fighting for "deregulation," knowing that in a deregulated market they will gain market share. When the firm batters down the doors, formal entry barriers go down, not up, and prices and profits for most firms may fall, not rise. Free-market economists consider that desirable, but the point is that the politicization of competition works both ways—to regulate and deregulate markets. An efficient firm also may seek to deregulate in order to institute a new cartel arrangement with other, more efficient firms. This situation occurred in France in the 1920s when Standard Oil broke down the previous cartel's regulatory protection only to institute a new cartel. To frame the matter in terms of "rents"

[25] Key 1967, p. 82.

[26] Vietor 1984 argues against the rent-seeking hypothesis. He shows that the classic rent-seeking model does not lead to permanent favoritism; but his evidence indicates that attempts to "capture" influence in a regulatory agency are like a game of king of the hill, where a coalition of losers is bound to try to unseat the smaller number of victors. He amply demonstrates that market competition is heavily politicized.

[27] E.g., a ten thousand dollar campaign contribution and a few lunches hardly constitute adequate reward for providing legislation potentially worth billions of dollars (Tullock 1989, pp. 11–12).

diverts us from the main objective—market control pure and simple. After several decades of oscillation between one cartel arrangement and another, in which new technical improvements and new sources of supply are introduced along with forced market concentration, the whole notion of a "normal rate of profit" is historically meaningless. There can very definitely be a struggle for control, to make sure the other party does not take away what one has regardless of the prevailing profit rate.

The fifth point is the size of the investment in political resources. As in investment in capital goods, political investment is motivated not by eventual real return, which can never be known at the time of the investment, but by the expectation of return. Political calculations are considerably more complicated than investment decisions. Contrary to rent-seeking theory, most political investment is not offensive in character, designed to secure privileges, but is defensive, designed to maintain what one has. The risks of "going to war" and trying to regulate a market in one's favor are great: even for the greatest monopolist there is no certainty of a return, and there is no telling what hornet's nest of opposition one might arouse. A firm does not usually decide, "It is time to do political battle and gain this or that advantage." It raises a large war chest when it has reached the conclusion, "If I do not engage in political battle to secure an advantage, my adversary will do so, and the costs of his victory are unknown to me." Thus Royal Dutch–Shell's World War I offensive against Standard Oil was provoked by the import policies that were tilting Standard Oil toward total domination of world oil markets and the ruin of the Royal Dutch–Shell combine. Standard Oil's brilliantly orchestrated counterattack from 1919 to 1921 had one objective: to stop Royal Dutch–Shell. Rome was obliged to conquer the world in order to defend itself from it: a similar logic starts business regulatory conflict. Expenditure on politics is not a fine probabilistic calculus but is akin to how nations determine what to spend on war: in normal circumstances they look around and try to match the competition, but when under a very great threat they devote extraordinary resources and energy to ensure their survival. Most of the time politicians are paid not to procure privileges, which is expensive and uncertain, but to warn clients when somebody else is threatening those privileges, which is far less costly.

Leading Theories of International Economic Regulation and Development

Many interpretations of state behavior that focus on economic interest groups outside the state will have something to say about transnational structuring. No one, however, has quite arrived at the scenario as I have described it. Strange comes tantalizingly close in suggesting "private bodies like industrial cartels and professional associations" as an alternative to regimes, but she does not pursue the idea.[28] In other work she also stresses the limited ability of modern states to cope with rapid capital flows.[29] This is still off the mark of how transnational structuring shapes states. Streeck and Schmitter (1991) have looked at the issue of European integration and described "transnational pluralism" as the successor to "national corporatism." At times they allude to parts of what I term transnational structuring, but nonetheless they neither explicate nor use the process as I do here. The works of Strange as well as Streeck and Schmitter reflect a convergence of concern about transnational structuring, and this book, by providing terms and a major empirical case, may further parallel inquiries that seek a broader level of generalization. Transnational structuring's distinctiveness and limits as an approach will appear best in comparison with other leading theories.

By international economic regulation I mean the adoption of state policies that affect the terms of international trade or in some way structure the pursuit of internationally distributed economic activities. I cannot be more precise than this: what theories of international political economy explain is a function of the theory itself. The theories below recognize many forms of economic regulation by the state and are not fully commensurable with one another. The prevailing explanations of international economic regulation fall into three broad categories.[30] Loosely described, the three are convergence theories, state-centric national interest theories, and world systemic theories. Transnational structuring is none of these.

Convergence Theories

In convergence theories, nations adopt similar regulatory policies because they are technically optimal.[31] The tariff as a revenue-raising device, independent of its role in fostering national industry, has roots in the comparatively easy enforcement of a tax on goods that enter a

[28] Strange 1983, p. 351.

[29] Strange 1986.

[30] "Postmodern" analyses of international relations in the past decade have made fashionable the extensive textual analysis of international relations theorists, putting many scholars on edge about summaries of their work. See, for example, Shapiro 1989, Grunberg 1990, Alker 1987, and Ashley 1984. The summaries here are not intended for such deep exegesis.

[31] Cf. Vogel 1986.

port.[32] When states are faced with similar problems, they adopt similar solutions. Repeated patterns of state behavior occur in a variety of circumstances. Convergence may occur, however, not because it is technically optimal but simply because a policy that looks like a good or workable idea in one place may be copied in another. An example is the adoption of alcohol fuel subsidy programs by some two dozen countries in the 1930s; this was due not to the operation of an international fuel alcohol interest but to a worldwide depression in agricultural prices. Purely local and national level agricultural lobbyists and their governments in a wide number of places found a similar way to increase demand.[33] "Technically optimal" and "copycat" causes of similar kinds of state intervention remain the simplest means to explain the adoption of similar policies by nations.

There are a large number of regulatory policies that "converge" simply because what works in one country may work in another. One theory of convergence along these lines is structural functionalism, which assumes that states are autonomous and that their internal development responds to the organic needs of an evolving system of "inputs" and "outputs."[34] Structural functionalism has Marxist variants,[35] but even these have no real theory of international economic regulation. The lack of emphasis on economics is characteristic of the orientation of the approach: if the evolution of "educational," "defense," "revenue," "legitimation," and other such "functions" of states is derived from the very fact of having a state, then the regulation of economic activity by some kind of institutional arrangement (taxation and revenue raising are among the acknowledged inputs) is merely one more institutional arrangement, but not necessarily one of special significance.

A more recent convergence theory focuses on states' unlimited appetite for funds, which leads to a search for more efficient ways to collect revenue. Levi's assumption that "rulers are predatory in that they try to extract as much revenue as they can from the population" is not like the realist-mercantilist power-maximizing state; she describes the state as having a "randomness of ends."[36] Though strongly influenced by the public-choice school of economics, her work is structural functionalist, with special attention to the politics of revenue. The worldwide development and adoption of tariffs and income taxes was not due to the operation of an international lobby or the explicitly state-created

[32] Levi 1988.
[33] Nowell 1985.
[34] E.g., Almond and Powell 1978.
[35] Poulantzas 1978 and Althusser 1970; see also criticism by Thompson 1978 and Gouldner 1970.
[36] Levi 1988, p. 3.

object of a "regime." Indeed, it would be nonsense to talk about an "income tax regime" unless states had somehow agreed to negotiate their levels of income tax with one another.

Yet tariffs, income taxes, and value-added taxes have an obvious and profound impact on economic behavior. The appearance of these taxes and the preference for them arise from a combination of technical collection ability and political palatability. Convergence theory can explain a number of important cases of government economic intervention, but it is far from identifying transnational structuring, which requires more than one state and the existence of international markets. Much of what Levi describes could occur even in a state that had no international competitors, whether military or economic; there is little in structural functionalism that requires us to look at the international system in order to understand one state.

Transnational structuring does not replace simple convergence theory or even realist-mercantilist or regime theories of economic regulation. The conundrum is more profound: Transnational structuring is a major practice that is often mistaken for other kinds of state intervention. Convergence theories work well for explaining the adoption of income taxes in many states or even the harmonizing of world driving standards on green for "go" and red for "stop." Nonetheless, a simple "technical optimality argument" disappears wherever we find partisan business interests pushing such policies for "nontechnical reasons," or for that matter where there is a strong "national security argument"—which might be construed as a higher-order technical optimality argument. Since both national and international oil interests as well as major national security arguments were part of the development of regulation in the pre–World War II period, our main focus should be on these classes of theory, not on convergence theories.

State-Centric National Interest Theories

The second explanation for international economic regulatory policies is "public interest" or "national interest" theories that focus on the state. Here I follow Beard and Smith, who identify "the national interest," "the public interest," and "the general welfare" as the same idea: the state represents general rather than particular interests of groups within society.[37] The state has responsibilities that stem from its unique capacity to represent general interests, but there is no a priori determination of whether an interventionist or a laissez-faire economic policy may be optimal.

[37] Beard and Smith 1934, p. 407.

The bias toward economic interventionism and protectionist regulation focuses on the state's obligations and its determination of its "needs." When a state's policies are designed to satisfy its needs at the expense of other states, we have a state-centric theory of public interest that can be identified as mercantilist, reflecting "realist" preoccupations with state power. When a state's economic policies are designed to further collaboration with other states, on the theory that less supervision of the economy in fact gives rise to greater economic growth, enhancing national wealth and the well-being of the nation, we have a liberal or "regime"-oriented explanation of economic policy. In theory, highly mercantilist, nonliberal protectionist states could negotiate "regimes" on a single issue or a spectrum of issues; but the general emphasis in the regime literature is on those "regimes" deemed consistent with a liberal economic international order.[38]

The realist-mercantilist theory and the liberal-regime theories alike hold that states intervene in markets to provide a public good. The theories agree that the primary locus of decision making is the state. They agree that the state's determination of needs and the "national interest" primarily determines policy. They do not agree on the means to achieve the ends, for lack of a general theoretical criterion by which "optimal" economic policy can be deduced: thus liberals sometimes favor mercantilist intervention,[39] and mercantilists sometimes favor liberal policies.[40]

State-centric national interest theories have no consistent prediction with regard to the time, nature, or extent of the appearance of international economic regulation, except as part of a general "new order" that results from the rise of a new hegemon. Nor can realists make reliable predictions about the success of nations' striving for power. The rational actor theory of power in international relations (realism) suffers from the same limitations as the rational actor theory of wealth in economics: the economic theory tells us nothing about whether wealth is to be had in the bond or stock market, in selling hamburgers or automobiles. It only predicts that individuals will pursue gain. In international relations, realism predicts that nations will pursue power, but we get no prediction of whether protectionist or interventionist

[38] Cf. Krasner 1983b, p. 357: "The most common proposition is that hegemonic distributions of power lead to stable, open economic regimes because it is in the interest of a hegemonic state to pursue such a policy and because the hegemon has the resources to provide the collective goods needed to make such a system function effectively."

[39] Cf. Adam Smith's defense of the navigation act: "As defence, however, is of much more importance than opulence, the act of navigation is, perhaps, the wisest of all the commercial regulations of England" (Smith 1976, 1:487).

[40] Cf. Heckscher 1935, 2:273–293, on imports needed by domestic manufacture and, on cotton, Hamilton 1968, pp. 311–313.

economic policies will be optimal for this, or whether laissez-faire or negotiated systems of trade (regimes) will be deemed optimal.[41] As Waltz writes, "Theory, as a general explanatory system, cannot account for particulars."[42] This point is not obvious: there have been some lamentable attempts to define the national interest in economic matters as some kind of definite thing whose existence can be deduced either from the realist theory of international relations, or from circumstances, or observed in the behavior of leaders. Krasner is typical: "Policies for securing military power are fairly straightforward. The state must not become dependent on unreliable sources of supply for raw materials that are necessary for the conduct of war. *Ceteris paribus,* this implies that the state should maximize self sufficiency through stockpiling, government purchase guarantees, subsidies, tariff protection for domestic industries, and similar devices."[43] This kind of deduction of specific interventionist policies is so endemic to the theory that even Waltz, who makes it plain why policies cannot be deduced from the theory, does exactly what Krasner does and makes nearly identical recommendations.[44]

Hirschman is more skeptical.[45] He observes that power-oriented economic arguments have been made on behalf of free trade and protectionism, and for mixed free-trade and protectionist regimes. There simply is no one economic policy attributable to the pursuit of state power, at least not beyond the homily that the state needs money in order to exist.

State-centric national interest theory cannot predict the preferred policies[46] of decision makers, nor can it predict the value of any given policy. When chess players sit down to play we can make no specific predictions about what will occur on the board, even though we know the rules that govern the actions. Knowing the rules, we can at least explain and analyze afterward how the decisions in the game were made. The state-centric theory of the national interest is really not so much a theory as a set of rules that allow us to interpret, after the

[41] E.g., Britain's 1848 decision to become more dependent on food imports by abolishing the corn laws, which was a mistake from a point of view that favors autarky but correct from one that values increasing national wealth (income) and hence national power. See Hirschman 1980, pp. 8–12.

[42] Waltz 1979, p. 118.

[43] Krasner 1978, p. 38. A general discussion is on pp. 35–58.

[44] Waltz 1979, p. 156.

[45] Hirschman 1980, p. 9.

[46] Timber for ships' masts is the preferred strategic commodity in the eighteenth century but not in the twentieth; realist-mercantilist theory predicts states' concern with each but has no means of predicting the shift from one to the other or the impact of each kind of technology on state capabilities and hence state strategies. Technology is exogenous to the theory.

fact, the appearance of strategically motivated economic regulation. We infer motivations from these rules, which leads to closed reasoning, because in the real world of international politics there is more than one game, and more than one set of rules, at work at any given time. Transnational structuring's rules differ from realism, but in the world both games are played simultaneously.

The Realist-Mercantilist Variant

In the mercantilist variant the state focuses on such narrowly defined "national interests" as strategic commodities, and it implements policies to procure or protect access to these goods, which enhances the state's military and economic power. Mercantilism is not too different from liberalism: both assume that the state is "motivated" to further the general good. There is tactical disagreement on what is best. For both, the state is a unitary actor. Let us imagine that a number of states enact protectionist policies in steel. They may perceive steel as a "technical necessity" for producing armaments; their assessment of the technical requirements of war leads them to adopt nearly identical steel policies. There is no "regime" per se, yet we have a kind of convergence: convergence derived from the states' need to protect citizens and territory. This state-centric approach emphasizes the dynamics of power in structuring the international political economy. Convergence theories based on technical optimality would attribute the repeated appearance of the same regulations to some kind of perceived efficiency. There would be no need to refer to the actions of other states. The realist-mercantilist variant's explanation of convergence must have as a starting point for one state's actions the existence of other states that pose a threat. This theory requires the existence of other states in order to explain the behavior of one state.

The realist theory of international relations focuses on the systemic constraints imposed on states that compete for power. Realism emphasizes individual actors, like the "convergence theory" identified above. But realism's isolated individual states are in a milieu of isolated individual states. Repeated patterns of behavior among states, which would include alliances, ideological functions, and economic regulatory policies, emerge in response to efforts to develop state power in response to other states that are developing their own power. If Germany wants to arm against Great Britain, it must have a steel industry; it enacts policies to favor the steel industry and proceeds to arm. That threatens other countries. Great Britain, France, and others respond to the growing German threat by similarly arming and implementing whatever policies they need to favor their steel industries. The arms race leads to similar kinds of state economic intervention, but the simultaneous appearance of that economic intervention is not a coordinated event: it emerges out of the dictates of a "system of states" in which the potential for armed conflict dictates the behavior of individuals, no one of which has any absolute control over the anarchic international system.

This action-reaction characteristic of the anarchic state system distinguishes realism from structural functionalist models of states originally developed for comparative politics. The structural functionalist descriptions of state development are extremely Spartan in their references, of whatever kind, to external or international influences and constraints on the development of nation-states.[47] Huntington's *Political Order in Changing Societies* (1968) offers an institutionalist perspective on development problems that is highly derivative of structural functionalism in explaining lack of institutional stability in the Third World, yet highly derivative of realism in explaining the development of institutions in Europe. His institutionalist pathology of the Third World is almost entirely free of references to imperialism, outside forces, and such, thereby maintaining the characteristic structural functionalist emphasis on the state as an isolated unit.[48] But he does refer to the evolution of military structures and government centralization in Europe as a result of the state system of competition: "The prevalence of war promoted political modernization. Competition forced the monarchs to build their military strength. The creation of military strength required national unity, the suppression of regional and religious dissidents, the expansion of armies and bureaucracies, and *a major increase in state revenues.*"[49] This capsule explanation of states' growing economic intervention as a part of the drive for power shares with simpler convergence theories a focus on the state's need to regulate businesses, rather than vice versa. Heckscher offers the strongest possible formulation:

> The state must have *one* outstanding interest, an interest which is the basis for all its other activities. What distinguishes the state from all other social institutions is the fact that, by its very nature, it is a compulsory corporation or, at least in the last instance, has the final word on the exercise of force in society; it has the "authority of authorities" *(Kompetenz-kompetenz),* to borrow the terminology of that eminent German constitutional jurist,

[47] Cf. Waltz 1970. Waltz, though a realist, offers structural functionalist "differentiation" (p. 207) for the different internal arrangements of states.

[48] Especially typical of the "input-output" models of comparative politics such as Almond and Powell 1978. There are structural functionalist exceptions; e.g., the potentially unifying effect of international institutions as analyzed by Haas 1964.

[49] Huntington 1968, p. 122; italics added.

> Jellinek. *Power* must therefore be the first interest of the state, which it cannot resign without denying its own existence. *La raison d'état* . . . is simply the claim of the state that regard for its power must, if necessary, precede all other considerations.
>
> Mercantilism would similarly have had all economic activity subservient to the state's interest in power. [M]ercantilist efforts at unification endeavour to secure the state's power *internally* against particularist institutions, and the question here is the external power of the state in relation to other states . . . it is only natural that all social forces must when necessary either serve or give precedence to the interest of the state.
>
> . . . .
>
> [M]ercantilism as a system of power was thus primarily a system for forcing economic policy into the service of power as an end in itself.[50]

The economy is subordinated to the state, and the explanation for international economic regulation lies in the power seeking of the state. The drive to power is dictated by the inexorable logic of competing with other states. This theoretical perspective does not recognize—at least in its most reductionist forms[51]—influences external to states as a constraint upon state behavior or upon the distribution of power in the international system.

The Liberal-Regime Variant

Realism depicts the Hobbesian war of all against all as the fundamental reality of international politics. Alliances and other cooperative behavior among states are caused by temporary advantages, combined with calculations based upon the existing distribution of military power in the international system of states. In war, states test whether the actual distribution of power corresponds to the perceived distribution of power.

Regime and interdependence theories of the past two decades solve the problem of international state cooperation in a system of anarchy in a manner analogous to the contribution of social contract theory in resolving the problem of the state itself.[52] Social contract theory assumes that individuals "in a state of nature" surrender individual rights in order to gain the benefits of living in a cooperative society; they

[50] Heckscher 1935, 2:15–17; all italics in original. Skocpol 1979, p. 29, offers a similar formulation.

[51] Morgenthau 1973 and Carr 1964, as realists, are more amenable to considering constraining factors on the power competition among states than, for example, Waltz (1959, 1970, 1979).

[52] For regime theories, see Krasner 1983; for interdependence, see Keohane and Nye 1977.

agree to do this in a fictional contract between themselves and the state. In regime theory, there actually often is a written contract: member states adhere to the General Agreement on Tariffs and Trade, or invited members sign on to the European Community, which is a regional free-trade regime. Sometimes cooperation occurs when states recognize that each depends upon others for essential needs. The basic notion is contractarian: states give up their anarchic state of nature to cooperate in one or several domains, gaining benefits they would otherwise lose.

The state remains the fundamental locus of decision making. Often there is a legal, publicly signed instrument—a treaty—that signals the existence of the regime. States may go to the bargaining table with other states, each lobbied by its own interest groups; but the fundamental calculus of interest lies at the level of the state and its officials. States are the centers of regimes.

Problematically, regime theorists contend that "regimes" can exist without formal documents or agreements. But it is difficult to affirm the existence of a regime for a specific commodity, which is hard to distinguish from virtually any kind of regularity in commercial exchange, in contrast to a pattern of enforced practices by which states regulate most commercial transactions. Simmons and Haggard correctly dispute Keohane's treatment of oil as a regime.[53] Strange describes "the efforts of states to interfere with market forces" as "puny" and would substitute for the "hegemonial and multilateral regime model" a "web of contracts model," which gets us away from the misguided attempt to specify regularity of commercial exchanges as a regime.[54] Lipson contends that commercial banking's control of international finance is an example of a privately organized "debt regime."[55] But this regime looks very unlike anything discussed by Keohane (1984) or Krasner (1983). Following Simmons and Haggard's objections (1987), I would contend that this is not a regime but an example of transnational structuring. The analytical problem is complicated because the stuff of international finance—money—is manufactured by states; states are intrinsically involved in anything that affects the manufacture, value, and distribution of this commodity, which sets finance apart from other international commodities and blurs distinctions between "private" and "state" interests to a higher degree than in any other branch of economic activity.

Leaving aside definitional problems, regime theory emphasizes

[53] Simmons and Haggard 1987, p. 494.
[54] Strange 1985, pp. 233–234.
[55] Lipson 1981.

actors within the state who, for reasons of psychological and institutional practicality, even inertia, adhere to an existing modus operandi. Ruggie defines regimes as "a set of mutual expectations, rules, regulations, plans, organizational energies and financial commitments, *which have been accepted by a group of states*."[56] This is rather expansive: NATO, which we used to think of as an alliance, becomes a "defense regime." Regime theory is probably at its best in describing multilateral negotiated agreements between nations; as the concept is extended, it gets more nebulous and becomes more inclusive at the expense of clarity.

The attitude of regime theory to hegemonic power is ambivalent. Keohane's *After Hegemony* gives a number of reasons why cooperative regimes might continue, once established, even in the absence of an uncontested hegemon to enforce rules. Without the prior period of hegemony, a regime would be exceptionally difficult to establish. Since the theory is state centric, it is best suited to explain the appearance of "world order" or "international regimes" with the dominance of a single power, a hegemon that enforces the rules. To have a regime when a hegemon is absent or in decline, the "leadership" of a state is necessary. As Keohane and Nye put it, "For international regimes to govern situations of complex interdependence successfully they must be congruent with the interests of powerfully placed domestic groups within major states, as well as with the structure of power among states."[57] Kindleberger's explanation of the disintegration of international economic agreements in the period between the two world wars is often cited: "The international economic system was rendered unstable by British inability and United States unwillingness to assume responsibility for stabilizing it."[58]

Transnational structuring can integrate states economically and add regularity to many aspects of production, consumption, and the power associated with the technologies that accompany production and consumption. State power can be either created or destroyed in this process, often at the instigation of business interest groups. Top state officials do not always realize the dimensions of their state's integration into a larger system, although some do. The world hydrocarbon cartel took form at precisely the same time that the international financial and trade practices entered a period of relative instability and disinte-

[56] Ruggie 1975, p. 570; italics added. Keohane concurs with Ruggie's definition (1984, p. 56). Krasner's definition (1983, p. 2) is similar to Ruggie's and frequently cited.

[57] Keohane and Nye 1977, p. 226.

[58] Kindleberger 1973, p. 292. The necessary stabilization or "regime" has to be provided by a hegemon; in Kindleberger's terms, "The main lesson of the inter-war years [is that] for the world economy to be stabilized, there has to be a stabilizer, one stabilizer." Grunberg's 1990 postmodern critique of hegemonic theory is strong corrective medicine.

gration.[59] Whereas regime theory is most problematic when applied to specific commodities, transnational structuring is most common in a commodity-specific form. Transnational structuring *is* affected, sometimes decisively so, by the competition for power in the realist sense, but it also shapes the world in which competition among states occurs and affects the endowments with which they compete for power. It does not require the existence of a hegemonic power.

NONSTATAL, SYSTEMIC THEORIES OF INTERNATIONAL ECONOMIC REGULATION OF STATES

Krasner has characterized pure realist models of the international order as a "conventional billiard ball model with its focus on zero-sum state interaction."[60] Regime theory, which focuses on modes of cooperative behavior among states, nonetheless shares "the fundamental assumptions of a realist structural paradigm: an international system composed of egoistic sovereign states differentiated only by their power capabilities."[61]

However, many international relations theories describe influences on the state system that are not part of the states themselves and that cannot be conceived, as regime theory is, as a product of the behavior of states and their institutions. Morgenthau describes the international system as composed of "autonomous units," reifying Krasner's billiard ball metaphor. But he also describes morality as restraining warfare, "consensus—both child and father, as it were, of common moral standards and a common civilization as well as of common interests—that kept in check the limitless desire for power, potentially inherent, as we know, in all imperialisms, and prevented it from becoming a political actuality."[62] Morgenthau's attribution of a genuine causal power to an unspecified system of cultural values is really an anomaly in realist thinking, one usually overlooked by his epigones.[63] I do not seek to explicate Morgenthau or to save realist theorists from their own inconsistencies. Morgenthau's reference to moral influences reminds us that not all nonstatal influences on state behavior must be economically grounded, even though that is the focus of this section. Ashley's analysis

[59] Cf. Keohane and Nye 1977, p. 131, on the instability of the international financial system in the 1930s.

[60] Krasner 1983b, p. 355; Keohane and Nye 1977, p. 226, criticize the metaphor.

[61] Krasner 1983b, pp. 355–356.

[62] Morgenthau 1973, pp. 173, 227.

[63] No trace of it, for example, is to be found in Waltz 1959.

of "technical rationality" advances another kind of ideational influence on actors within the system of states.[64]

The principal nonstatal theories of the international system focus on economic agents. Polanyi, for example, affirms that in the nineteenth century haute finance restrained competition among nations: "The Pax Britannica held its sway sometimes by the ominous poise of heavy ship's cannon, but more frequently it prevailed by the timely pull of a thread in the international monetary network." This financial system had its own autonomous logic: "Only a madman would have doubted that the international economic system was the axis of the material existence of the race. Because this system needed peace in order to function, the balance of power was made to serve it."[65] Subordinating the balance of power to the operations of haute finance, Polanyi stresses the evolution of a world market economy as a fundamental social transformation of the human race; it is a line of reasoning foreign to realism. "The Great Transformation" was an international social change based on the transition from feudalism to capitalism, and it happened as much through states as by means of them. For Polanyi, realism's world of unitary, power-maximizing states resulted from the subordination of social relationships to market relationships.

Polanyi does not show how nonstatal actors influence the international system. He refers to "unconscious" influences on nations' attitudes toward the international gold standard and the market system. These unconscious influences are difficult to interpret concretely.[66] Still, his description of the evolution of values and ideology of liberalism is more articulated than Morgenthau's references to moral factors. Polanyi identifies the Rothschilds and the Morgans of the world as major international influences but does not specify what they do and how they do it.[67] Hannah Arendt also sees a role for finance in the European balance of power, but she restricts herself to international Jewish banking.[68] She argues that the internationalism of the Jewish banking houses was historically linked to absolutist monarchies, and that with the rise of the nation-state, mass politics, and democracy, this influence on the state system disappeared. These powerful nonstatal actors were not a necessary feature of the modern capitalist system but a phase of the prior absolutist monarchical system.

[64] Ashley 1984. See also Shapiro 1989 on the role of language in international relations.

[65] Polanyi 1957, p. 18.

[66] Cf. Polanyi 1957, p. 27: "In the struggle to retain it [the world gold standard], the world had been unconsciously preparing . . . to adapt itself to its loss." Or again (p. 77): "Eighteenth century society unconsciously resisted any attempt at making it a mere appendage of the market."

[67] Polanyi 1957, p. 27.

[68] Arendt 1966, pp. 13–54.

The ideas of Polanyi and Arendt are highly idiosyncratic in comparison with the world systems theories of international relations.[69] In modern world systems theory the "world-economy"[70] is stratified by class—ownership of the means of production—and by space, for the world-economy has an outer boundary (the periphery), an intermediate zone of development or de-development (the semipheriphery, which can include states on the rise or in decline), and a developed "core." A state's economic, military, and political behavior and development are determined by its placement in the world system, which evolves over time and structures the ability of nations to maximize power in the realist sense.

Modern world systems theories have a Marxist lineage, going back to Lenin, Hilferding, and Luxemburg.[71] There is an important and often overlooked liberal theory of imperialism, linked to the development of underconsumption theory, which in some respects is closely allied: Hobson, in particular, is heavily cited by Lenin[72] and had some influence on Max Weber.[73]

Marxist and non-Marxist economic systemic theories generalize about the world system far beyond transnational structuring's more limited effects. The world-encompassing theories explain state behavior with regard to whole classes of actors—as, for example, capitalists in a competitive versus a monopolistic economy. They make claims about a larger level of world organization and other generalizations that go beyond anything in this book. Kautsky sees in "ultra-imperialism" a potential for peace and unified action among industrialized nations;[74] Lenin sees in monopoly capitalism the cause of war and

[69] See discussion in Gilpin 1987, pp. 67–72.

[70] "World-economy" is hyphenated by these theorists (e.g., Amin 1974; Wallerstein 1980, 1982, 1984; Frank 1982) to emphasize the single economic system of which states are constituent elements, as opposed to an aggregate of national economies. See also a critical analysis in Brewer 1980.

[71] Lenin 1977, Hilferding 1970, and Luxemburg 1980. "Dependency theories" often mix the Marxist critique of capitalism with the mercantilist critique of laissez-faire; classic mercantilists include List 1966 and Hamilton 1968. In some cases, as with Cardoso and Faletto 1979, the dependency thesis of lack of peripheral development is inconsistent with traditional Marxist notions about the spread of capitalism: significantly Cardoso and Faletto are not included in Brewer's 1980 survey of "Marxist theories of imperialism."

[72] As a theory of imperialism, Hobson (1902) influenced the Marxist tradition via Lenin (1977). As significantly, he influenced Keynes (1964), who acknowledged Hobson's importance (1964, pp. 364–368) in a chapter (chap. 23) on mercantilism that is all too often ignored by today's theorists of political economy. Fichte's *Closed Commercial State* (1980) is one of the earliest works linking capitalist activity to international conflict.

[73] Cf. Weber 1968, 2:910–921.

[74] Kautsky 1970. Some Marxists (e.g., Hymer 1972) see in the multinational corporation the element from which supranational authority will be built; but transnational structuring sees the use of local authority as the "stuff" out of which international market regulation is achieved.

conflict; modern world system theorists may make claims about the development or lack of development among broad categories of nations in the world-economy.

Transnational structuring does not explain capitalism but assumes it. A world systems theorist such as Wallerstein could absorb transnational structuring as part of the politics of the world-economy without doing great violence to his larger theoretical structure. Wallerstein's world system is "constituted by a cross-cutting network of interlinked productive processes which we may call 'commodity chains' such that for any production process in the chain there are a number of 'backward and forward linkages' on which the particular processes (and the persons involved in it) are dependent. These various production processes usually require physical transportation of commodities between them, and frequently transfers of 'rights' to commodities in a chain are made by autonomous organizations, in which case we talk of the existence of 'commerce.'"[75] Implicitly, each commodity chain has its own associated politics, but this particular kind of political struggle is not foremost in his analysis, and it is unclear how he would reconcile the politics of the "hydrocarbon commodity chain" discussed in this study with a world system based on *states* divided into core, periphery, and semiperiphery. The world systems theorists have not described transnational structuring as set out here.

Because transnational structuring might fit in with some variants of Marxism, the question arises whether it is a form of Marxism in disguise. The answer is that it is not: I offer no dialectical notion of history, no reference to class conflicts as part of what drives history, no attempt to build up a complete picture of "the world" or human history in the world. I *am* saying that a complete picture of the world will have to take into account transnational structuring, and that some theories of international relations will find this less of a problem than others. But transnational structuring does not "refute" realism: the point is far more subtle. Much of what is passed off and understood as realism is in fact transnational structuring, and this must perforce affect our understanding of states—how they are constructed, how they make decisions, and how capabilities are distributed among them. In fact, transnational structuring admits the possibility that in some cases realism reflects objective circumstances and decision making; I do not try to interpret all actions of the state as beholden to "capital"; but it is true that all actions of the state are legitimately open to question on those grounds.

Since many of the elements, institutions, and capabilities within the

[75] Wallerstein 1984, pp. 2–3.

state system are derived from transnational structuring, states are not by themselves "fundamental"; realism's posited "anarchy" of their systemic relation to each other simply has no meaning from the point of view of transnational structuring. Realists would see a dozen or so states with highly protectionist and interventionist oil policies as reflecting the fear engendered by systemic anarchy. From transnational structuring's point of view, these dozen states represent a chain of regulated markets consciously created with an eye toward establishing order in the world marketplace. That is not anarchy but a true transnational institutional construct. It certainly looks mercantile enough, especially when compared with a "free market." And who would say the mercantilism is "not real"? The interventionism and control exist, and the state may even on occasion use its powers for "the national interest," however one gets around to defining it. The seemingly objectively dictated fear-driven interventionist policies reflect "reality" to the realists, who thus name themselves precisely because of this. But their reality is only a useful illusion to the firms that help them along: the world of mercantile states is also a mercantile illusion.

In another dozen states, international firms may reach a consensus to maintain "free markets" because that is advantageous; they may be unable to impose their will because of local resistance; or both may be true. The free market is an alternative form of institutional construct. Some might call it a free-trade regime, but it is inconsistent to do so unless one also sees the protectionist circumstance as a regime, which to date no regime theorist has. In yet another ten or twenty states, the outcome is perhaps mixed. In fact all three outcomes are part of the same process of transnational structuring, and the true world "regime" is the one that encompasses within a single view the totality of these seemingly contrary outcomes.

Since each of these dozens of states will justify its actions in "the national interest," it is clear that this is not a useful theoretical point with which to approach the problem; we must begin by looking at the state system from the point of view of business interests. Whether we choose to call that Marxist or Schattschneider-style pluralism or an expansion of public choice, rent-seeking theory has more to do with labels than the reality and implications of transnational structuring in the world system.

This book in fact has no strict antecedents, but there are many literatures that are allied or close. Transnational structuring is not a version of the "power of multinational corporations" literature. Vernon's "sovereignty at bay" model of corporations and state conflict focuses on the relation between corporations from a "developed" country and the state in an "undeveloped" one.[76] Vernon also discusses relationships between firms from developed countries and host governments of other developed countries. Vernon sees the "autonomy" of governments as challenged by the growth of multinational corporations' power;[77] he posits that a corporation might "blackmail" a country into cooperating with it by threatening to move its operations somewhere else. His other potential source of conflict is between the home country government of the multinational firm and the host country government. Such considerations do indeed hold "sovereignty at bay," but Vernon does not consider what might occur when the multinational attempts to control political outcomes in its home *and* its host country *and* still other host countries.[78] The multinational's control of technology and its access to capital allow it to "negotiate" with a host government; this view typically sees multinationals as having a global business strategy, but seldom as having a global political strategy. This approach emphasizes the host government as a thing apart, which "negotiates with" a multinational as one bureaucracy to another. It ignores the means by which multinational corporations gain political representation in the foreign host government and strive to make sure the agents they "negotiate" with are their own or are sympathetic to them. Sovereignty at bay theory ignores how the goals of a government may be politically defined by the multinational corporation.[79] The theory also does not consider that drastic increases in regulatory power, or "sovereignty," may also result from political competition among multinational firms.

Gilpin accepts a core/periphery distinction in the world economy and worries about how multinational corporations affect national power by transferring wealth and investment.[80] Home and host governments retain sufficient autonomy to reverse long-term disinvestment trends. He makes no systematic reference to state economic intervention caused and inspired by private interests versus state economic intervention on behalf of some "national interest." The government's ability to define and pursue an objective national interest is axiomatic. Thus multinationals "have stimulated both home and host governments to

[76] Vernon 1971. Vernon's model is based on the product cycle, where production is internationalized from the home country to another country with lower labor costs.

[77] Vernon 1977.

[78] The attempt to control host as well as home country is not typical of the multinational literature. See, for example, Morse 1976, Encarnation and Wells 1985, Martin 1985, Biersteker 1988.

[79] Evans 1979 describes the interaction among state, national, and international interest groups in Brazil. The book does not examine what this process of political competition looks like on a global scale.

[80] On the core periphery relationship, see Gilpin 1975, pp. 47–63; on the need for neomercantilist intervention to halt the United States' decline, see pp. 198–214.

utilize industrial and other policies to make these powerful institutions serve what each perceives to be its own national interest."[81] He does not mention attempts by multinational corporations to shape government definitions of "the national interest" to suit their particular interests, in a manner that makes the governments reflect the presence of multinational business interests as a constituent element of their articulated "national interest."

As Gilpin and Vernon present the multinational corporation, transnational structuring does not exist. Gilpin refers to an amorphous larger process, the "international trading and investment regime," which is the sum of all trade practices wherein multinational corporations do business with governments all over the world. These practices constitute "a complex pattern of relationships among corporations, home governments, and host countries that has increasingly politicized foreign investment both at home and abroad."[82] Gilpin's catalog of diverse and contradictory acts does not describe the historical process by which corporations leave their imprint on the world order.[83]

The Problem of Intentionality

State-centric theories of market regulation contend that intervention furthers "the national interest." Rent-seeking critiques of public interest arguments contend that regulation is really a means of getting favors for private interests at the expense of the public. Schattschneider is explicit: "Unsupervised conduct in pressure politics means that the few will control the process at the expense of the many. The inertia of the masses is so great that not even the strong incitements of economic interest arouse them, especially if they are affected obliquely and indirectly."[84] The veil of public-interest rhetoric is most opaque where the national security is invoked. No matter what their genesis or who their instigators are, policies adopted to favor the national interest, once in place, are touted by the full power of the state and its functionaries. The commercial interest behind the scenes collaborates to project this image; many functionaries in the state itself will adopt the official view as their own, and many of those who think otherwise will remain silent

[81] Gilpin 1987, pp. 236–237.

[82] Gilpin 1987, pp. 254, 262.

[83] A weakness shared by some Marxists. See Wendt 1988 and Duvall and Wendt 1987. These essays leave few hints as to what "in the world" corresponds to the abstract categories.

[84] Schattschneider 1935, p. 287.

or be reluctant to commit any dissenting opinion to paper: Why argue against a victor?

The popular press, some and possibly the whole of the industry press, and many internal government memoranda will reflect the national interest perspective. The quantitative weight of evidence aligns with those who prefer state-centric interpretations of state policy. Realist-mercantilists always find abundant confirmation of what they choose to believe, so much so that to confirm the theory is uninteresting, especially since the existence of the confirmatory evidence for realism-mercantilism is predicted by rent seeking and transnational structuring. It is always more difficult to prove that there are special interests at work behind the mask of the state's public interest rhetoric.

Yet the question of the intentions behind policy is not confusing because of the empirical question of how state regulation occurs. The confusion arises over whether the regulation, once in place, serves "the public good." This normative concern has a curious retroactive effect on our assessment of the people and interests that put a policy in place. If we approve of a policy, we want to believe that state decision makers enacted it as best serving the public interest. Rent-seeking theorists, muckrakers, and other critics of regulation reason from the reverse stance: showing how selfish interests of the private sector manipulated the state to regulate the market "proves" the state policy is illegitimate.

In fact, those who see the state "defending the national interest" as it "ought" share a blind spot with muckrakers and other critics of state regulation. The muckrakers see "harmful" state regulatory practices as the distortion of the true purpose of the state by selfishly minded interest groups. Realist-mercantilists defend state intervention, convinced there truly is a "national interest," or at least that there truly are people who believe in it and act on their belief.

The shared blind spot is, What if policy measures are enacted to pursue the "national interest," which we later discover were put in place by selfishly motivated private groups, whose only conception of the national interest is as a cynical instrument to make themselves rich? Does that mean the policy is in fact bad? Does that prevent the policy from being good? A firm that bribes representatives to land a tank-building contract may still build excellent tanks; and another firm may try similar bribes and produce inferior goods. But the performance of the tanks, the value we put on them as the concrete products of policy, has nothing to do with the empirical-historical question of the politics behind the awarding of the contract. Perhaps builders of good tanks do not need to bribe anyone, but this seems unlikely, for builders of good tanks may still need to contribute money to counter the money from builders of bad tanks, and both resort to influence purchasing in order to further their business. The normative judgment about the value of a policy is thus a problem distinct from describing the process that led to that policy's adoption.

Realist-mercantilists never consider that a "sound" state policy that they approve of as "defending the national interest" may in fact have been motivated wholly by private and selfish interests who controlled the state in exactly the manner described by the rent-seeking theorists, except that the outcome was "good." In fact, business interests have one primary goal in their relationship with the state: to make money, perhaps by restricting market entry, perhaps by selling products to the state for any of a number of purposes. These rationales are not opposed, they are orthogonal, and much confusion occurs because those who prove business influences in arms, strategy, or other policies seem to think they have necessarily shown a detriment to the public interest and those who defend the public interest also assume that autonomy from private sector influence is required in order to carry out good policy. This leads to a tremendous confusion over empirical consequences and normative judgments, and it is one reason this issue has gone around and around in political science analyses without ever reaching a conclusion.

Realism-mercantilism, qua rational actor theory, can no more specify the correct power-maximizing behavior of a nation than economic theory can specify the best place to invest. There are many possible strategies for national power, and many for making money. There may be no "objectively best," and what is more, that there is no way to determine the objectively best is what opens up the whole policy process to intense debates about what the "best" is. This is the chink in the armor of state logic, through which rent-seeking interests eagerly armed with lobbyists, campaign contributions, bribes, and technical reports insinuate their private goals into the body politic. To observe that this happens is not the same as saying that it is always deleterious; and even the discussion of what is "good" and "bad" gets caught up in the routine of legitimation or delegitimation of state activity, which is definitely a part of politics but nonetheless detracts from the empirical problem of what *is*.

The bias toward state-centric logic leaves business influence theories, and transnational structuring, with the burden of proof: in the absence of documents showing that private interests played a determining role in the adoption of an economic policy, the common opinion will hold to the contrary, "public interest" view. The bias inherent in state officials' justifications of state policy, combined with the covertness of the special interests and the generalized lack of a hard documentary record about business lobbying activities (seen from the domestic level) or transnational structuring (as a product of world politics), is the reason for this "default setting" of interpretation in the public interest mode.

Obviously many in political science would prefer that business elites not be given pride of place in *any* political decisions. Nonetheless, business influence is clearly a part of local, national, and world politics. It is pointless to argue about whether business influence does exist. The only pertinent issue is, Just how far can it go? Business influence has a profound effect on the behavior of states: it can control the resources and technologies with which states play such realist games as the balance of power and the struggle for hegemony. There are occasions when the cartel-like behavior of internationally coordinated interests acts as a kind of international commodity government, in which states are used by business interests as instruments for enforcing trade and market shares. The power-maximizing goals posited by realism as determining state behavior do not cease to exist, but they can be profoundly modified by business behavior. The world system as such is not "fundamentally" composed of states, but also comprises the economic interests who see those states as parts of the world chessboard on which they move.

Private interests do not always win over *raison d'état,* but "realist" goals of the state do not always prevail over private interests. The integration of France into the world hydrocarbon cartel is a good case to consider: France's "etatist tradition," in which the goals of the sovereign power are considered stronger than private interests, is a recurring theme in the literature, including Krasner's "neomercantilist" analysis.[85] There are, moreover, a number of specialized studies on oil, and France's policy on oil, that repeat the myth of the French state's control of this portion of the economy "in the national interest."[86]

The "strong state" argument about France is well established in comparative politics.[87] Mény and Wright exemplify the consensus when they observe, "France clearly has the longest and strongest statist tradition in Western Europe."[88] But the issue of French statism is of limited interest here for two reasons. The first reason is the inherent limitation

[85] Krasner 1978, pp. 58–61; see also Gourevitch 1978, p. 906, and Shonfield 1965.

[86] The principal secondary sources on French oil history that adopt a state-centric view are: Murat 1969, Nayberg 1983, Touret 1968, Labarrière 1932, Melby 1981, Marchand 1937, André 1910, Thomas 1934, Rondot 1977, Faure 1939, Kuisel 1967, and Nouschi 1970. Adelman 1972, pp. 234–235, differs, describing protectionism as drawing "detested foreigners into the French market like bees to the jam pot."

[87] Traditional statist views are in Green 1983, pp. 161–192, and Hall 1984, pp. 29–33, 1986, pp. 246–247, for the postwar period; Shonfield 1965, pp. 71–87, surveys the entire topic of French state interventionism. See also Chazel and Poyet 1963; Cole's studies of Colbert (1939, 1943) are classic.

[88] Mény and Wright 1987, p. 42.

of a case study. Proving France's oil policy was dominated by private interests does not seriously challenge realism or French statism. Alternative arguments can save both: France is strong in all areas except oil; or France is generally strong, but not during the Third Republic, when weak institutions and fractionated interests prevailed; or France's "weak policy" of accommodation to American interests was actually a clever "strong" alliance-fostering policy, based on a realistic assessment of possibilities in the wake of the First World War; or France's oil policy was a mistake. Generals make errors on the battlefield; politicians can make errors in policy.

All these possible explanations provide a one-state explanation for France's decisions. To call French oil policy "weak" when it was supposedly "strong," to characterize it as "statist," is to adopt the language and values of realism-mercantilism. This theory cannot be falsified based on any one case, or even many, just as thousands of business failures do not falsify the theory of the economic rational actor. Accepting these criteria guarantees no serious challenge to realism or to the theory of French statism.

This brings us to the second reason the French "strong state" argument is of little use: it binds us to the incorrect notion that, in oil matters, there is something special about France. Let us go back to the original discussion of transnational structuring based on the analogy of indoor smoking regulation and the nationwide struggle between the American Tobacco Institute and the American Lung Association. The discovery of Draconian anti–indoor smoking laws in California is interesting because of what it tells us about the balance of power in that state between pro- and antismoking forces. The discovery of weak regulation of indoor smoking in Virginia is not surprising; but we might find that Virginia's laissez-faire indoor smoking exists side by side with highly interventionist or "strong" measures designed to support tobacco prices; in fact, a laissez-faire policy on indoor smoking, given the weight of the evidence on health, might actually *be* a "strong state" measure to boost demand for tobacco. That "strong" and "weak" can be so easily twisted according to context, then embedded in an even more difficult debate about cultural preferences on market intervention, suggests that the terms themselves are misleading. If we know that a confrontation over indoor smoking occurred between two well-organized national lobbies in dozens of state legislatures, this is the main event, not California's or Virginia's particular cultural traditions, which are contributing rather than causal variables.

The spectrum of oil regulatory measures introduced in France had counterparts in legislation introduced in Great Britain, Japan, Spain, Germany, Italy, and even parts of the United States. The interesting issue is how France fits into this larger whole. This book may provoke some scholars of France to reconsider their generalizations about etatism, because the oil intervention case is one of the defining archetypes of the statist theory, and if the archetype is wrong, one wonders what else may be too. Proving that would require additional studies, of which this one is only the model. It may turn out that many cases of both regulatory control and alternating phases of French liberalism have resulted from transnational structuring. That certainly is a strong implication of this work, but the importance of the conclusion is not limited to France.

If we consider the French oil case essential to the notion of the etatist tradition, then this book strikes at the heart of the scholarly consensus on the supposed autonomy of the French state. Some might consider that a sufficient objective. But if this work were limited to France alone, it could not make the argument it does. It rewrites part of French economic history in order to rewrite the history of the world oil market. Damage to the etatist tradition as such is collateral. The use of French "statist traditions" to explain the country's oil policy is part of the larger confusion about the evolution of the world oil market; and the confusion over the oil market is part of the immense conundrum of capitalism's role in the international system.

This book will arrive at a definitive picture of neither world capitalism nor the French statist tradition. It will show this: *There was one world political struggle over the oil market, and France's "nationalist" regulatory policies were one local outcome of that struggle.* Strong states, weak states, and national interests were all secondary to the world political struggle to rationalize markets.

Three Principles of Realism-Mercantilism and Transnational Structuring

Of the three major explanations for international economic regulation (convergence, realist-mercantilist state-centric, and world-systemic), only the realist-mercantilist perspective offers an explanation of international economic regulatory policies that directly contradicts transnational structuring. As I argued above, there can be no "economic prediction" derived from realism; yet many realists persist in tying a state's pursuit of power to particular forms of economic interventionism. State-centric explanations are recurrent features of international relations theory, of current and past politics (how officials justify their actions), and of studies of the oil industry.

Krasner's general criteria for mercantilist economic regulation are consistent with a number of other studies of oil policy in France and elsewhere: "What do mercantilist prescriptions look like in modern garb? The distinction between policies designed specifically to maintain the sinews of war and those related to achieving general foreign policy goals and maintaining general economic activity remains useful. Policies for securing military power are fairly straightforward. The state must not become dependent on unreliable sources of supply for raw materials that are necessary for the conduct of war. *Ceteris paribus,* this implies that the state should maximize self sufficiency through stockpiling, government purchase guarantees, subsidies, tariff protection for dometic industries, and similar devices."[89] In the oil industry, France did exactly as Krasner predicts: it sought access to foreign supplies; it tried to bring the refining industry to its own soil and to protect it through market sharing, stockpiling, and preferential tariffs. The government took a partial equity participation in a state company, a kind of subsidy and a means of sharing risk with private investors.

Many authors see the French state's oil initiatives as having met Krasner's mercantilist standard. Shonfield describes the French oil policy as forming "a fighting company which would involve the national prestige in a struggle for position with the established giants of the oil industry."[90] Melby contends that even though self-sufficiency was "impossible" for France, national independence could be realized by cooperation with the great oil multinationals.[91] Five goals dominated French policy, writes Labarrière: first, the protection of national production and encouragement of prospecting; second, the improvement of national defense through acquisition of refining capabilities, which would allow France to address producing countries rather than go through intermediate refiners (such as the Standard Oil Company); third, the building up of defense stocks without having the state pay for them; fourth, regulation of the industry, without diminishing treasury revenues and without raising prices to consumers; and fifth, providing a stable market that would incite companies to invest in France.[92] "The 'future' was a bright one for the [state oil company] and the French effort to gain oil independence," concludes Kuisel, reviewing the various state actions.[93] French policy was the "fruit of the national will," according to Nayberg.[94]

The great weight of scholarly consensus favors a statist interpretation

[89] Krasner 1978, p. 38.
[90] Shonfield 1965, p. 82.
[91] Melby 1981, pp. 104–105.
[92] Labarrière 1932, pp. 115–117; Thomas 1934, pp. 71, 124, provides a similar list.
[93] Kuisel 1967, p. 43.
[94] Nayberg 1983, p. 106.

of French oil policy and sees the French statist policy as fitting in with a realist-mercantilist worldview in which internal economic policies are driven by the need to maintain national security. By refuting this view, however, we do not "refute" realism. If we show that a firm went bankrupt trying to make a profit, we do not demonstrate that the economic theory of profit seeking is false. We show it did not work in one instance. If all we do is show that France's supposedly "strong" realist-mercantilist policy was in fact "weak," we have added nothing to the understanding of the world system beyond a footnote about French oil policy in a particular period.

The fault of realist-mercantilist theory is not that it is totally wrong: it is incompletely descriptive. Transnational structuring is orthogonal to the operations of the "realist" international system. At times there can be a clash between the rival logics of realpolitik and transnational structuring. Though a more subtle objective than trying to "refute" realist-mercantilist theory, elucidating transnational structuring's development and its interaction with the realist goals of state actors is ultimately a more significant contribution to our understanding of the world system.

Transnational structuring at a local level looks like what economists call "rent-seeking" behavior, which political scientists classify more broadly as business interest group conflict. We distinguish transnational structuring from purely domestic business conflict and rent-seeking behavior only by looking at the local market as a part of the world market.

Few who have written on French oil policy challenge the enshrinement of *les raisons d'état* as the state's prime mover. Some important studies oppose the general view; on the whole, a certain sensationalism combined with an unwillingness to publish thoroughly documented studies has kept the anti-etatist interpretations on the margins of serious historical analysis.[95] Even where these anti-etatist studies are based on leaked documents or, in the case of the Schvartz parliamentary report, subpoena power, there are no in-depth, systematic studies that relate the problem of specific oil regulatory measures to a theoretical perspective that challenges the state's claim to preserving the national

[95] Some French researchers have claimed that this is due to the limited French market for academic publications. Studies have to be "sensationalist" to sell even modestly, and footnoting is dropped so as to lower costs and not discourage the lay public. The market forces of the publishing industry combine with the publicly promulgated discourse of state interest to minimize the potential for a serious scholarly challenge to the dominant view. This "oppressive" interpretation is offered with a grain of salt, however; the French government footed about 50 percent of the bill for the research in this book, and the reason there are few anti-etatist studies of any quality may simply be that the good ones have yet to be written.

interest. Though often conspiratorial or sensationalist, the anti-etatist studies are not inaccurate. They are underdocumented; most, when checked against historical records, prove reasonably reliable.[96]

The most authoritative "anti-etatist" analysis of French energy policy is a 1975 parliamentary report written by deputy Julien Schvartz. The deputy neither was nor is anti-etatist; but his report, written from a pro-etatist point of view, documented failings of the French oil industry during the 1973–1974 oil crisis. The industry ought, from the Schvartz commission's point of view, to have represented the "national interest" better than it did.[97] Had France's oil policy been as mercantilist as described above, there would have been no need for a parliamentary report on that policy's failure in 1975. Schvartz's investigation was necessary because the idea that oil policy was "motivated" by commitment to the national interest was false—false in 1975, false in 1928, and false earlier. It has always been false. What has been taken for "national policy" is in fact a prime example of transnational structuring, which operates orthogonally to the requirements of a purely state-driven system. Lacking a vocabulary with which even to look for transnational structuring, the sophisticated theorists of the state and the more focused authors of monographs and parliamentary reports have been unable to do more than express occasional frustration at seemingly inexplicable contradictions.

We can sum up these contradictions in this way. The realist-mercantilist view of economic regulation has three underlying principles of interest:

1. States are the primary actors in the world system; all other entities are ultimately subordinate to them. Because firms hold both wealth and know-how, states use them to pursue power.
2. International competition among states creates an objective need for access to strategic resources. Interventionist policies to secure supplies develop where the "free market" is unreliable. Public discourse about "national security," the "public welfare," and "the national interest" largely reflects objective threats and efforts to deal with them.

[96] In this category we can put the works of Fontaine (1960, 1961, 1967), Bergier and Thomas 1968, Péan and Séréni 1982, and Simonnot 1978. These studies do not cover the formative period of the French oil industry in any detail. Feigenbaum's 1985 study is brief but an advance over many of its French predecessors.

[97] Schvartz 1974 does not focus on the origins of French oil policy. The commission assumed that the 1928 legislation ought to have guaranteed the state's interests. That it did not, surprised many French people in the 1970s but certainly is consistent with transnational structuring. There are no in-depth historical studies of the origins of the French oil industry that challenge the accepted canon.

3. The "free market" is "natural," and "strong interventionism in the market" is an artifact of state power. State power is therefore measurable by the "degree of intervention."

In going their own way, the firms that pursue transnational structuring may or may not adhere to the world of realist-mercantilist principles. The three main traits of transnational structuring are:

1. Transnational firms view individual states as resources to provide conditions amenable to business. Large firms may pursue an agenda for rationalizing world markets across many states at a time. Tactics in each state will vary. Deregulation (weakening the state) and regulation (enhancing state power) are possible tools in a larger strategy. Of the many ways to influence state policies, the most effective is to become one of the players in that state's domestic politics.
2. States routinely engage in wars and other struggles for advantage with other states. These struggles provide good opportunities to gain special privileges, dressed in the rhetorical garb of supporting the "national interest."
3. Like cartel agreements, the coalitions of influence that weaken or enhance state power are ephemeral. State authorities, once created, may take on an institutional life of their own or be captured by other interest groups.

It may be useful to distinguish yet another dimension to the world system, one that is opened up for study by transnational structuring but that may have a deeper logic than these few rules suggest. These rules apply most easily to the relatively simple circumstances of similar international firms, all in the same business, in conflict with one another on a world chessboard of states. The analytical problem is much deeper when the conflict among firms includes a true shift in the productive base of society from one era of dominant technology (such as coal) to another (such as oil). The historical circumstances of that conflict are described in this book, but they suggest some additional considerations about the evolution of states and technologies in the world system that I shall discuss in the conclusion.

The chapters that follow retell the economic history of France, and of the world oil market, in the first four decades of this century. They afford the opportunity to look at transnational structuring in the world oil market before the First World War (chapter 2). The war allows us to follow the rivalry of the principal world companies as they fought each other over world oil supplies even as the world's governments fought over territory (chapter 3). We then proceed to the dynamics of the French market as affected by the turmoil in Russia and the continued antagonism between international oil giants, leading ultimately to

France's integration into the world oil cartel (chapter 4). We consider the collective rivalry of the oil interests against the coal interests, and how the struggle over hydrocarbon technology led to the world hydrocarbon cartel, affecting the distribution of national capabilities in refining, synthetic fuels, and synthetic rubber (chapter 5). The conclusion reconsiders the multiple implications of transnational structuring.

Chapter Two

Transnational Structuring before World War I: The Pattern of State Monopoly Attempts and Market Interventions

Before World War I many states tried to intervene in the oil market. These attempts constituted the first great effort at transnational structuring. The development of a state-owned oil company in Great Britain, state monopoly proposals in Germany and France, and negotiations in Constantinople over the development of oil in Mesopotamia were all part of the same process. This process explains a nonevent: the indifference of the French government to the acquisition of an overseas oil source. From a realist-mercantilist perspective this failure is incomprehensible on the part of the world's second largest empire; but the great French financial house with oil projects in Russia—the Rothschild Bank—had allied itself with the international projects of the Royal Dutch–Shell Oil Company and the German Deutsche Bank. This group had no need for a distinctly "national" policy, and so neither did France. Moreover, domestic French interests were hostile to the development of a petroleum-based economy. Even though the linked German and French oil monopoly projects never reached fruition because of the war, parts of the plan were preserved and used in the international arrangements of 1928.

Origins of the French Oil Industry

In the nineteenth century oil was primarily a fuel for lamps. In the towns it competed against gaslight. By the mid-nineteenth century coal gasification plants had become common in major European cities, but the cost of the gas left lamp oil the fuel of the poor. In the countryside, where the low population density could not support the investments necessary to make coal gas, oil lamps were even more generally used. The lamp fuel market was competitive, attracting producers of alcohol, shale oil, liquid coal by-products, and vegetable oils, which were all commonly used for lighting.

France was an early starter in producing shale oil and vegetable oil for lamp fuel; its shale oil industry was so prosperous that for a brief time the country was even an oil exporter, before the American Civil War.[1] After the war the Pennsylvania oil fields underwent rapid development, and the whole of the French lamp fuel industry faced tremendous competition as American products began to make their way to foreign markets.[2] Before 1863 tariffs on oil products imported into France had been assimilated into tariffs on coal products. They rapidly became a separate political issue, with wild gyrations in policy that followed the political fortunes of the country, of the free traders and the protectionists, and of the evolution of the oil industry itself.[3]

The French refining industry grew out of the French vegetable oil industry.[4] Purifying vegetable oil for lamp fuel, usually oil made from rapeseed *(colza),* involved agitating it with sulfuric acid, allowing it to separate, and then "washing" the oil with soda to neutralize the acid. Identical processes were used in the shale oil and petroleum industries, but shale oil works operated near the source of the raw material, chiefly in Autun, to avoid the prohibitive cost of transporting the rock to a processing plant. Vegetable oil was typically processed in the major towns, since the seed pulp waste could be sold and easily transported as fertilizer.

The vegetable oil refiners refined crude petroleum from the United States. Relative to American refiners they were inefficient, and about 20 percent of the volume was lost in the refining process in France: it made more sense to refine in the United States and ship only the usable product. That course would have put the vegetable oil refiners out of the petroleum business, so they asked for, and received, a tariff to compensate for the disadvantage. American refiners were then forced to split the difference with their French competitors. They did the first distillation of oil, burning out most of the waste product and thus

[1] The whole of the French oil market in the nineteenth century is presented extensively in Nowell 1988, pp. 102–147. Citations have been deleted from this summary, but relevant materials are listed in the bibliography.

[2] See Tarbell 1904 and Johnson's *Development of American Petroleum Pipelines* (1956), whose suffocatingly dull title conceals a lively history of *capitalisme sauvage.* Other standard references on the United States include Williamson et al. 1963, White 1962, Gibb and Knowlton 1957, Hidy and Hidy 1955, and Yergin 1991.

[3] André 1910, pp. 89–97, 136. On free trade and protectionism, see Smith 1980 and Golob 1944.

[4] Faure 1939, pp. 63, 137, wrongly argues that the oil refining industry did not come to France until the late 1920s.

saving on shipping. The crude was purer and less heavy but still too dirty to burn in lamps. As such it qualified to enter France not as a "refined" product but as a lower-tariff "crude"; the French refiners did the final distillation before marketing. The one-distillation petroleum exported to France came to be known in the trade as "French crude," and from 1885 until World War I a tight cartel of French oil companies dominated the national lamp oil market, moving gradually into automotive gasoline as this market developed. Because their product arrived in predigested form, the French oil refining industry stagnated technologically.

Though domestic and foreign petroleum ultimately gained the advantage, the protected market left room for rival lamp fuels: shale oil, vegetable oil, and alcohol were the most important. By the beginning of the twentieth century France, nearly bereft of petroleum deposits, had a "national oil industry" whose members mostly could trace their origins to the vegetable oil industry. Indeed, one of these companies, Desmarais Frères, maintained its dual vegetable oil and petroleum business until 1961, a century after its founding. Other companies had equally venerable roots: Alexandre Deutsch's company, Alexandre Deutsch de la Meurthe et Ses Fils, was founded in 1843; Fenaille et Despeaux, named after its owners, dated from the same period.[5] These companies led the ten principals of the French industry, known familiarly as the Cartel des Dix (Cartel of Ten).

The French national firms were dependent on oil multinationals for their supplies. At first that meant Standard Oil, which was growing rapidly in the United States. Later, however, with the development of Russian oil in the Transcaucasian region, the French Rothschild firm followed the Nobels into major oil production. The Rothschilds backed a major tariff adjustment in 1893 that equalized Russian oil's prices with those of oil coming from the United States. In 1902–1903 the oil industry came under a jealous attack from alcohol interests that wanted to monopolize the lamp fuel market.[6] Alcohol proponents reasoned that without protection the French national oil refiners would be put out of business; they stripped away the tariff favoring French refining. But the move proved healthful for the French industry: no longer

[5] Menjaud and Dion 1961, p. 5; Faure 1939, p. 62, puts A. Deutsch's beginning at 1845; Peyret 1952, p. 211.

[6] The alcohol industry's fascinating maneuver was a limited national rent-seeking ploy that was not part of transnational structuring. See Nowell 1988, pp. 128–134. The alcohol industry returned in force after World War I (Nowell 1985). Déb.Ch., 6 March 1902, p. 1195; Hubbard 1903, pp. 626–627; Fin. B34034, 3 March 1902, "Projet de loi de finances pour 1902: Monopole des pétroles"; B34034, 10 August 1903, letter signed by Brunet; B34034, 9 January 1904, "Monopole des alcools dénaturés et des pétroles."

protected from Standard Oil and the Rothschilds, the French national firms sought vertical integration as a defense against the increasing competition and began to invest in production abroad. A number of them also clustered protectively around either Standard Oil or the Rothschilds.

At the turn of the century, therefore, the French oil industry as a whole received the great bulk of its supplies from Standard Oil, which was unable to break the refiners' cartel. The French national firms were the gateways to the French market. The Rothschild banking family had begun large oil operations in the Transcaucasus and was beginning to make its presence as an oil multinational felt not only in France but throughout Europe. Finally, some of the French national firms had begun to invest in direct production in Russia, Romania, and Galicia (now Poland). In 1904 a group comprising Desmarais Frères, Lilles-Bonnières, the Raffinerie de Pétrole du Nord, the Compagnie Générale des Pétroles, and Paix et Lesieur started syndicated operations in both Russia and Romania.[7] French capital also established one of the largest firms in Galicia, Limanowa.[8]

Before the abolition of the protective tariff in 1903, French firms could timidly collect monopoly rents; after 1903 they had nothing to lose and everything to gain by going abroad to secure independent supplies. They feared the eventual penetration of large multinational firms such as Standard Oil, whose principal tariff obstacle to direct operations in France had been removed, though the dominant members of the Cartel des Dix continued to be able to negotiate for supplies. French retail capital also entered the business of importing refined products from independent concerns in producing areas abroad. Large store retailers bypassed the French cartel and imported directly from producers in new oil regions that were opening up in Romania and Galicia. The tariff repeal, intended to curtail the oil industry's strength, instead forced significant readjustment and opened the market to more vigorous growth. By 1905 the market was still cartelized, but at least its members were trying to get alternative sources of supply. Legally the regime was a free market, but a major effort to put an end to that would come as the international oil interests carried out their grand maneuvers.

THE WORLD STRUGGLE AGAINST STANDARD OIL

Standard Oil dominated the American market by concentrating on refining and marketing. Geared to supply the continental American

[7] Menjaud and Dion 1961, p. 33.
[8] Lender 1934, p. 105.

market, the firm also flooded world markets with refined oil exports. From the nineteenth century until World War II, Standard Oil was "the firm to beat" in the world oil industry.[9] The American giant had tremendous technical, marketing, and capital advantages. Buying from many different American oil fields meant its supplies were steady regardless of fluctuations in production from a given field.

There is a characteristic pattern of challenger international oil companies before World War I. They tended to grow up around oil fields at "the periphery" or in out-of-the-way places where Standard Oil's influence was relatively small. The additional advantage of remoteness was that it added higher shipping costs to Standard Oil's retail prices. A challenger producing firm needed a local market that had not been penetrated by Standard Oil or that enjoyed some kind of protection. The Nobels at first specialized in the Russian market, Royal Dutch first sold Dutch East Indies oil in Asia, far from Standard Oil's home base, and the Burmah Oil Company produced and sold in the protected markets of British colonial India.

A firm producing much oil in a region with a small market had to find buyers before mounting an operation of any consequence; at the same time, remoteness exacerbated start-up costs, which increased the need for buyers. So a British capitalist in Persia, whose crude was unsuited for lamp oils and not well suited for gasoline, but just possibly passable as a fuel oil, would look for a guaranteed buyer in virtually the only consumers of consequence who happened to pass through—the British navy. A German capitalist looking at the enormous costs of developing Mesopotamia would want a protected market for the eventual product, to be sure the intrinsic risks of a major project in a foreign, precapitalist environment were not compounded by the risks of competing against a corporation of Standard Oil's formidable resources and abilities.

Such were the problems of developing oil in out-of-the-way places while Standard Oil dominated the world's largest markets. The search for a solution drove the policies of the international companies that challenged Standard Oil in the pre–World War I era. Another problem for these firms, linked to the first set, was how to compete with each other. Frenetic competition among rival Russian producers only weak-

[9] The firm was broken up in the wake of a successful antitrust prosecution in 1911. It became thirty-two separate corporations, the largest of which were Standard Oil Company of New Jersey (also "Esso," derived from the initials SO; sometimes abbreviated SONJ or referred to familiarly as "Jersey"); Standard Oil Company of New York (Socony; later Mobil); and Standard Oil Company of California (Socal; later Chevron). In this book "Standard Oil" in any post-1911 reference means Standard Oil of New Jersey. Many oil company names (e.g., Sunoco for Sun Oil Corporation, Socony for Standard Oil Company of New York, Texaco for Texas Oil Company) were derived from Western Union's six-letter address abbreviations for large firms.

ened them collectively against Standard Oil, and the same principle held in the other oil-producing regions of the world.

THE VANTAGE POINT FROM RUSSIA: THE ROTHSCHILDS, THE NOBELS, ROYAL DUTCH, AND SHELL

The major "Russian" oil producers, the Swedish Nobel family and the French Rothschild banking firm, maintained two objectives.[10] One was to increase their market share at the expense of Standard Oil; the other was to divide the world oil market with the American giant. These objectives, seemingly contradictory, were not so in fact. Cartels are most advantageously negotiated from a position of strength, and increasing one's own market share is a sure way to bring an opponent to the bargaining table. The Rothschilds and Nobels for a time backed a government-sponsored Russian consortium, which would have allowed them to unify and negotiate with Standard Oil. They needed to regulate the production, exporting, and market share of the smaller Russian firms, which would otherwise have expanded production to take advantage of the entente. Count Sergei Witte, Russian minister of transportation, backed the objective of a Russian consortium in 1893 and again in 1901, but proposals foundered on the inability of the larger firms to agree on sharing export facilities with the smaller firms.[11]

Russian oil exports had two possible routes to world markets, one from Baku to Batum, which meant through the Black Sea and to the European markets, the other from Baku south to the Persian Gulf, which would have dramatically improved the position of Russian producers in Far East markets. It was common knowledge in the pre–World War I oil industry that anyone who could load oil in the Persian Gulf, from whatever source, would have a tremendous competitive advantage in Asian markets over oil exported from the Black Sea through the Suez Canal. As early as 1892 the American chargé d'affaires in Constantinople wrote to John D. Rockefeller about using Mesopotamian oil for this purpose, saying that "in one or two interior districts of Mesopotamia in the desert oil has been found" and that "this gov't

[10] "Russian" firms invariably means firms owned by foreign capital. The Nobels were Swedish and the Rothschilds French; the major "native" firms were owned by Armenians, of which the largest, Mantascheff, could perhaps count as a third Russian "major." There were also small British firms and the German Spies Petroleum. Apostol and Michelson 1922 list the oil firms producing in Russia in their appendices. See also McKay 1983.

[11] See Gerretson 1955, 2:148, 233.

is inclined to be liberal and there may be a chance to pipe oil to the sea and wipe the Russians out of all this Eastern trade."[12]

The Transcaucasus–Persian Gulf pipeline route was backed by the Nobels and Rothschilds, and it is worth noting that these powerful European interests may have been backing Russia's "drive to Persia," which has so often been attributed specifically to Russia's imperial ambitions.[13] The Batum route prevailed because the pipeline, which replaced a less efficient rail tank-car system, still lowered the cost of exporting to the Far East via the Suez Canal and also allowed exporting to Europe from the Black Sea, all without antagonizing the British, who blocked the way in Persia.

The decision between the two routes was not easy, however, and the Baku-Batum pipeline's construction was slowed as a consequence. Though the pipeline was planned in 1895, only the first leg was opened by 1900. Subsequent authorization to complete it to Batum was not given until 1901.[14] The reason for the delay may be surmised from Kazemzadeh's study of Persia: in 1901 a report was filed with the czar urging the Persian pipeline option, contending it would cut the cost of reaching Asian markets to one-third the rival Baku-Batum route that also used the Suez Canal. The Persian route would have dealt a decisive blow to the position of United States exports in Asian markets. Count Witte backed the Persian route, and in January 1902 he made a loan to the Persian government contingent upon a concession for the pipeline.[15]

Witte's nationalist economic policies were in the tradition of Friedrich List.[16] A promoter of industrialization, he welcomed the efficiency of the Nobels and Rothschilds, but he wanted the foreigners under the control of the state, merged with the other "independent" producers in order to reduce squabbling and strengthen the Russian position. The Rothschilds and Nobels were willing to consider a consortium in order to consolidate their control of the Russian oil industry, but not if it meant losing even partial control over their own capital investments.

[12] Rock.Arch.Rockefeller Family, RG 1, John D. Rockefeller, Business Correspondence 1879–1894, box 65, folder 483, 19 June 1892.

[13] Whigam 1903, p. 267, and Feis 1964, pp. 366–381. Feis deals extensively with Mesopotamia but curiously leaves out any mention of the Baku-Persian Gulf pipeline.

[14] Leo 1978, p. 114, erroneously lists 1905 as the completion date of a *railroad* already in service. The *pipeline* was finished in 1906, according to Darby and Fullard 1970, p. 175. Gerretson 1955, 2:231, is vague on the date but gives the details of its history repeated here.

[15] Kazemzadeh 1968, p. 360.

[16] Gerretson 1955, 2:148. List's *National System of Political Economy* (1966) is a classic in the mercantile tradition.

Witte's ambitions were defeated by the united opposition of the major foreign companies. The Rothschild firm had, since 1892, exported most of its oil through a specialized oil carrying firm, Shell Trading and Transport. British Shell had a large fleet that sailed east of the Suez to Japan, succeeding where the Rothschilds had not. Before 1892 oil had been exported in the inefficient five- and ten-gallon tin cans once used to distribute refined products. Standard Oil had introduced tankers in 1890, but a permit to run them through the canal was denied because of the opposition of the Welsh tin industry, which would have lost its kerosene-related trade to tanker ships.[17] British Shell built its own ships, and new canal regulations allowed only the new Shell-designed tankers through, without admitting Standard Oil's tankers.[18] Strong in transport, Shell was weak in marketing, which was the strong point of Royal Dutch in the Far East. Relative to the producing potential of the Rothschilds, both companies were short of crude to meet their needs. The union of the three was a good combination, bringing together the disparate elements of the industry that Count Witte was trying to unite, but in a way that avoided the interference of the government and the smaller producers.[19] The smaller Russian producers were drawn into the orbit of the new combine, whose production, transport, and marketing network offered them an outlet under terms dictated by the members.

The new company was called Asiatic Petroleum. It was a jointly held and operated subsidiary of the Rothschilds, Royal Dutch, and Shell, all of which retained their autonomy. The new general manager of Asiatic, Henry Deterding, used his new combination of resources to invade the Indian market with a low-quality oil that could undersell Burmah's even though this company enjoyed the special protection of the Indian colonial government.[20] The threat posed by Asiatic was potent enough that the Burmah Oil Company, which had a near monopoly of the Indian market and was the British Empire's second largest company after Shell, was forced to make a cartel accord with the upstart Asiatic Petroleum in 1905.[21]

Hostile to consolidation schemes under the Russian government, the members of Asiatic still favored the Persian pipeline. Before the forma-

[17] Gerretson 1953, 1:217.

[18] Henriques' defense (1960, p. 85) of Samuel against charges of using bribery to obtain authorization for his tankers is not entirely convincing.

[19] Henriques 1960, p. 393.

[20] See Corley 1983a, p. 234. Lord Curzon, refusing in 1902 to see Standard Oil's representatives, explained, "It is not desirable for an American company or subsidiary to work the petroleum fields of Burma" (Gerretson 1953, 2:340).

[21] Gerretson 1955, 2:342.

tion of Asiatic this had been primarily the pet project of Count Witte and the Swedish and French interests represented by the Nobels and Rothschilds; after the consolidation British Shell threw in its weight, trying to persuade the British government to give up its opposition to the pipeline route to the Persian Gulf. Shell had succeeded in gaining authorization for its ships to cross the Suez Canal where Standard Oil had failed, so there were hopes for success. Asiatic Petroleum, already encroaching on the Indian market, would have its costs cut by two-thirds, threatening ruin for Burmah Oil, which looked for a way to block the Persian pipeline route.

Origins of the Anglo-Persian Oil Company

Russia's project for the Persian Gulf pipeline and domination of Asian oil markets was blocked by the oil concession of William Knox D'Arcy, a British subject who, having made a fortune in gold in Australia, at the turn of the century tried his hand at oil in Persia. In 1901 D'Arcy had obtained from the shah the rights to a vast oil concession covering most of southern Persia.[22] Article 6 of the concession agreement provided that no other company could build a pipeline across the concession territory to the Persian Gulf; so long as D'Arcy fulfilled his contractual obligations to explore for oil, the Russian pipeline project was stymied.[23] D'Arcy, however, was running an expensive exploration project in territory that was only nominally under the control of any administrative power in Tehran; tribal chieftains retained considerable influence in the suspected oil-bearing areas. This, combined with dubious prospects for oil, added up to a costly, high-risk operation.

D'Arcy's genius lay in finding the strengths behind his apparently weak position. Strapped for cash, he went to the Rothschilds to negotiate backing. Consolidating his interests with the Rothschilds' was a nightmare for Burmah Oil: it would have sped the trans-Persian pipeline or, worse, brought Rothschild-owned Persian production onto the market. The negotiations were dropped in February 1904, but they convinced Burmah Oil that it needed to play along with D'Arcy. In November 1904 the company backed D'Arcy's exploration efforts.[24]

Still, Burmah Oil was not enthusiastic about Persia. The company had two strategies to meet the new challenge from Asiatic. It kept

[22] Kazemzadeh 1968, p. 358, calls the concession agreement "one of the more significant documents of the twentieth century."
[23] Ferrier 1982, p. 44.
[24] Ferrier 1982, pp. 60–70. Ferrier does not stress the Persian pipeline.

D'Arcy afloat to buy time so that the Persian Gulf pipeline would be blocked. It would then use the time to expand production on its home turf in Burma. D'Arcy's exploration team surprised everyone by striking oil in May 1908, and the Anglo-Persian Oil Company was incorporated in 1909 to work the fields. At that time it was Britain's third largest oil company by capital, after Shell and Burmah, its single largest stockholder; but technical problems made the company a dubious prospect,[25] since the poor quality of the crude left it with limited potential in the lamp oil and gasoline markets.

Persia's low-quality oil required difficult decisions. Lack of funds kept D'Arcy's Anglo-Persian under constant pressure to sell out to the Royal Dutch–Shell combine.[26] The director of the Anglo-Persian Oil Company fought to maintain its independence. That required finding a market for the low-quality Persian product: the only possibility was fuel oil. Anglo-Persian's director, Charles Greenway, correctly ascertained that fuel oil was the "market of the future" for the oil industry and that one of the largest potential buyers was the Royal Navy.[27] In addition, he wished to expand his available supply of oil, and the nearest probable source lay in Mesopotamia.

Greenway found in the British government the most ready solution to Anglo-Persian's problems: the government could provide capital, a huge potential market, and political leverage to compete with other interests in the Middle East, particularly in Mesopotamia. Royal Dutch–Shell attempted to go over Greenway's head, motivated as much by the desire to bring in a pipeline from Baku as by wanting to develop the poor-quality oil in Anglo-Persian's concession. Royal Dutch–Shell offered to buy Anglo-Persian from Burmah Oil. Greenway represented the move to the Royal Navy as an attempt to monopolize potential oil supplies to the British fleet. He even pointed to the Royal Dutch–Shell group's participation in the successful 1905 cartel agreement in India, sponsored by his mother company, to support his argument. To be sure, there were inconsistencies in accusing the Royal Dutch–Shell group of monopolistic practices, since the protectionist policies of the colonial government favored cartelization, but Greenway did not need to dwell on these in his appeals to the British Admiralty. In October 1912 he proposed that the Admiralty guarantee purchases from his company, that the British government exert maximum influence to help his company gain control of the Mesopotamian concession, that

[25] Ferrier 1982, p. 107.

[26] Ferrier 1982, pp. 158–62.

[27] D'Arcy traded his Anglo-Persian shares for Burmah Oil stock at a profit and retired from direction of the company. Charles Greenway became acting head of the concern. D'Arcy died in 1917.

the Indian government exempt his company's oil from import duties, and that his company receive a subsidy to help it raise capital.[28] To show he was willing to represent "British" interests—his main selling point over the vastly more extensive Royal–Dutch Shell combine—he offered the government "one or two" representatives on the board of Anglo-Persian.

As Ferrier points out, these proposals originated in "company policy, not national policy."[29] The Anglo-Persian Oil Company eventually would offer the Admiralty participation as a stockholder rather than direct subsidies. The search for capital of the first national oil company was "hardly the image of glorious imperial prescience mirrored in many commentaries."[30] The latter are guilty of not having looked behind the veil of Winston Churchill's ringing nationalist rhetoric when he proclaimed that the reason for government participation in Anglo-Persian was to ensure that Britain would be dependent "on no one process, on no one country, and on no one field."[31] Three years earlier in 1910 he had affirmed, "You have got to find the oil; to show how it can be stored cheaply; how it can be purchased regularly and cheaply in peace, and with absolute certainty in war."[32] Behind this splendid rhetorical ambition lay a different reality: when the government bought shares in Anglo-Persian, this instrument of national policy was on the brink of bankruptcy; it was in no position to supply the Royal Navy in adequate quantity, it was entirely dependent on one field in one country, and an Admiralty chemist, reporting on field experiments with the oil in the North Sea, summed up its quality thus: "All Persian cargoes of Fuel Oil were unsatisfactory."[33] During the First World War Anglo-Persian focused its efforts on expanding commercial, not military, production. Its role as a naval supplier was minor.

From a realist-mercantilist perspective, the navy would surely want to ensure its supplies of oil as the Admiralty contemplated conversion to the petroleum age. But of the companies for sale, Anglo-Persian was a poor choice, especially since one of the largest oil companies in the

[28] Ferrier 1982, p. 168.
[29] Ferrier 1982, p. 168.
[30] Ferrier 1982, p. 72.
[31] Ferrier 1982, p. 182.
[32] Ferrier 1982, p. 165.
[33] Ferrier 1982, p. 199. Earle 1925, p. 270, quotes Churchill as contending that the British had "despatched a mission to Persia to survey the Anglo-Persian Fields and to report upon their availability as sources of fuel for the fleet. The report of the commission was favorable." And, I may add, a whitewash of what was going on in Persia. Earle did not have access, in 1925, to Ferrier's documentation.

world—the largest oil company in Britain, an oil company that was actually making money rather than losing it—was also on the block.

The Admiralty Refuses to Buy the Shell Trading and Transport Company

British oil policy before World War I protected Burmah Oil from competition. In 1902, when offered controlling interest in the (then entirely British) Shell Trading and Transport Company, one of the largest oil concerns in the world and far more capable of meeting Admiralty needs than the future Anglo-Persian, the Admiralty declined.[34] Four years later, when Samuel was forced to sell off some prized tankers from his fleet, the London *Times* criticized the government for "the folly of parting with a vital war interest."[35] No national policy kept Samuel's fully developed Shell Trading and Transport, the largest British oil concern, from amalgamating with Royal Dutch in 1907. Nor was government participation sought for British oil operations in Mexico. Alternatives were not explored because the British government had no oil policy other than protection of Burmah. To this end it put resources into the fledgling Anglo-Persian, whose stockholders were unwilling to accept the necessary risks.

Had Shell been under Admiralty control, its greater resources might have allowed a swifter, more deliberate development of Persia, and the company would have retained greater, perhaps total, autonomy with regard to Royal Dutch. As Ferrier summarizes it, the British government was "ultimately without a positive and consistent . . . commitment to the overriding priority of an oil policy and its effective implementation."[36]

Royal Dutch–Shell before the European Petroleum Union

The Anglo-Persian Oil Company was therefore the first oil concern to get the national interest defined in a manner synonymous with its

[34] Henriques 1960, p. 401. Shell's connections through Admiral John Fisher were apparently no match for the India Office and Burmah Oil. More on Shell follows in the next section.

[35] Cited by Henriques 1960, p. 492.

[36] Ferrier 1982, p. 260. Ferrier never adequately explains how a new company, with no production record, on the verge of bankruptcy, would find the clout to gain the advantages it did. Burmah Oil had both the size and the motivation to try to get someone else to pay for its protection. The Admiralty's refusal to buy Shell, which, when offered, was in difficulty but in far better shape than Anglo-Persian when it made an equivalent offer,

private interest. There were attempts in other countries to achieve a similar goal. These more directly concerned the development of the Royal Dutch–Shell combine.

Royal Dutch and Shell Trading and Transport both entered the oil business in the Far East in the 1890s.[37] Competition with the Standard Oil Company forced these two companies to converge on Russian production as a crucial supplement to their own holdings in the Far East. The Asiatic Petroleum Company, formed in 1902, in which the Rothschilds, Royal Dutch, and Shell Trading and Transport were principal stockholders, helped to pool the diverse resources of these three concerns. Shell was strongest in transport, Royal Dutch was strongest in marketing, and the Rothschild company, known by the Russian acronym Bnito, was strongest in production.[38] Samuel delayed agreeing to participate in Asiatic, however, until after his offer to allow the Admiralty control of his company had been refused. He had hoped the Royal Navy would offer a secure, lucrative market that would salvage Shell's independence.[39] Disagreements between the three groups were not infrequent, and by degrees Henry Deterding, leading Royal Dutch, gained control of the principal operations of the group. By 1907 Marcus Samuel had made crucial errors in marketing and production, and he sold out to Deterding. Even so, Samuel preserved his company's nominal independence. The Shell Trading and Transport Company and Royal Dutch each became a holding company for the combined operations of the two. Royal Dutch retained 60 percent of its own operations and gained 60 percent of Shell's. The Shell Oil Company retained 40 percent of its own operations and also received 40 percent of Royal Dutch's. Each parent company remained independent. The legal provisions that allowed for this were extremely complex but workable: the arrangement has lasted without fundamental modifications to the present.[40]

---

makes no sense from a strategic point of view. Henriques (1960, p. 610) cites a postwar Board of Trade memorandum: "The full importance of petroleum and its products for war purposes and the vital necessity for maintaining adequate stocks and supplies does not seem to have been fully appreciated until the latter half of 1916."

[37] Gerretson's (1953–1957) four-volume history of the company before World War I is also a survey of world economic history at this time.

[38] The initials were the Russian translation for the Société Caspienne et de la Mer Noire.

[39] Henriques 1960, pp. 399–401. Samuel proposed Admiralty control on 13 March 1902. He received the letter of refusal 17 June 1902 but held off signing an agreement to form Asiatic until after the sea trials of the *Hannibal,* the success of which perhaps might have persuaded the Admiralty to change its mind. Samuel's company was in difficulty owing to price wars with Standard Oil. It was still turning a profit, and in no way in a situation comparable to that of the Anglo-Persian when it sought government participation a decade later.

[40] Gerretson 1955, 2:345–348.

Shell's founder, Marcus Samuel, was knighted in 1898 and became lord mayor of London in September 1902. His knighthood rewarded Shell's salvage of a grounded British naval cruiser, HMS *Victorious*, in the Suez Canal. Socially and politically prominent, he observed the high Jewish holidays in state and used his position and his many ambassadorial contacts to intervene actively on behalf of Jewish causes around the world. At his inauguration he antagonized the Foreign Office by refusing to invite Romania's ambassador, his protest against pogroms there.[41] He actively urged the adoption of fuel oil by the Royal Navy beginning in 1898, trying to capitalize on the goodwill his rescue of HMS *Victorious* had earned him. He was a friend to Admiral John Fisher, a leading proponent of British fleet modernization who had invested most of his life savings in Shell stock. The two vigorously lobbied for British naval sea trials of an oil-powered vessel, but they were disappointed in 1902 when the first experiments ended in smoke-filled failure. The Admiralty remained uninterested in oil and in Samuel's company, forcing Shell to become increasingly bound to the affairs of Royal Dutch.

The Asiatic Petroleum Company, founded in 1902, foreshadowed the full amalgamation of Shell and Royal Dutch in 1907. The Royal Dutch–Shell combine's size dwarfed all companies save Standard Oil and the Nobel interests in Russia. The united actions of the Royal Dutch and Shell companies, in combination with the Rothschilds, had political consequences all over the globe. The group defeated an unfavorable state-sponsored cartel proposal in Russia; it scared the Burmah Oil Company into helping D'Arcy in Persia; and it intimidated Anglo-Persian into appealing for a government bailout. The growth of the combine had consequences in the European market as well.

In Europe, Standard Oil was as everywhere the dominant supplier, and competition was intense. The chief non–Standard Oil groups included Royal Dutch, Shell, the Nobels, and the Rothschilds, with one important addition: the German Deutsche Bank's participation in Steaua Romana, a Romanian oil-producing company. Shell's marketing and production strategy between 1902 and 1906 suffered multiple buffets. In Russia Standard Oil forced prices up by buying all available independently produced oil at Batum; in Europe it forced prices down by marketing at a loss. In Texas, where Shell bought from Guffey Petroleum (which later became the Mellon family's Gulf Oil), the decline of production in the Spindletop region caused serious supply problems.[42] Shell was not sufficiently integrated vertically, and it foundered because a pure "trading and transport" company was vulnerable

[41] Henriques 1960, pp. 228–229, 421–422, 456.
[42] Henriques 1960, pp. 460–462, 490.

to normal production fluctuations that were deftly exploited by Standard Oil.

The Deutsche Bank allied with Shell in the European market to take business away from Standard Oil. It was forced to retreat from its aggressive position in the ensuing price wars, leaving its partner, Shell, burdened with contractual obligations it could not meet. In 1906 Shell had to withdraw from the European retail market and sell some of its prized freighters to the Deutsche Bank at a loss. This final blow forced Samuel to sell out majority control of his worldwide operations to Henry Deterding and Royal Dutch.

Development of the European Petroleum Union (Epu)

With Shell out of the way as a retailer, the Deutsche Bank sought the initiative in the European market. It united the principal Russo-European competitors of the Standard Oil Company into a single marketing organization that would divide production, sales, and profits. This company, formed in July 1906, was called the European Petroleum Union (Europäische Petroleum Union), or Epu. First hailed as a means to compete with Standard Oil, Epu quickly settled for a cartel accord that gave Standard 80 percent of the European market to Epu's 20 percent. Epu's 20 percent was divided up 36.19 percent to the Nobels' oil production, 29.61 percent to the Rothschilds' Bnito, 28.2 percent to the Deutsche Bank's Steaua Romana, and only 6 percent to the declining Shell. Epu also purchased independent production in Romania and Russia, to keep that oil from reaching the market directly and disturbing prices. Gerretson calls Epu "a peak in the development of the world market economy which was interrupted by the Great War."[43]

Indeed, the market agreements between Epu's members and the Standard Oil Company, on 1 May 1907, mark the date of the first successful worldwide oil cartel. International companies finely coordinated the distribution of a resource whose importance was rapidly increasing. British, German, French, Russian, and American capital was united in a common attempt to control production and prices. Since the Asiatic's members of the cartel (the Rothschilds, Royal Dutch, Shell, and the major Russian firms) also cooperated in non-European markets, the impact of the Epu agreement covered the entire Eurasian landmass (see fig. 1).

[43] Gerretson 1957, 3:88–94.

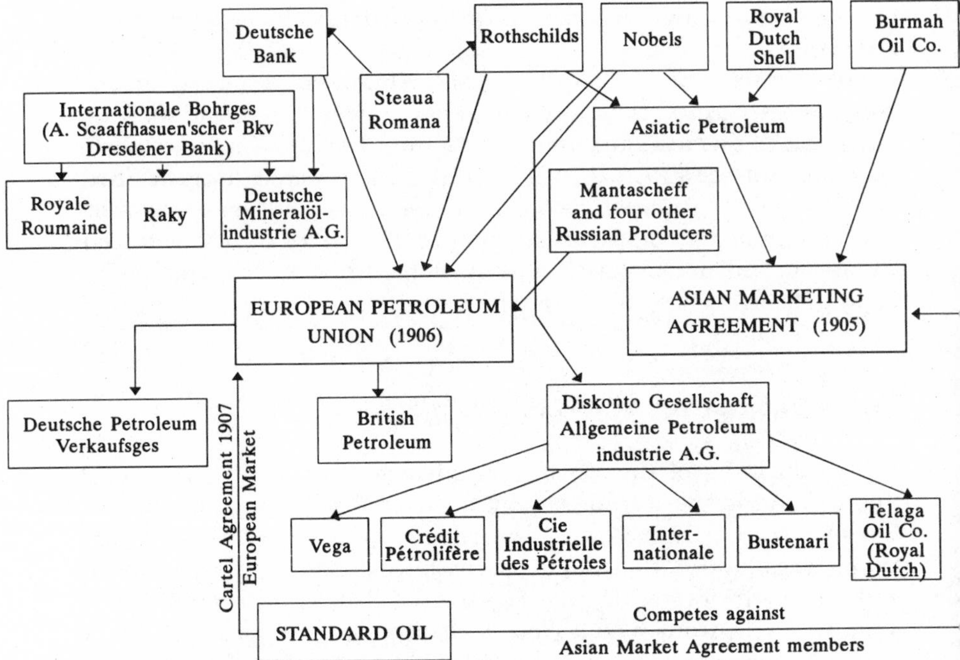

Figure 1. The first world oil cartel, 1907. *Source:* Muffelman 1907, p. 142, Gerretson 1955, 2:342, and this chapter.

This global reach may at first escape the eye. Epu, after all, was limited to Europe. The 1907 creation of Epu worked in tandem with the 1905 cartel agreement, however, including many of the same members, which applied to the Asian subcontinent and the Far East. The combined effect was that of a world cartel uniting smaller producers against Standard Oil. The main production areas of the anti–Standard Oil companies were the Caucasus, Romania, and the Dutch East Indies. The agreement coalesced over time and was not so much the result of a grand plan as pieced together under competitive pressures that repeated themselves in market after market. The stability of the 1907 world cartel came from Epu, which served centralized production, purchasing, and marketing, and enforced market shares for its participants. The resulting unity allowed the diverse interests of Epu to negotiate en bloc with Standard Oil. Epu's chief defect was that Shell was treated as a weak fourth partner, but after its amalgamation with Royal Dutch later that year, this no longer reflected reality. In the new combine, Deterding pursued the conquest of the European market that had proved too much for Samuel's Shell Trading and Transport.

Epu agreed with Standard Oil to maintain market shares as they were. Epu could not sustain a price war with the Standard Oil Company because, in the wake of the 1905 revolt in Russia, production in that country was restrained. Czarist troops were withdrawn from the Baku area, causing the Russian governor to fear Armenian independence movements. As a counterbalance he armed some of the Azerbaijani population. Many of the leading "Russian independents" were owned by Armenian Christians, and since their badly underpaid workers were the Moslem Turkic Azerbaijani, the region exploded with class, ethnic, and nationalist tensions.[44] Oil continued to flow, but rapid expansion of output was out of the question.

Members of the Epu–Standard Oil accord also sought to use state interventionism in the oil industry to limit competition. Producers wanted to fortify the protection the accord offered with government regulation. This regulation, once made available, would become the means to increase market share while still adhering to the letter of the cartel accord. Cheating on the accord by price competition in the market would have meant a swift disintegration of Epu. Government regulation allowed for the letter of the accord to be respected while each partner sought to create zones of privileged status within which it might not just defend itself, but also expand. The Burmah Oil Company protected its hold on Asian markets by checking the drive of the Caucasus producers to reach the Persian Gulf with a pipeline. As a result, Anglo-Persian grew under the umbrella of Burmah Oil and the Admiralty.[45] In Germany the Deutsche Bank's efforts to expand oil investments into the Ottoman Empire would be paralleled by an attempt to regulate shares in the domestic market. In France a similar effort would be made, without the link to Mesopotamia, for the main regulatory effort would be directed at limiting the growth of small importers of refined products.

The German Monopoly Attempt: Its Link to Mesopotamia

As in the British creation of Anglo-Persian, in Germany the drive for state interventionism had ample commercial as well as imperial motivation. In the British case low-quality Persian oil, unsuited for the kerosene market, could be unloaded only on the Admiralty, leading to government equity participation as the particular form of state intervention in Britain. In Germany the Deutsch Bank's Steaua Romana had good-quality kerosene to sell in competition with Standard Oil. But even within the protection of Epu it was hard to get a competitive advantage over the American adversary. Once the cartel was in place,

[44] Gerretson 1957, 3:105, 133.
[45] See above, p. 53.

the Deutsche Bank sought to accomplish through political competition what it could not achieve in the market.

In 1908 Dr. Leo Muffelmann, director of the Berlin Bureau of the oil journal the *Moniteur du pétrole Roumain,* published an article analyzing the obstacles facing Epu and made some startlingly accurate predictions. "Some people desire to oppose the Standard Oil Company with a state oil monopoly. . . . To be sure, the next tendency of the oil market will be an effort toward state monopoly. In all countries this goal is becoming apparent as all other means to combat the Standard Oil Company come to nothing. Up until now Standard has gone from success to success in Europe. . . . Will it be possible to stop the triumphant march of the Standard Oil Company in Europe?"[46] Muffelmann also noted that the German finance ministry had begun preparation for a lamp oil monopoly project.

The Deutsche Bank was the inspiration and chief political force behind the project. Various sources concur on this: Muffelmann's articles, the version of the monopoly project in Standard Oil's company history, and a special report written by a member of the French finance ministry, who was sent on a mission to Berlin to study the project in May 1914.[47] The monopoly project was officially proposed in spring 1911 in the Reichstag. Through the use of propaganda leaflets the Deutsche Bank aroused the support of the small shop-owning class that sold lamp oil and that resented Standard Oil's mushrooming home deliveries by horse-drawn carts.

The German monopoly plan was linked to German progress in Mesopotamian negotiations. There was an important but indirect consequence for the French market: the Shell group forged ahead in Mesopotamia in alliance with the Deutsche Bank, and French capital, in the form of the Rothschild bank's holdings in the combine, was brought along in tow. The largest French oil group pursued its interests through Germany. Hence the French government had little to say about Mesopotamian oil until World War I. For the same reason, it was natural for the combine to turn to its French stockholders for help when the war sundered the Germans from Mesopotamian development plans: Royal Dutch–Shell deftly turned to France for diplomatic assistance, and almost overnight France developed an oil policy.

[46] Muffelmann 1908, pp. 94, 96.

[47] Muffelmann 1908, 1914; Gibb and Knowlton 1957, pp. 205–220; Fin. B27305, "Mission à Berlin de M. Schweisguth, Inspecteur des Finances, du 13 au 26 mai 1914," dated 8 June 1914, and the supplemental letter of the same date. In Berlin Schweisguth interviewed seventeen members of the Reichstag, as well as directors of the major banks and officials of the principal oil companies. According to Coston 1975, p. 516, Pierre Schweisguth was a nephew of the influential Mirabaud family, of the Banque Mirabaud.

Deterding's newly formed Royal Dutch–Shell combine supported the Reichstag monopoly project. In 1912 he claimed to have had foreknowledge of the monopoly project "five or six years before." This places the monopoly's earliest planning somewhat before or during 1907, at about the time of the Royal Dutch–Shell amalgamation and approximately when Muffelmann gathered material for his article, which appeared in early 1908.[48] Deterding wanted his new combine to have a larger share of the European market, but because of his close association with the Rothschilds through Asiatic, he could not expand Royal Dutch–Shell's market share at the expense of Epu, of which the Rothschilds were an important member. An assigned share in a German monopoly would allow him to expand at the expense of Standard Oil instead.

The Deutsche Bank wanted to enlarge the market for Epu and its own Steaua Romana. The proposed state kerosene monopoly would have reduced Standard Oil's share of the market from 72 percent in 1912 to 57.6 percent for 1915–1920, and then to 46.1 for 1920–1925.[49] This aggressive move on Standard Oil's market share was not part of the Epu agreement then in force. It was a means to break the cartel agreement by being "forced" to accept government-assigned market shares, which of course would favor the German firm and its allies. State control of kerosene had no military or strategic consequences and was aimed purely at enhancing the position of the Deutsche Bank and its allies vis-à-vis Standard Oil. To ensure the support of small shop owners, the monopoly protected their participation in the oil trade and ensured their representation on the proposed government company.

Standard Oil reacted vigorously to the threat of the government monopoly. The director of the German subsidiary (the Deutsche Amerikanische Petroleum Gesselleschaft, or DAPG), Heinrich Riedemann, led the defense; his family, according to the Standard Oil history, "stood high in the councils of the Church and of the Imperial State."[50] The French inspector of finances expressed the same opinion less delicately. His sources said Riedemann put a stop to what had been, in 1911, favorable conservative press coverage. The Catholic Center party's votes for the oil monopoly project were reversed because he was "known for his generousness towards the works of the party," which

[48] Muffelman 1908; Gibb and Knowlton 1957, pp. 211–212.

[49] Fin. B27305, "Mission à Berlin de M. Schweisguth, Inspecteur des Finances, du 13 au 26 mai 1914," dated 8 June 1914.

[50] Gibb and Knowlton 1957, p. 205.

had "earned him great sympathy."[51] Riedemann's political counterattack on the monopoly project vexed the Deutsche Bank and Arthur von Gwinner, one of its directors, who offered Riedemann an outright bribe of 250,000 Reichmarks annually for life if he would withdraw as director of Standard Oil's subsidiary.[52]

The monopoly project failed because no one knew whether enough oil could be found for Germany that was not already controlled by Standard Oil. Adding up available tonnages not controlled by Standard, this looked difficult. Two independent American companies, Gulf Refining and Union Petroleum, both of which had dealt with Shell before its amalgamation, had made presentations to the German parliament in Berlin; but the enthusiasm of these "independents" did not assuage genuine doubts about their ability to provide sufficient quantities of oil.[53] To defeat the plan, Standard Oil tried, in the United States and Germany, to buy up surplus supplies of oil.[54]

In addition, Standard Oil threatened to embargo virtually all oil deliveries to Germany if a monopoly attempt were passed. In the Standard Oil account, the German military was so frightened by the embargo scare, which would have included fuel oil, that it put pressure on the government to withdraw the plan. The government did in fact dissolve parliament in May 1914; Gibb and Knowlton as well as Schweisguth describe this as a measure designed to kill the oil monopoly legislation.[55] Schweisguth also believed that amendments to the monopoly plan to the general advantage of small shop owners may have caused the Deutsche Bank to lose interest. The legislation was supposedly destined to be reintroduced, but the ambitious project fell prey to the exigencies of war.

THE REICHSTAG PROJECT'S LINK TO MESOPOTAMIAN OIL

The Deutsche Bank did not need a parliamentary committee to tell it that there were only limited world oil supplies beyond Standard Oil's reach. The Deutsche Bank had an expanded, long-term procurement strategy that consisted of extending the Baghdad railway into Mesopo-

[51] Fin. B27305, "Mission à Berlin de M. Schweisguth, Inspecteur des Finances, du 13 au 26 mai 1914," dated 8 June 1914.

[52] Gibb and Knowlton 1957, p. 214. Allegations of bribery in the German monopoly project are also discussed in Leo 1978, p. 168.

[53] Muffelman 1914, pp. 20–21; Fin. B27305, "Mission à Berlin de M. Schweisguth, Inspecteur des Finances, du 13 au 26 mai 1914," dated 8 June 1914.

[54] Gibb and Knowlton 1957, pp. 209–10.

[55] Gibb and Knowlton 1957, pp. 212, 220; Fin. B27305, "Mission à Berlin de M. Schweisguth, Inspecteur des Finances, du 13 au 26 mai 1914," dated 8 June 1914.

tamia, a concession that included the rights to oil on both sides of the tracks for twenty kilometers. Muffelmann quotes the first rumors of the oil monopoly plan in Germany in early 1908, which indicates that his "news" dated from late 1907. That same year the Royal Dutch–Shell combine, under the guidance of Calouste Gulbenkian, opened an office in Constantinople.[56] The Deutsche Bank was simultaneously trying to develop new oil in Mesopotamia and create a regulated market for it in Germany. The secure, competition-free market would guarantee a stable return for the immense capital investment in Mesopotamia, which in fact was beyond the means of German financial markets. Projected Mesopotamian development was so costly that help from British capital in the City was needed to go forward;[57] British investors would feel more confident about the worth of German oil stocks and bonds if they knew the investments would not face devastating competition with Standard as soon as oil began to flow from Mesopotamia.

The Young Turk nationalist revolution of June 1908 delayed the Mesopotamian oil and the German monopoly projects; with the Young Turks came a recrudescence of British influence in the Ottoman Empire.[58] Abdul Hamid, trying to preserve his power against the Young Turks' new parliamentary regime, attempted a coup d'état in 1909 that failed; Mohammed V became the new sultan. The pro-British atmosphere did not last, since the loss of Tripolitania (Libya) and Thrace in 1911 and 1912 brought the new nationalist government around to more complex power balancing maneuvers that ultimately tipped Turkey toward the German alliance of World War I.[59] In the interim, an alliance between British and German capital looked to be the best way to make progress in the Mesopotamian project, and this justified establishing a major new British-controlled bank, the National Bank of Turkey.[60] The bank combined British and German nationalities in one oil project and thus helped the Turkish government resolve its ambivalence between the two powers, also solving the German capital shortage problem by bringing in British participants.

Calouste Gulbenkian

"British control" was only figurative, however. Although the National Bank of Turkey had British members and backing, it was formed at

[56] Kent 1976, p. 22; Earle 1925, p. 267.
[57] Kent 1976, p. 96.
[58] Gerretson 1957, 3:24, 243–245.
[59] See also Feis 1964, p. 354, who says the switch from Britain and France to Germany was in part a result of the French and British loan policies toward the new regime.
[60] Gulbenkian 1945, p. 7; Gerretson 1953, 3:244.

the initiative of Calouste S. Gulbenkian and his uncle, Nubar Pasha.[61] Gulbenkian, son of an Armenian oil importer, had spent his early career in the service of the Mantascheff Oil Company, a Baku concern with close ties to Asiatic Petroleum. The major participants in this firm were Royal Dutch, Shell, and the Rothschilds. In London Gulbenkian had worked for Frederick S. Lane, chief coordinator of the Rothschild family's oil operations.[62] In Mesopotamia Gulbenkian's father had also been a tax collector for the sultan. The younger Gulbenkian's experiences with Byzantine politics in the Ottoman Empire were his training ground for international politics. In his memoirs Gulbenkian asserts that he was responsible for several of the oil combinations described in this chapter, particularly the original oil concession for Mesopotamia and the amalgamation of Royal Dutch–Shell. Other histories do not credit him with such a preponderant influence in oil affairs; but in his lifetime he negotiated as an equal with the principal companies of the great nations of the world and became one of the world's wealthiest men with his 5 percent royalty on Middle East oil production, never losing influence or wealth in such major political upheavals as the Russian Revolution, the collapse of the Ottoman Empire, and the Nazi occupation of France. We may thus conclude that he did not stumble into his fortune by chance. He was a genius of international politics and finance, an oil factotum whose relative obscurity stems from his never acting on behalf of any nation or corporation, but always negotiating for himself. What solitary glory he enjoyed was buried with him. His fortune was pounced on by the Portuguese government and survives as the Gulbenkian Foundation, headquartered in Lisbon, his last place of residence. This, as well as an Armenian church in London, is perhaps all that survives of him.[63]

Gulbenkian owned 30 percent of the National Bank of Turkey, through which, as an original participant in the Mesopotamian oil concession, he would emerge with a 5 percent holding of the Iraq Petroleum Company in the interwar period.[64] Sir Henry Babington-Smith, a British civil servant, was governor of the bank, and Sir Ernest Cassel was also a member of the board of directors. The capital was privately subscribed by Cassel and by Lords Revelstoke and Farringdon. The National Bank of Turkey's attempt to gain the Mesopotamian oil concession was stymied by Arthur von Gwinner and the Deutsche Bank;

[61] Gerretson 1957, 3:244.

[62] Gulbenkian 1945, p. 5.

[63] See Hewins 1957 for his biography; Gulbenkian's memoirs, dated 16 September 1945, are in the U.S. State Department library, Washington D.C., and are referred to here as Gulbenkian 1945.

[64] Kent 1976, p. 80.

Cassel and Gulbenkian determined to remove the obstacle by bringing the Germans into a united effort to gain the rights to the concession.[65] British capital thus came to be allied with German capital in the extension of the Baghdad railway and its oil concession. Together the National Bank of Turkey and the Deutsche Bank incorporated as a British limited liability company, called African and Eastern Concessions, Ltd., in January 1911.[66] Gulbenkian's 32,000 shares were reserved for "myself and the oil group, the choice of which was left to me."[67] Gulbenkian's pride aside, his only partner capable of developing Mesopotamia was Royal Dutch–Shell. This group was represented through Asiatic Petroleum.[68] Through Asiatic, Frederick Lane, representing the Rothschild interests, and Henry Deterding of the Royal Dutch–Shell combine became members of the new African and Eastern Concessions, Ltd., in September 1912, and the company was rebaptized the Turkish Petroleum Company. Gulbenkian's services on behalf of Royal Dutch–Shell in these negotiations explain his appearance as an oil ambassador in France during World War I, the subject of the next chapter.

There was intense commercial rivalry, especially in Romania, between the Deutsche Bank and the Royal Dutch–Shell combine, as well as personal antagonism between the Deutsche Bank's Arthur von Gwinner and Royal Dutch–Shell's Henry Deterding.[69] The Deutsche Bank would have preferred to exclude Royal Dutch–Shell from Mesopotamia but could not do so because of "contractual obligations."[70] These obligations are not hard to surmise: the Rothschilds were a major participant in Epu, and in 1911 they had sold virtually all their Russian oil interests to Royal Dutch–Shell in exchange for stock. The French bank became one of the largest shareholders in one of the world's largest oil companies.[71] When the Rothschilds sold their oil operations to Royal Dutch–Shell, the combine became a powerful de facto member of Epu. The Deutsche bank had to tolerate Royal Dutch–Shell as a participant in Mesopotamia because Epu was the centerpiece of its European marketing strategy.

[65] Gulbenkian 1945, p. 9. Cassel "enjoyed the confidence of the Turkish Government and the good will of the Deutsche Bank" (Earle 1925, p. 268).

[66] Gulbenkian 1945, p. 9; Ferrier 1982, p. 165.

[67] Gulbenkian 1945, p. 10.

[68] Kent 1976, p. 34. Gulbenkian mentions the participants by individual names, Lane and Deterding. Kent identifies the company as Asiatic. Differences between the oil operations of the Rothschilds and Royal Dutch were disappearing as the two were being merged.

[69] Gulbenkian 1945, p. 11.

[70] Kent 1976, p. 71.

[71] Tolf 1976, pp. 189–190; Déb.Ch., 26 April 1920, p. 1370. Tolf writes that the Rothschilds got 40 percent of the British and Dutch sides of Royal Dutch–Shell, that is, 40 percent of Shell's 40 percent, and again 40 percent of Royal Dutch's 60 percent.

The Deutsche Bank's monopoly project in Germany caused parallel developments in Turkey. News of the monopoly project reached Standard Oil's German offices in the spring of 1911, shortly after the Deutsche Bank and the National Bank of Turkey had registered as African and Eastern Concessions, Ltd. The Royal Dutch–Shell combine joined this company on 25 September 1912, and just over a month later, in November, the Deutsche Bank–inspired monopoly project was introduced in the Reichstag. Progress in Mesopotamia and the Deutsche Bank's political assault on Standard Oil's position in German markets were linked. The monopoly project would have allowed the Deutsche Bank amicably to divide up markets with the growing Royal Dutch combine, at Standard Oil's expense.

This illustrated a basic principle that would drive Standard Oil's interest in Mesopotamia after World War I: even if one has no particular need for an oil field, one does need markets, and to hang on to the market, one must control the oil fields that might flood that market. The German market-sharing legislation would have been the keystone of heightened market power for Epu all over Europe and would have protected new Mesopotamian production from severe price wars. The protected, dependable rate of return for the bond issues for the Baghdad railway was a key selling point in the capital markets of Germany and Britain. The joint participation of British, French, and German capital also provided a guarantee against new railroads and new oil developers in Mesopotamia.[72]

Thanks to Standard Oil and Heinrich Riedemann, the 1912 German monopoly project encountered a series of parliamentary delays. As the legislation languished in Germany, the Deutsche Bank's Ottoman consortium was attacked by Anglo-Persian. In its third year of legal existence, Anglo-Persian sought to protect its precarious status as the first producer of Persian Gulf oil. The growth of the well-financed Royal Dutch–Shell combine, prepared to produce and market oil that might compete in the Asian market against Anglo-Persian's products and those of its parent company, Burmah Oil in India, put the small company on the spot. To counter the growing Royal Dutch–Shell and Epu, it had to come up rapidly with the cash to enter into development it was not prepared for.[73] Obtaining participation in the Mesopotamian concession was ultimately part of the overextension that pushed the company to propose Admiralty participation as a stockholder.

[72] Cf. Buell 1925, p. 455: "No foreigner will build a railroad at his own expense in a backward region without assurance that no other railway will be built in the same territory which his road is designed to serve. In the case of certain 'natural' monopolies, competition becomes disastrous."

[73] Kent 1976, pp. 98–99.

Anglo-Persian had a strategic as well as a commercial card to play. Its growth would protect the interests of Burmah Oil, but also block the expansion of German interests. The Deutsche Bank wanted British capital for the expensive railway and oil development project, which would in geopolitical strategy have the effect of opening up continuous land transport for troops from central Europe to the Persian Gulf. The British knew that the extension of the railroad would bring German influence to their very doorstep in the Gulf, for the Baghdad railway was seen to have a "natural" terminus in the harbor of Kuwait, over which the British had established a protectorate in 1898.[74] The development of the railway was linked to the development of the oil, and Anglo-Persian could hope that, by selling itself as the "chosen instrument" of British imperial policy, the defense of its commercial interests would also be the means by which contending British and German commercial ambitions would reach a negotiated compromise.

But the National Bank of Turkey had already sponsored its own plan for the coexistence of German and British capital in Mesopotamian oil: in 1912 the bank and Sir Ernest Cassel bought into 35 percent of the Turkish Petroleum Company.[75] The terms of coexistence had been worked out, but they did not include Anglo-Persian and, indeed, threatened its viability. The Turkish Petroleum Company neglected not so much the strategic considerations of British policy as the needs of Anglo-Persian, and behind it Burmah Oil, for the Germans had agreed not to extend the railway beyond Basra without British consent.[76] The British oil companies did not want any Persian Gulf oil development that was not theirs. Oil piped out of Mesopotamia into the Persian Gulf, with or without a railroad, was just as much of a threat to these interests as Witte's schemes to export Transcaucasian oil through the same waters.

The British members of the National Bank of Turkey were convinced by the British Foreign Office to sell their shares to Anglo-Persian, and Gulbenkian came under pressure to do the same. But he took the matter to Henry Deterding, his chief oil partner in Royal Dutch–Shell, who "became wild" and "got into a state of frenzy" at the thought of losing one of the great undeveloped oil fields of the world just when it seemed within his grasp. Deterding told Gulbenkian to "sit tight, do nothing, and that he himself would take the challenge against Sir Charles Greenway and his friends' [Anglo-Persian and Burmah] machinations and intrigues."[77] Deterding promised to take up the mat-

[74] Martin 1959, pp. 81–83.
[75] Gulbenkian 1945, p. 10.
[76] Feis 1964, p. 357.
[77] Gulbenkian 1945, p. 13.

ter directly with the Dutch government, through which he pressured the British; moreover, the "Shell" part of his company *was* British, and this part of his political influence could also be brought to bear.[78] Consulting daily with Gulbenkian, Deterding carried out the negotiations.[79]

In the meantime Anglo-Persian and the Foreign Office had stopped the progress of the Turkish Petroleum Company's concession. The Turkish government, under the rule of an international treaty informally referred to as the "capitulations," was unable to raise tariffs on imported goods, and thus to deal with its chronic insolvency, without the agreement of the treaty signatories. Britain refused to allow tariff increases unless the Turkish Petroleum Company was denied its concession. Gulbenkian, as usual well placed to monitor his interests, was senior financial advisor to the Turkish government and passed information to the Royal Dutch–Shell combine; the Deutsche Bank looked for a compromise among the competing interests.[80]

To break the impasse, Gulbenkian offered to reduce his holdings in the Turkish Petroleum Company; the British interests that had originally been represented in the National Bank of Turkey withdrew and were replaced by the Anglo-Persian Oil Company, which ended up with 50 percent of the concession, in which the Deutsche Bank and Royal Dutch–Shell were the other two partners. Gulbenkian's share was reduced to a minority participation in Royal Dutch–Shell and Anglo-Persian, a move to reduce his influence that he would later successfully combat. The result was that on 19 March 1914 an agreement for the development of the Mesopotamian concession was reached and signed in the Foreign Office.[81] The signatories were the representatives of Anglo-Persian, Royal Dutch–Shell, the Deutsche Bank, and the German and British governments. The Burmah Oil Company, as a guarantor of Anglo-Persian but a nonsignatory of the Foreign Office agreement, was brought into the deal in such a way that "the Burmah Oil Company would have all the profit and the Anglo-Persian all the risk."[82]

Ferrier remarks that by "curious coincidence" the settlement on the Mesopotamian concession was reached at about the same time that the details for Britain's government stockholding in the Anglo-Persian were also reached, which along with the Burmah guarantee were essen-

[78] Deterding himself would become a naturalized British citizen in 1915 (Earle 1925, p. 273).

[79] Gulbenkian 1945, p. 14.

[80] Gulbenkian 1945, p. 13.

[81] Ferrier's date (1982, p. 197).Gulbenkian 1945, p. 14, gives the date as 24 March 1914.

[82] Kent 1976, p. 100; see pp. 24–100 for one of the best diplomatic accounts.

tial if the overextended Anglo-Persian was to meet its new commitments to oil development.[83] There is nothing curious about the coincidence. The Admiralty approved participation on 11 May 1914; the cabinet approved it on 13 May; and it was approved in Parliament on 17 June. In Germany the Deutsche Bank's state monopoly project had been excessively modified in the Reichstag by pressure from small shop owners, and to resurrect it in a form closer to the original intent of the Deutsche Bank the entire Reichstag was suspended on 20 May 1914.[84] The German plan was to be revised and resubmitted. The close concurrence of dates shows how two forms of government market regulation—state participation in Anglo-Persian and monopoly regulation of the German market—were linked to oil in Mesopotamia, even when on other matters these governments were drifting toward war.[85]

Looked at from a global perspective, the spring 1914 Admiralty participation in Anglo-Persian, the German government's withdrawal of one monopoly plan in preparation for another, and the accord on Mesopotamia appear no more coincidental than that the musicians of an orchestra should conclude a symphony at the same moment. The major oil interests sought to resolve a central problem with several interconnecting themes. The problem was the development of Mesopotamia; the themes were dividing up markets, finding capital, and guaranteeing an adequate return on a difficult development project. The Deutsche Bank and Royal Dutch–Shell, their interests concerted by participation in Epu, alike supported a monopoly project in Germany that would ultimately allow for the expansion of their interests at the expense of the Standard Oil Company. The history of the monopoly project in Germany closely parallels related developments in Mesopotamia, the vast oil field that would allow the conquest of the European as well as Asian markets. Similarly, Anglo-Persian and Burmah Oil, faced also with competition from Standard Oil but also directly concerned about the growing vigor of the Royal Dutch–Shell combine, had to find the means to get a rapid infusion of capital and the freedom

[83] Ferrier 1982, p. 197.

[84] Fin. B27305, "Mission à Berlin de M. Schweisguth, Inspecteur des Finances, du 13 au 26 mai 1914," dated 8 June 1914.

[85] On the drift toward war see Montgelas 1925, pp. 41, 53, 85, 92–95, 118–119. Germany's willingness to compromise with British interests, both financial and imperial, does not support the notion that the Baghdad railway and oil in Mesopotamia "caused" the war, though the oil certainly became a primary objective once war broke out. As Fischer observes (1967, p. 41): "The agreements on the railway, the ports on the Persian Gulf, and the navigation rights of Mesopotamia show how strongly Germany was endeavouring to reach an understanding—and ultimately an "alliance"—with Britain, even at the cost of sacrifice. In these fields Germany was unmistakably willing to play the junior partner to Britain as a world power."

from competition that a guaranteed customer would provide: they found both in the Admiralty. The German oil monopoly was never reintroduced because the costly project could not be undertaken during World War I. As Lord Curzon observed afterward, "Had no war supervened, and had Mesopotamia remained until now under Turkish rule, the exploitation of these oil deposits would long since have begun."[86] Germany's defeat and the breakup of the Ottoman Empire rendered the question of a German domestic monopoly moot: the oil for it was gone. German strategy switched to synthetic oil from coal, which would be a factor in the great accords of 1928.

THE 1914 FRENCH MONOPOLY PROPOSAL'S LINK TO MESOPOTAMIA

In 1902 French alcohol producers had introduced a plan to create a state monopoly in the lamp fuel market.[87] That attempt failed, and only half the alcohol lobby's plan was put into action: protective legislation for French refining was killed, but no monopoly ensued. Internationalization of French oil capital followed, creating the conditions for a new attempt to regulate the market. The 1914 legislation was in most respects similar to the legislation introduced in Germany.[88] It proposed a *régie interéssée:* one company would hold a monopoly on oil sales in France. The state would own 20 percent of the company, and private oil firms would own the remaining 80 percent. Their profits would be proportional to their shares. Those choosing not to join the *régie* would be expropriated with compensation. The idea was to create a state-sanctioned division of the market along proportions roughly similar to those enjoyed by the members of the dominant cartel. It brought a vigorous protest from the growing sector of small importers: "The majority of our members are direct importers of oil and gasoline from producing countries," complained the Féderation des Syndicats et Groupes de l'Epicerie en Gros, "and if from one day to the next someone takes away from these merchants the right to import oil, the fruits of their labor will be quickly wiped out."[89]

Remembering that by 1914 the Rothschilds had become part and parcel of Royal Dutch–Shell, which was actively part of the British-Turkish-German plans for Mesopotamia, the introduction of the French legislation in January 1914 is hardly surprising. Under this

[86] Earle 1925, p. 271.
[87] See above, pp. 47–48.
[88] Doc.Ch., annexe 3382, 15 January 1914, pp. 18–19.
[89] Fin. B34034, "Féderation des Syndicats et Groupes de l'Epicerie en Gros–Projet de Loi de M. de Monzie sur les huiles et essences minérales."

legislation France too would become a secure, regulated market for Mesopotamian oil, protecting costly investments against a future price war with Standard Oil imports. The French project also resolved the conflict between Royal Dutch–Shell's Henry Deterding and the Deutsche Bank's Arthur von Gwinner over how to accommodate, simultaneously, the plans for European market expansion that Mesopotamian oil would open up for both the German bank and Deterding's company: the Deutsche Bank would dominate Germany, and through the Rothschilds, Royal Dutch–Shell would have greater preponderance in France, where in any case a major German marketing effort would not be well received in the chauvinist decades after the Franco-Prussian War.[90]

The French monopoly would have set the state up to buy oil produced in Romania and the Transcaucasus by the "independent" French houses of the Cartel des Dix, and also the Rothschild group. The French monopoly would import from them first, getting whatever else was needed from Standard Oil. As the only oil company without ties to a major French domestic banking group, Standard would lose the lion's share of the market when the Mesopotamian oil started to flow. The French 1914 project preceded by five months the comparable British project and German Reichstag revision. In all three countries, major regulatory initiatives sprang from the development of Mesopotamian oil.

Even though after 1911 the French Rothschilds were major stockholders in Royal Dutch–Shell, and even though the French ministry of foreign affairs was deeply involved in all aspects of negotiations over loans to Turkey and railroad development, which of necessity meant some knowledge of the oil concession, we see virtually no formal French "oil policy" in the period before World War I.[91] A French finance ministry official wrote in February 1914 that it was "too late" to envision gaining access to any significant oil deposits that were not

[90] Cf. Feis 1964, p. 200: the French Banque de Paris et des Pays Bas and Banque Imperiale Ottomane were favorable to rapprochement with German capital and to the introduction of German securities on the Paris exchange because they "wished to share in the German development in Turkey." Importantly, their means of "sharing" was in the financing of the operation, not as stockholders like the Rothschilds; the Banque de Paris et des Pays Bas would emerge after the war in the opposite camp of Royal Dutch–Shell. The interest of the Banque de Paris et des Pays Bas in financing was part of the general rapprochement of some German and French capital groups following a 1913 agreement on the division of Syria and Turkey into spheres of influence for railroad development (Feis 1964, p. 357); the Banque de Paris et des Pays Bas' rival, the Rothschild group, got "its" share by participating through Royal Dutch–Shell.

[91] On the French ministry of foreign affairs, see Feis 1964, pp. 355–357.

already in the hands of the oil multinationals, because the "monopolization had gone too far."[92]

Given the oil industry's growth in areas such as Mexico, Romania, Galicia, and Texas, and given Britain's purchase of shares in Anglo-Persian less than five months after that sentence was penned, this official must have been seriously misinformed. The navies of Britain and Germany were moving to oil: Why did France do nothing? Had the French government been pushed as hard as the British government by Anglo-Persian and Burmah Oil, or the German government by the Deutsche Bank, such misinformation would have been impossible. The largest French oil producer, the Rothschild bank, and the national company that marketed its products, Deutsch de la Meurthe et Ses Fils, were both closely tied to the Royal Dutch–Shell group. They did not need to pursue their oil interests through the French government. Royal Dutch–Shell's European marketing was handled through Epu by the Deutsche Bank; France's "national policy" was to allow the Mesopotamian deal to go through and let the national monopolies divide up the profits. The Rothschilds owed their entry into Mesopotamia before World War I to the actions of the Royal Dutch–Shell combine and the German Deutsche Bank, not to the French government. In spite of the rising international tension, the German Deutsche Bank and Holland's Royal Dutch–Shell combine were the principal spokesmen of French oil policy.

In other oil-related strategic concerns, the French government was at best lackadaisical. The navy had built small oil tanks for storage at Cherbourg in 1909, to supply a few submarines and torpedo boats. An oil tanker was purchased in 1910 and for two years was used as a floating storage tank at a naval base. Two more tankers were bought in 1912 and 1913; the navy wished to import its own oil and had in mind a reserve stock of oil that would last for at least nine months of normal usage. Because of funding considerations these reserve stocks were cut to three months.[93]

The French navy was interested in modernizing but did not pursue it as vigorously as happened under Admiral Fisher in Britain. British oil supply policy lurched forward, more or less in parallel to fleet development, not as a result of the same kind of strategic planning that went into the fleet, but in response to the exigencies of the evolution of the commercial market. Because the Rothschild interests were locked to the fate of Royal Dutch–Shell, the French lacked even this. Nor was there any French counterpart, governmental or otherwise, to the vigor-

[92] Fin. 34034, 9 February 1914, "Proposition de loi no. 3382, etc.," pp. 8–9.
[93] Nayberg 1983, pp. 6–10, 18.

ous pursuit of oil development by the German Deutsche Bank, whose commercial policy, at least in the Ottoman Empire, in most regards could not be differentiated from imperial policy. An assessment of the French 1914 monopoly project in the ministry of industry did not even mention strategic considerations but worried that "the refiners, as masters of the state company, will procure hidden advantages, such as in the purchase of raw materials, thereby reducing their apparent profits and thus what they will have to share with the state."[94]

France's oil lethargy might be attributed to its traditional status as a land-oriented power, though this does not square well with the world reach of the French empire. France had passed up the chance to be one of the earliest oil-fueled seafaring nations: as early as 1869 the navy had experimented with oil fuel for a boiler-powered river gunship.[95] According to France's Admiral Dégouy, objections about oil's safety and availability were "vigorously exploited by the coal people—owners of and stockholders in the mines, as well as the mines' directors, engineers, and miners." This opposition "paralyzed" the development of French naval oil policy. Dégouy explains France's lack of an oil policy thus: "Fifty years of stagnation, or at best extremely slow progress, progress continually fought by the representatives of particular interests, people in most other respects quite respectable: the owners of mines, the builders of mining equipment, the manufacturers of boilers for steam powered vessels, and so on."[96]

Dégouy's explanation raises the question of why the coal industries in Britain or Germany did not mount a similar resistance. The short answer is that they did. Coal-allied interests headed by I. G. Farben helped drive Germany toward its successful synthetic oil policy, which will be discussed in chapter 5. Ferrier writes that in Britain "professional engineers at the Admiralty" in 1906 were "advising the head . . . not to commit themselves too deeply to Fuel Oil, as in their opinion the next few years will probably see a development in internal combustion engines worked by coal gas."[97] Marcus Samuel, founder of Shell, was a trader in coal before he got into the oil trade, and his biographer gives good but brief accounts of the coal-oil controversy.[98]

The introduction of France's monopoly bill fulfilled the prediction made by Leo Muffelmann in 1908, that state monopolies would be the trend in the world oil market. His forecast was based on world oil

[94] Min.Ind. 12892, "Note sur le monopole des pétroles," by Georges Bexton, March 1914.

[95] Dégouy 1920, p. 661. Also cited by Leo 1978, p. 159.

[96] Dégouy 1920, pp. 662, 685.

[97] Ferrier 1982, p. 82.

[98] Henriques 1960, pp. 96, 277–280, 610–611.

competition and said nothing about strategic considerations. As the case studies of Germany and Britain show, he was right. He was also right concerning France, for in France as in Germany the problem was how to compete against Standard Oil's superior resources. The merging of the Rothschild interests and the Royal Dutch–Shell combine had changed the circumstances prevailing at the time of the last tariff revision in 1903. In 1903 the Rothschilds linked up with French national refiners who marketed the Rothschild's Bnito oil. After 1911 the Royal Dutch–Shell combine wanted to enlarge that share of market at the expense of Standard Oil. This would make room for the expanded production of Royal Dutch–Shell, which in 1903 had been two independent companies; in 1911 the production capacity of the combine included all the previous Rothschild production plus what the Shell and Royal Dutch had each brought to their now united operations.

In France as in Germany, Mesopotamian oil seemed on the horizon. The larger firms also wanted to contain the small producers and importers of Romanian oil that had mushroomed in the post-1903 market. Lowering prices to compete against these firms, and against Standard Oil, would have scared off bondholders from investing in the Mesopotamian dreams of Royal Dutch–Shell, the Deutsche Bank, and the Rothschilds. Legal allocation of market shares would guarantee access to capital and a market for the product. The time for new regulation seemed propitious too, because Standard Oil was being reorganized under the Supreme Court's antitrust ruling of 1911: "The dissolution decree had been widely publicized in Europe, and much of the public, at least, believed that the Standard Oil empire was now crashing in ruins."[99]

TRANSNATIONAL STRUCTURING BEFORE WORLD WAR I

Transnational structuring does not invariably prevail over the logic of state power and international competition; it is an element that exists in addition to that dynamic. I may say flatly, with no fear of contradicting the general tenets of transnational structuring, that much of what might have been in the oil market as a result of the agreements of 1914 was swept away by the "realist logic" of World War I. But the war originated in other concerns: *all the private interests and relevant governments of Europe had assented to the 1914 proposals for the Turkish Petroleum Company.* The agreements were impossible to execute with the outbreak

[99] Gibb and Knowlton 1957, p. 204.

of war, but significant fragments, or pieces of the effort, remained and became the underlying premises of what was to occur in later decades.

We can best see the point by imagining what the "strong version" of transnational structuring might have been in the circumstances described in this chapter. In the strong version we would have had simultaneous passage of domestic oil monopoly laws in 1914 in Germany and France and simultaneous approval of the British government for equity participation in Anglo-Persian, even though at the time it appeared to be a dubious commercial and military prospect, as a means to fund participation in what was not a dubious commercial prospect, the oil of Mesopotamia. We would have also had the simultaneous passage of concession plans in Turkey for the development of Mesopotamian oil.

These groups were a part of even larger world agreements. Through the Nobels, who were linked to Epu and major producers in Russia, and the Royal Dutch–Shell interests, Russian output would have been tied to the vast operations of this transnational structure for the control of oil, this pan-European and anti–Standard Oil cartel. The passage of cartelistic monopoly legislation in France and Germany would also have affected the two largest markets for Romanian exports. These producers too would have found themselves enmeshed in the same logic of controlled markets. The oil revenues of the producing regions in Romania, Galicia, and Russia; the prices and economic activity in the oil industries of Britain, France, and Germany; the rate of development of these industries and thus their ability to serve "the national interest"—all of this would have been enmeshed in the same structure, seamless in design and intent, from the oil fields of Mesopotamia to the great capitals of Europe's major powers.

The markets of Asia would also have been included: Burmah Oil was tied, through Anglo-Persian, into Mesopotamia; and Royal Dutch–Shell controlled the production of the Dutch East Indies. In scope and vision, the ambitions of 1914 were a test run for the accords of 1928. A long road had to be traveled before that would come true. The Great War and the Russian Revolution eliminated the Nobels and brought turmoil to major world producing areas. The ambitious scheme would require laborious reworking before it could be tried again, and Standard Oil was aware that Mesopotamia was the means by which its opponents hoped to chase it from Europe.

So much for "what might have been." What was accomplished, and what changed? To begin, let us observe that even though a tradition of "etatist intervention" in markets may have prevailed in France and Germany, it certainly did not prevail in Britain; and the Admiralty stake in Anglo-Persian demonstrates a point I shall make repeatedly:

local traditions and customs count for very little in this process. The weak Third World country of Turkey, the French, the Germans, the British—all were integrated into the same transnational logic. We see no reference to the strategic interests of the state in the French case, and in the German case the military opposed the Deutsche Bank's projects, timid about how these grand transnational schemes might pan out after a threatened embargo from Standard Oil. In the British case we see some attempts to mollify security concerns about the potential threat posed by the Baghdad railway; but as Ferrier observes, the decision to back Anglo-Persian is hard to justify in light of the poor-quality oil and poor prospects of that company. Moreover, the far-seeing empire had managed to pass up the chance to own a first-class oil company in declining to buy Shell, demonstrating that the oil interests of the state were nonstrategic.

Were we to follow realist-mercantilist logic to explain the growth of these oil policies, we would have to say that each country, observing the development of oil policies that posed a potential strategic threat by enhancing the strength of rival powers, initiated its own oil policies. We still have the problem of the form, timing, and content of each country's oil efforts; and the fact is that even were we to say that some kind of "oil policy race" had miraculously broken out in 1914, it stretches credulity to say that these tightly interwoven collaborative efforts, needed to manage a world-scale marketing and production project, somehow drew their essence from the dynamics of power competition, when in fact the commercial incentives are plain to see.

The politics of transnational structuring before 1914 left four durable elements in world oil politics:

The creation of the Anglo-Persian Oil Company. The initial goal was to block the Baku producers' drive to the Persian Gulf. Heavily backed by Burmah Oil, which was Britain's number one purely British oil firm (Royal Dutch–Shell was not), this company passed development and risk capital costs on to the Admiralty. Burmah Oil reaped profits at no risk whatever.

The Mesopotamian Concession agreement, known as the Turkish Petroleum Company. The piece of paper giving title to the oil in what became Iraq became the basis for negotiations in the post–World War I period. It had an enduring effect on the world oil market and even contained the first version of what would later be known as the Red Line agreement. The sharing of oil resources by the major powers in the Middle East, the internal development of Iraq, and even Franco-Iraq relations in the latter half of the twentieth century were affected by this document.

The creation of Royal Dutch–Shell. Royal Dutch–Shell was not created by international interests who lobbied various governments. The politics and economics of doing international business in many countries forced a particular structure upon Royal Dutch–Shell, making Britain face the conundrum of a "British" company that worked suspiciously closely with German interests, that is, the Deutsche Bank in Mesopotamia, but that also had high French and British equity.

The absence of a French oil policy. France began World War I with no oil policy or objectives. This was due to the nature of the capital groups that dominated the French oil industry before the war. France did come up with a highly "interventionist" policy in 1914, but it had no strategic content. The development of a coherent, stable French policy would be delayed until 1928.

There were also what we might call the "lessons learned," which would become part of the underlying assumptions that governed the actions of the major international oil companies in the coming decades. The first was that the introduction of Mesopotamian oil into world markets would have major effects that required an agreement to control production at one end and to apportion consumption at the other. That was what we see general agreement on in 1914, and what we see becoming the operating principle for the oil producing and marketing part of the world hydrocarbon cartel in 1928.

Standard Oil also learned a lesson from fighting the monopoly project in Germany: it is not enough to have oil to sell, since Standard Oil had plenty. Without a hand "on the spigot" of new oil sources, the American colossus would be run out of entire national markets by rivals wielding specially fashioned regulatory policies. Since Germany was the largest market outside the United States and France was the third after Britain, the message was clear. The company's strong postwar drive to acquire overseas oil production derived from this, not from the self-serving myth of declining American production.

The Great War rearranged the international chessboard of states on which the oil interests sought to control world production and marketing. The international oil companies struggled mightily to adapt to those changes. While ships were lost and men slaughtered, the companies fought their own war for the control of oil resources. It is to that commercial war, waged through rather than by nation-states, that we now turn.

Chapter Three

The Great War and the Struggle for Commercial Advantage, 1914–1921

The state should most vigorously promote the "national interest" in time of war. During the First World War German armies came within a dozen miles of Paris: modern French oil policy is often dated to the alleged oil crisis of late 1917, when scarce supplies forced the French government to take an interest in this strategic commodity.[1] The realist-mercantilist interpretation of state interventionism focuses on increased petroleum demand for the new gasoline-powered weapons of war: tanks, airplanes, trucks for transporting troops and supplies, and ships. It focuses on the effects of the German submarine campaign on shipping. These factors gave France an overwhelming justification for organizing and overseeing petroleum deliveries.

This summary of events is erroneous. This "world war" version of the genesis of French oil regulation, which combines German submarines with the malfeasance of Standard Oil, shows how a realist-mercantilist approach can be led astray by its own bias to accept state rhetoric. A telegram from Clemenceau to Wilson, reproduced below, is the fundamental document in this tradition. Though a source document, it obscures the essential dynamics of the world oil market during the Great War. Yet the possibility for a renewed realist-mercantilist interpretation is discernible. That is because the *essential* "realist" events stand out from an analysis of commercial interests and realities: in oil the chief strategic fact of the war was not submarines but control of the Dardanelles. Without the operating assumption that state rhetoric veils the pursuit of private gain, we would accept the *official* "realist" version of events. Transnational structuring focuses our attention on

[1] Cf. Nayberg 1983, pp. 130–260; Murat 1969, p. 18; Thomas 1934, p. 41; Touret 1968, p. 23, to name but four. Most authors consider the strategic rationale so obvious as not to need thorough examination.

the worldwide corporate struggle for the control of the means to produce, transport, and distribute oil and on how companies sought to preserve or enhance their positions by supporting or opposing government policies. It assumes that whatever is readily available—such as Clemenceau's telegram—may be misleading.[2] Archival sources reveal a different layer of meaningful realist "fact." At this deeper level transnational structuring and realism, while still having radically different emphases, show a close interdependence: rent seekers must structure their goals in light of the realist limits of their circumstances, while a realist perspective is inaccurate unless it follows the goals of the companies that, after all, produce, transport, and distribute the strategic commodity.

The breakup of Epu, the closing of the Dardanelles, and the military competition for control of Mesopotamia and the Caucasus are distinctly "realist" in flavor. Organized warfare is the province of states and a defining characteristic of the realist theory of international relations. The "realist effects" of war will be explicitly discussed in this chapter, but the general point is the intertwined nature of the geopolitical (realist) military struggle for power and the attempts of international private interests to harness government policies for their own advantage.

The First World War's victors and losers—the principal territories lost and gained—constitute the history of states and can be found in any encyclopedia. From the point of view of transnational structuring, World War I is considerably more of a challenge. Who won, who lost, and how is not "obvious." The information is in no encyclopedia, for two reasons. First, encyclopedias assume the preeminence of states in international conflict and have not recognized international business interests as major players. Second, the way international business interests react to wartime competition is a function of both economic pressures, which can be understood at a theoretical level, and industry-specific circumstances, entailing knowledge of the specific firms. Neither a theory of the constraints on business competition in a wartime economy nor the specific regulatory measures because of them are as obvious as the outcome of a military battle. Transnational structuring gives a view of a different kind of world war, whose main outlines nonetheless depended on the course of the military conflict.

Not one but two French oil crises occurred, one during and one shortly after the war: the first, generally considered the more serious, was in fact of little significance; the second crisis, usually ignored,

[2] Clemenceau's telegram was published, for example, by L'Espagnol de la Trameryе 1921, pp. 104–105; Touret 1968, p. 24; and Bérenger 1920. Henry Bérenger probably is the telegram's real author.

caused a greater threat to the French oil supply than had ever been posed by the kaiser's submarine fleet. Both crises were key events in the international politics of oil. In each crisis the French government sought to help Royal Dutch–Shell at the expense of Standard Oil. The first crisis was a ploy to divert Standard Oil's fleet from Asian waters. The second resulted from the blundering attempts of the French government to evict Standard Oil from the French national market, a plan whose strategic significance, if it had any, lay in the utter foolishness of antagonizing France's principal supplier.

The specific regulatory moves of 1914–1921 originated in the attempts of international groups to use the state. Their goals were determined by changes in the oil market, and they influenced policy. The private sector's desires weighed more heavily than strategic requirements, even during this period of mortal combat. Yet many changes in commercial strategy were dictated by developments on the battlefield. French oil policy emerges as an instrument of companies that, in a world of warring states, did their best to survive according to the criteria important to them: profits, access to new and old oil production, product transport, and market share. The struggle for commercial advantage was orthogonal to the world military conflict, yet it shared crucial points of intersection with it.

The need for the state to regulate the wartime economy may seem obvious by realist-mercantilist standards. But war also increases the incentives for, and means to accomplish, the private sector's use of the state. War increases both realist-mercantilist dynamics and business instrumentality. It is not true that realism-mercantilism describes economic regulation in war while transnational structuring describes economic regulation in peace. The wartime transnational structuring described in this chapter follows logically from the nature of a wartime economy: France is only a case in point. The worldwide military and commercial struggles extended to the remotest corners of the world. To clarify the main outlines of the historical process, I have provided a map of the world showing oil movements along with a numbered commentary (see fig. 2 below).

WARTIME PRODUCTION, SCARCITY, AND INFLATION

The business of war is also the business of peace: the principle is epitomized by the Roman Vegetius's famous dictum *Qui desiderat pacem praeparet bellum* (Who desires peace should be prepared for war). Procuring strategic materials in peacetime is part of the state's preparation for conflict, which in war becomes a matter of survival. In war the

Figure 2. The struggle for world oil, 1914–1922 (see map, overleaf)

Britain blockades the North Sea (1); Turkey and Germany close off the Dardanelles (2). Standard Oil commits its Baltic and German fleets to Asian markets (3). Royal Dutch–Shell is cut off from its main sources of oil in Romania and the Caucasus (4, 5); to supply the European market the combine must stretch supply lines to Dutch East Indies (6). In Britain (7), Royal Dutch–Shell's efforts to create rapprochement with the Admiralty meet opposition of Anglo-Persian, which seeks control of Mesopotamia (8). Disruption of Asian markets by the increased presence of the Standard Oil companies, especially the market of Anglo-Persian's parent company, Burmah Oil (9), forces a partial truce between Anglo-Persian and the Shell combine. The Admiralty confiscates Socony's Hong Kong fleet and puts it into the service of the Shell combine's Dutch East Indies trade route to Europe (6), but also awards rival Anglo-Persian control of the Shell combine's confiscated marketing company, British Petroleum (7).

Royal Dutch–Shell is doubly hard pressed because war conditions and the Russian Revolution cause the loss of markets, including Russia and the European Petroleum Union (10, 11), as well as supplies. The company increases its search for new supplies, including in Venezuela (12) and in Mexico (13), where revolutionary factions receive money from competing oil groups. The Royal Dutch–Shell offensive includes France (14) where the government is convinced by the company to request the German share of Mesopotamia (8) as war booty, to preserve the Shell group's share in a concession that Anglo-Persian had sought to invalidate.

French policy evolves into an alliance with Royal Dutch–Shell as its designated company to receive oil booty won during the war. The combine also benefits by receiving favorable allocation of shipping and markets: state-enforced market sharing in the form of an oil import monopoly is projected to last well beyond the war, promising to ease competitive pressures on the Shell combine due to its stretched supply lines and loss of oilfields. Backed by the Shell combine, France negotiates terms with Britain after the war for an alliance covering the Caucasus, Romania, and Mesopotamia, as well as Algeria (5, 4, 8). In France, Standard Oil counters the French government's alliance with the Shell combine by forging its own alliance with the Banque de Paris et des Pays Bas, suspending oil deliveries until it is exempted from the oil import monopoly (14).

Standard Oil extends its counteroffensive, challenging Anglo-Persian and Royal Dutch–Shell in other parts of the globe. In the United States (15) legislation is introduced, aimed primarily at the Shell combine, to retaliate against policies enacted abroad against Standard Oil.

Standard Oil applies diplomatic pressure in Holland (16). In Persia (17), Standard Oil bribes the Majlis for North Persian concessions, thus pressuring Anglo-Persian to come to terms so that the British firm can maintain its supremacy in Persia. By the early 1920s Standard Oil has forced France to deregulate its oil market and cease favoring the Shell combine; it has gained access to oil concessions in the Dutch East Indies (6), and it has forced Britain and France to agree in principle to American participation in Mesopotamia (8). The British abandon the Caucasus to the Bolsheviks (5), favoring the development of rival Persian oil (17), while French efforts to gain control of suspected oil deposits in Armenia (18) have fallen prey to Turkey's nationalist revival under Mustafa Kemal.

Source: Gerretson 1957, vol. 3, foldout facing p. 44, and this chapter.

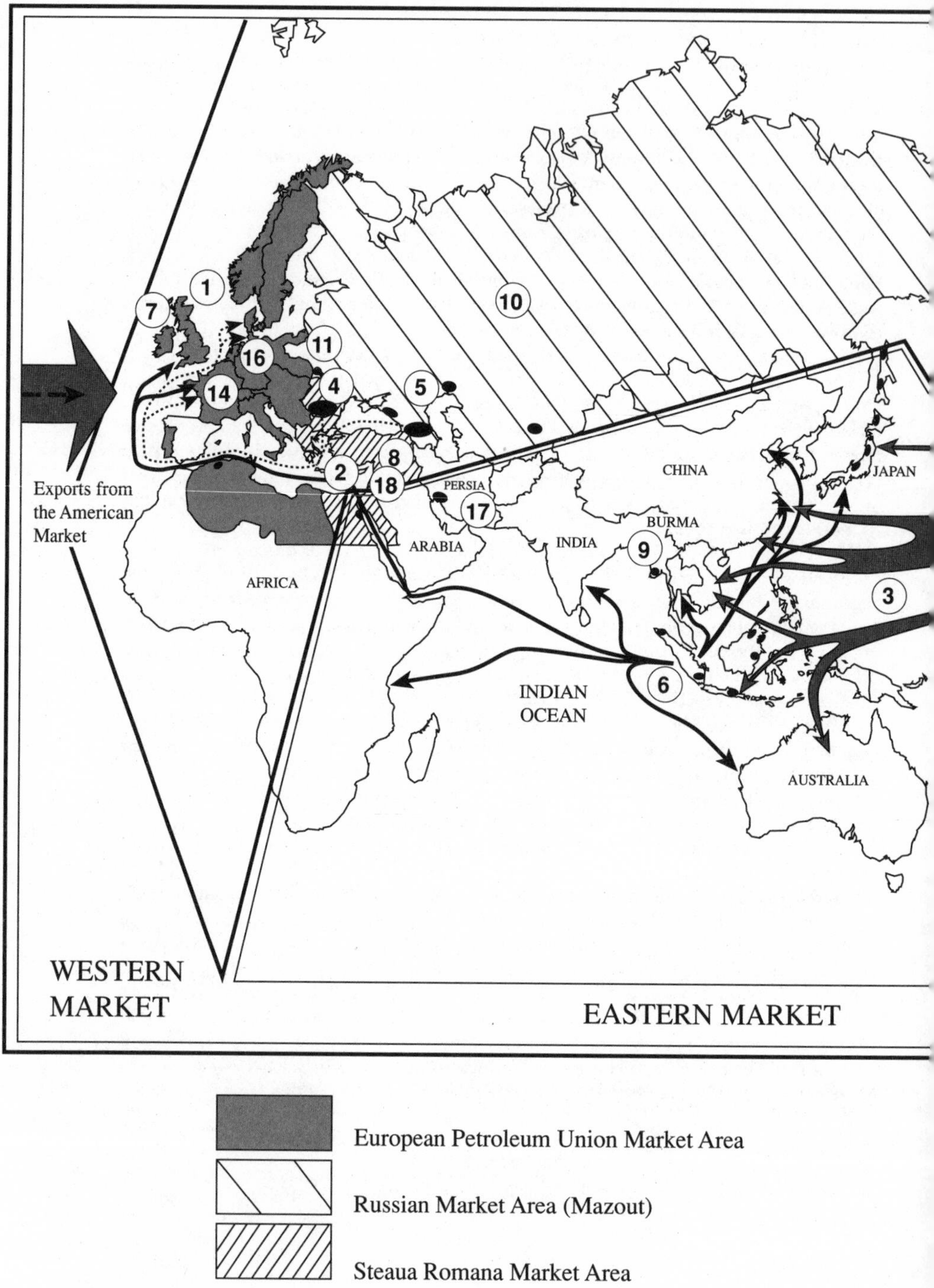

1
7
10
11
16
4
5
14
8
2
18
Exports from the American Market
PERSIA
17
CHINA
JAPAN
BURMA
ARABIA
INDIA
9
AFRICA
3
6
INDIAN OCEAN
AUSTRALIA

WESTERN MARKET

EASTERN MARKET

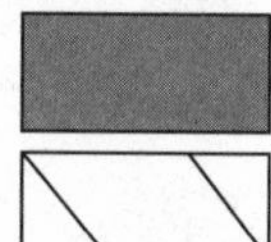

European Petroleum Union Market Area

Russian Market Area (Mazout)

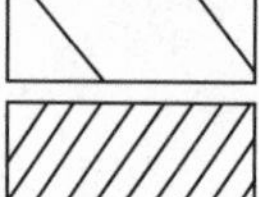

Steaua Romana Market Area

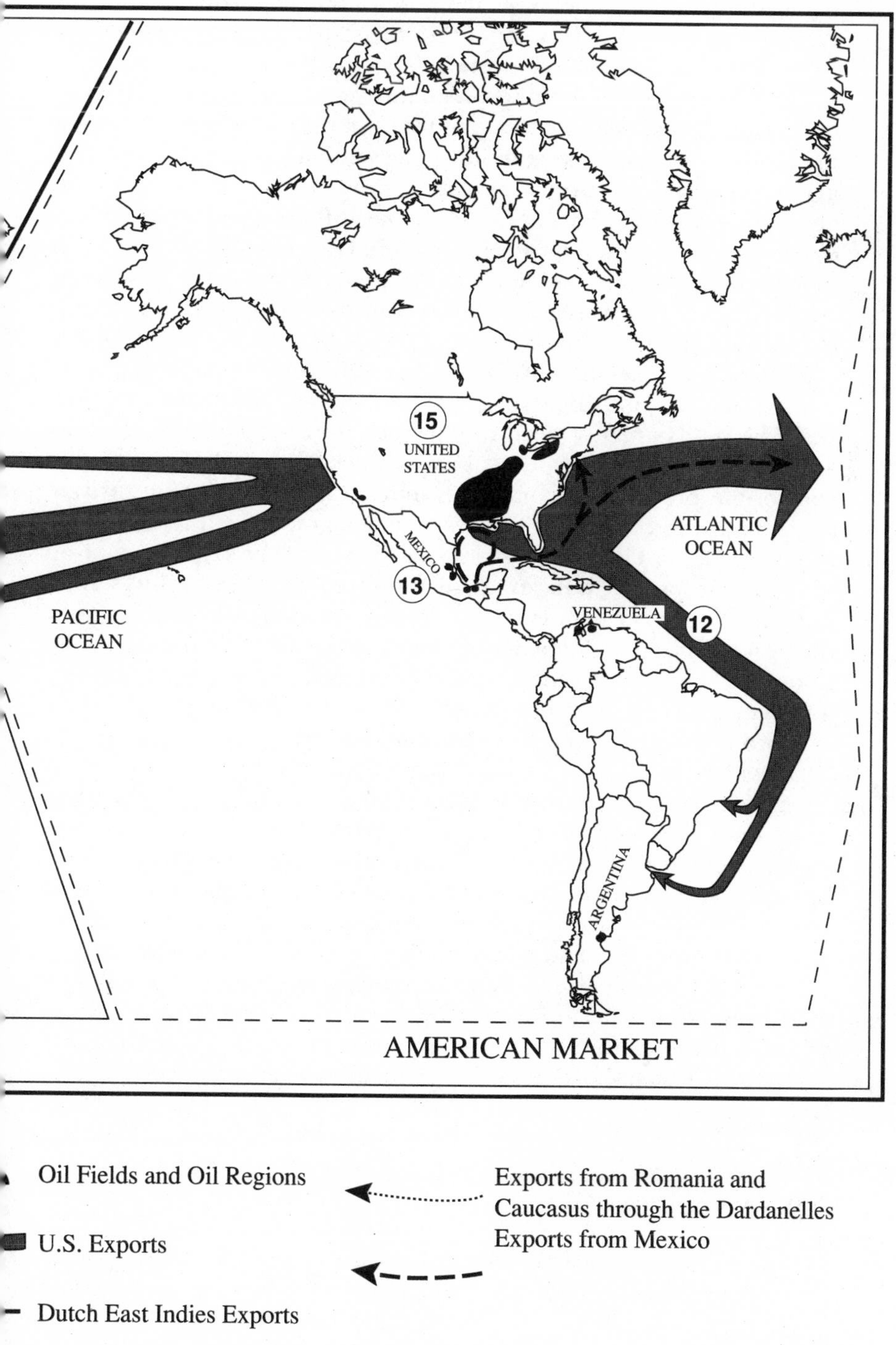

15
UNITED STATES
MEXICO
13
VENEZUELA
12
ATLANTIC OCEAN
PACIFIC OCEAN
ARGENTINA

AMERICAN MARKET

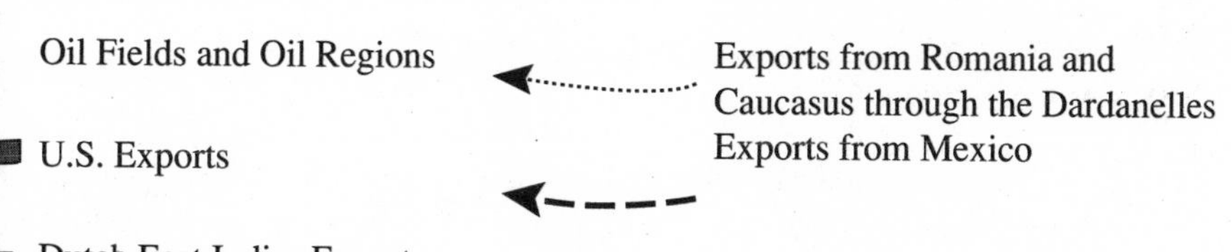

Source: Gerretson 1957, Vol.3, foldout facing p. 44, and this chapter.

propaganda of the state proclaims an unambiguous message about the national interest and mobilizes the population. The battle deaths, the genuine dangers of defeat, legitimize the state's claim to regulate the economy in the "national interest."

The government of a market economy never appears so directly in control of its economic interest groups as in time of war. In the era of mechanized warfare, mercantilist intervention to produce needed goods has been a feature of even such traditionally noninterventionist governments as the United States. The need for intervention seems so obvious as scarcely to require explanation. France required the utmost mobilization of industrial and military power to avoid the capture of Paris. Every economic regulatory action of a wartime government proclaims the need for self-sacrifice.

At the same time, the normal conditions of commercial struggle intensify; war catapults the struggle for commercial advantage into politics. The state's greatest need to intervene in the economy occurs just as business lobbying reaches its highest pitch: the need of the state to control the private sector increases, but no more so than the private sector's need to control the state. A wartime government is vulnerable to manipulation by interest groups, as shown by the regulation of oil in France. If wartime policies are manipulated by interest groups, then state policy is probably never free from such influences. I begin with the reasons business lobbying increases in wartime, then pass to the historical example provided by the oil industry, first explaining why governments must intervene in the economy in the face of war's "clear and present danger" and why that intervention fosters an increase in business lobbying.

War increases demand beyond normal levels, and prices rise. The price rises are intensified by labor shortages that result from conscripting the working population. Parts of the economic infrastructure become a target of enemy attack and are destroyed, causing more scarcity and more upward pressure on prices. Laying aside widespread attempts to "corner" critical commodities, even a "fair minded" capitalist has to adjust his prices upward in anticipation of his own constantly increasing costs. The more prudent he is, the more he will have to raise his prices, because war's uncertainties never allow him to anticipate exactly which shortages will develop or to what extent costs will increase.

The Excess Profits Tax

Price controls and excess profits taxes are regular features of capitalist economies at war and may seem to favor government rather than business. This is false. Before entrepreneurs produce, they buy—they purchase labor and products such as tools and raw materials. The upward trend of prices causes intolerable uncertainties for entrepreneurs in their capacity as consumers. A good, X_1, priced at the cost of its materials plus a reasonable profit, say 10 percent, may appear to cover the cost of raw materials and tools used to make it; but when those materials rapidly increase in cost, a business must make enough money on X_1 to cover not just original expenses, but also the cost of making the *next* good, X_2, at higher prices. To have the income necessary to produce X_2, the manufacturer may charge what looks like an excess profit on X_1. Since it is impossible to foresee which prices will rise most, and how they will vary from sector to sector, one cannot distinguish who is honestly staying abreast of price hikes and who is profiting from increased scarcity to get rich. In the wartime economy charges of speculation abound, demoralizing the collective effort. Business may gain relative to workers, but any given industrialist feels that suppliers are unfairly hiking their prices while he himself is only realizing a fair profit in difficult conditions. Unchecked, wartime inflation makes goods impossibly expensive and halts production. As buyers, capitalists are the first to realize this.

Officials of the wartime government also face economic disintegration directly, as buyers. Contracting suppliers either raise prices or fail to deliver goods. From these members of the private sector, the government cannot fail to learn of the economy's bottlenecks. Government contractors also find, in government purchasing agents, ready ears for their complaints about the unfair price hikes of other producers. Price controls and an excess profits tax emerge by consensus, because the capitalists need them to continue production and the government needs to have production continue. The excess profits tax may seem qualitatively different from price controls, but it is functionally equivalent. An excess profit is defined by the cost of production plus some "reasonable" profit; as we have just seen, when a good X_1 is extravagantly priced in order to cover producing X_2, what may appear as "excess profit" in fact compensates for rising costs. An excess profits tax, universally applied, stops the pricing of today's goods to meet tomorrow's costs. Along with price controls, it cools the whole inflationary process. In war, capitalists and government alike hasten to apply these controls.[3]

In France and other belligerent states, wartime regulation emerged

[3] In peacetime inflation, dominant industries oppose price controls and excess profits taxes. Inflation has a more predictable character, and costs are easily passed on to consumers.

for the reasons given above. Businesses joined government-sanctioned consortia to allocate production and set prices. The testimony before parliament of the French minister of commerce, Etienne Clémentel, illustrates the point: "I only wish you had been able to hear these industrialists yourselves. They told me that, in spite of these constraints [i.e., price regulation and excess profits taxes], which were imposed more because of the situation than because of my desire, they were happy to enter into the consortia. For they said that they were overwhelmed by the constantly increasing demands of their suppliers, and that they would have been forced to mark up their own prices beyond anything that had been foreseen. The disturbances in the market were such that they would have suffered much more from a free market than in the organizations to which they consented. *They preferred to be regulated.*"[4]

The United States also asked the belligerent nations to consolidate their purchasing into consortia. The purchasing agents of foreign firms bid up prices in the United States; purchasing consortia helped offset inflation with government sponsored monopsonistic practices.[5]

WHY WARTIME REGULATION INCREASES POLITICAL BUSINESS ACTIVITY

Regulation in peacetime often emerges from businesses' desire for predictable environments in which risk is minimized and profits are stabilized. Individual capitalists compete through a number of tactics. They compete in pricing; they can seek regulation that makes such competition unnecessary; or they can do both. Price competition is "nonpolitical" for firms, who judge their own prices and competitiveness by watching who is selling what for how much. But regulation is politically competitive and requires producers to monitor prices and watch who is influencing whom in the enforcement of rules and taxation that affect businesses unequally. Firms' patriotic eagerness to do government work at nominal pay reflects their need to monitor their competitors and their desire to gain the upper hand. During World War I, executives "sat in the morning as 'foremost patriots' directing mobilization, while in the afternoon they conducted their own business

[4] Arch.Nat. $F^7$7662, Commission du Budget, 17 May 1918, Audition de M. Clémentel; italics added.

[5] On United States consortia, see Clémentel's testimony (Arch.Nat. $F^7$7662, Commission du Budget, 17 May 1918, Audition de M. Clémentel). Consolidation of purchasing was unnecessary in the oil sector, where Standard Oil formed a virtual private monopsony. Contemporary observers and later historians have sometimes erroneously thought that the French oil monopoly in 1918 resulted from United States pressure. The causes were quite different, however. See Murat 1969, p. 16; Nowell 1983, p. 238 (where the error is repeated); and Doc.Ch., annexe 989, 3 June 1920, p. 1555.

with newly acquired inside information and often with government contracts."[6]

In wartime the government is forced to monitor the economy to a degree unnecessary in peace but is hamstrung by its lack of expertise in the myriad details of even the simplest industries, each a complex orchestration of procurement, marketing, and labor. The government must rely on information obtained from the private sector, and must invite industrialists to help. Coercively obtaining information, as by judicial procedures and hearings, takes too much time. Nominally committed to being "fair," the government must accept slanted information from a select number of industry officials. It gets advice from major producers, who have the most goods for sale; political risk is minimized because only the least significant firms will be offended. By definition, "wartime efficiency" favors one group of producers at the expense of another.

Under economic controls, the "normal" outlet for capitalist competitiveness is closed off. In addition to price controls there are production quotas, quality requirements, and regulation of all other important aspects of industry. Yet the goals of capitalist enterprise remain intact. There is still a need to get a return on invested capital, to hold or increase market share. When "normal" competition through price and quality is controlled by regulation, the entire competitive energy of the firm is concentrated on the regulatory process, either to secure an advantage or to make sure competitors do not. All the factors that drive a capitalist enterprise to compete remain in place. In peacetime competition is both in the market, in terms of prices, and in politics, in the struggle for favorable regulation. Under wartime controls only the possibility of political competition remains. In securing information from industry sources, the state is particularly vulnerable to manipulation. There is no time when business competition is so heavily politicized as in war, which for business means a covert struggle for advantage under a publicly emphasized desire to serve the national interest. The bitterness of this political struggle is mitigated only by the increased wartime demand, which increases business volume without eliminating the desire of individual firms to achieve commercial supremacy.

Additional Factors Exacerbating Business Lobbying in Time of War

The elimination of competitive pricing increases competition for favorable regulation, but it is not the sole cause. The elements of produc-

[6] Bates 1963, p. 99.

tion also become increasingly scarce. Access to production facilities, transportation equipment, and raw materials—all part of competition in peacetime—becomes even more urgent in time of war. War increases demand, but the strategic goal of the enemy is to destroy the capacity for production. More scarcity results. The loss of a firm's ship entails not only the lost capital of the ship (depending on insurance regulations), but more important, the market share sustained by that ship's operation. Should a rival firm enjoy the competitive advantage that its larger, still intact fleet affords it, or does some criterion of "fairness" require it to lend a ship to its competitor? What if no ship of equivalent tonnage is available for loan? These are hard questions for rival firms of a single country; when rival firms come from different, allied countries, each enjoying the support of its respective government, the result is intense squabbling.

The return of peace does not resolve these issues. A firm is forced to lend a ship to a second firm: if the second firm is forced to return the ship immediately after the armistice, it will be left ill equipped for the rigors of competition in the unregulated peacetime market. Wartime regulation therefore guarantees a fight over the timing and extent of peacetime deregulation. The two phases of regulatory struggle are so closely bound together that to dissociate them would be misleading: by its nature wartime economic conflict does not stop when the guns fall silent. This chapter on World War I therefore covers both the war years and the regulatory conflict that followed.

World War I and Royal Dutch–Shell

Royal Dutch–Shell was severely stressed by World War I. The company's aggressive reaction to the war's effects and Standard Oil's dominance are at the heart of the world struggle over oil. France's first round of interventionist state oil policy was only one part of this world struggle, which is diagramed in figure 2.

Large-scale commercial disruptions caused by the war set off an intense struggle among the world's oil firms. Antecedent commercial accords were suddenly useless. Royal Dutch–Shell's largest single source of oil was cut off by Turkey in September 1914. Exports from the rich fields of the Caucasus through the Dardanelles were cut off; the Russian Revolution three years later brought chaos to the producing fields. Royal Dutch–Shell's Romanian production was also blocked at the Dardanelles. While Romania stayed neutral, exports to either side were officially banned; when Romania sided with the Allies in August 1916 it sequestered corporations in which German nationals held shares,

Table 1. Impact of World War I on Russian and American oil supplies to France (in hectoliters)

	1911	1912	1913	1914	1915
			From Russia		
Crude	966,484	473,510	356,868	159,872	4,939
Refined oils	3,209	4,214	1,984	309	0
Gasoline and kerosene	496,797	432,184	626,684	393,115	13,275
Fuel oil	480,271	553,726	614,412	358,743	44,421
TOTALS	1,946,758	1,463,634	1,599,948	912,039	62,635
			From the United States		
Crude	934,206	1,104,023	1,127,632	985,564	211,578
Refined oils	1,667,529	1,643,438	1,841,880	1,781,363	2,651,173
Gasoline and kerosene	489,514	925,759	708,063	1,246,983	2,519,968
Fuel oil	679,902	822,294	707,908	579,726	963,796
TOTALS	3,771,151	4,495,514	4,385,483	4,593,636	6,346,695
			Russia: United States—Ratio of Totals		
RATIO	0.516	0.326	0.365	0.199	0.009

Source: Calculated from Doc.Ch., annexe 2533, 28 September 1916, p. 1491. This table compares only Russian and U.S. sources of France's oil. All other source countries represented about 15 percent of total French supplies.

even minority shareholdings. Royal–Dutch Shell's Astra Romana, which had a 4.11 percent Deutsche Bank shareholding as part of the arrangements that held Epu together, was sequestered by Romania in a maneuver that was part of a larger confrontation between national and multinational capital.[7] Then Germany invaded, and Royal Dutch–Shell lost the oil along with other allied companies.

Royal Dutch–Shell's remaining large source of oil was in Borneo, forcing the company's supply lines to Europe to stretch halfway around the world.[8] The increased transit time meant more tanker tonnage was needed to keep oil flowing to Europe. The company could hardly stay even, much less increase its supplies in accordance with rising demand. Robert Waley Cohen, a Shell officer, lamented the inability to take advantage of rising demand: "We not only can't dream of going into new trade, but can't even approach maintaining our existing one."[9] The shift in market supply forced by the loss of Russian production can be seen by the increase in market share of American compared with Russian oil on the outbreak of the war, as shown in table 1.

[7] See Pearton 1971, pp. 75–79.
[8] See map, figure 2.
[9] Henriques 1966, p. 222.

The war disrupted marketing and cartel agreements as well. The loss of Russian and Romanian exports destroyed the pan-European German, French, Russian, British, and Swedish oil interests that had enabled Epu to compete with Standard Oil. Standard Oil stood to increase its domination of European markets. With Russia's civil war in 1917 not only the Transcaucasian production was lost, but also the huge internal market in which the Shell combine had played a major role. The company was short of oil, short of tankers, and short of places to sell oil.

Royal Dutch–Shell's Problems in Great Britain

The Admiralty and Royal Dutch–Shell had an antagonistic relationship from 1907 to 1912. The Admiralty had declined to keep Shell Trading and Transport British by buying it and saving it from merging with Holland's Royal Dutch and France's banking interests. Now it accused the group of being "foreign." In addition to a Dutch-Gallic taint, the combine seemed German too. Its prewar collaborator, the Deutsche Bank, played the leading role in Epu, in which Royal Dutch–Shell participated. The Royal Dutch–Shell and Deutsche Bank interests had fought hard against the "purely British" Anglo-Persian for the Turkish oil concession. The oil concession in turn would make the Baghdad railroad a reality, a prospect dreaded by influential British imperial and commercial interests.[10] On top of this, the prominent role of Marcus Samuel and Robert Waley Cohen, both Jews and both directors of the combine, as well as the Rothschilds, may have made the corporate group a target of anti-Semitic sentiment: there is no direct evidence, but it would be naive to think otherwise. The Admiralty had a veritable potpourri of reasons why the combine was "not British" and therefore "suspect." The British navy bought no oil from the combine from its merger in 1907 till 1912, when Admiral Fisher was appointed chairman of the Commission on Oil Fuel and Engines.[11] The policy was nonsensical: the combine controlled so much oil that the Admiralty, to maintain the boycott, had to buy oil from Royal Dutch–Shell through intermediary companies.

Admiral Fisher supported the combine, but Churchill—hostile—alleged Royal Dutch–Shell's monopolistic schemes to justify in Parliament the Admiralty's stake in Anglo-Persian.[12] The combine courted

[10] Lord Kitchener opposed German influence in the Ottoman Empire; see Antonius 1939, p. 129.

[11] Henriques 1966, p. 184.

[12] Henriques 1960, pp. 282–287, 534–536, 575–586. Fisher was forced from his position as First Sea Lord in 1910; from 1912 to 1913 he chaired the Royal Commission on

the Admiralty for its business; the Admiralty required ardent efforts to "prove" the combine's fidelity to Britain. At the war's outbreak, the combine leased much of its commercial fleet to the Admiralty. The agreed rates did not reflect the higher shipping costs that soon resulted from increased demand and the damage done by German submarines. The company had prematurely concluded that the loss of German markets freed up additional shipping tonnage for itself, and it tried to use these vessels to get on the Admiralty's good side. But the decision proved poor, for the company was forced to lease new freighters elsewhere at higher prices, subsidizing the Admiralty's use of its own tankers. The Admiralty used the tankers to buy oil from the United States, which meant mostly from the combine's archrival Standard Oil.

Royal Dutch–Shell thus began the war subsidizing the use of its prized commercial fleet to bring in oil from its commercial nemesis.[13] Standard Oil meanwhile had transferred twenty-two of its thirty-eight ships under German registry back to the United States, since the British blockade of German commerce rendered them useless for that market.[14] Because the German market was the largest outside the United States, the tankers set free by the blockade released for Standard Oil a fleet to be used worldwide. On top of this, the Admiralty's delivery contracts put Royal Dutch–Shell's fleet into service for the American firm. So surfeited was Standard Oil that it leased some "extra" oil vessels to carry cotton. It also put what amounted to a new fleet to use in taking over Royal Dutch–Shell's Asian markets, which the combine had neglected in favor of the Allies. As Henriques writes, Standard Oil "exploited to the utmost all the conditions of hostility which could help them to capture the Shell's markets." Royal Dutch–Shell kept detailed records of its competitor's ship movements in the Far East, urging the Admiralty all the while to appropriate some of Standard Oil's tonnage.[15]

Admiralty favoritism for Anglo-Persian created further difficulties for the combine. Anglo-Persian wanted to use the war to get the whole of Mesopotamia for itself, about which more below. The British government also helped Anglo-Persian acquire retail outlets in Great Britain at the combine's expense. The retail firm British Petroleum was controlled by the Deutsche Bank as the British marketing outlet for Epu:

Oil Fuel and Engines, and he was reappointed First Sea Lord from 1914 to 1915. He consistently supported Royal Dutch–Shell's commercial ties to the Admiralty (Ferrier 1982, p. 690).

[13] Henriques 1966, p. 199.

[14] The remaining ships were used on the Danube between Germany and Romania ("Oil for Germany," 1916).

[15] Henriques 1966, pp. 200, 212–214, 230.

it was a large vendor of Shell products. When war came, it was confiscated as enemy property; Royal Dutch–Shell wanted it back, but Anglo-Persian director Charles Greenway wanted it to create a vertically integrated commercial oil company.[16] Greenway won the ensuing struggle and even managed, in the midst of war, to persuade the government to capitalize his company's purely commercial downstream operations.[17] Anglo-Persian absconded with Shell's former marketing firm on 2 April 1917.

Summary of Royal Dutch–Shell's Situation

Loss of markets, loss of shipping, and loss of supplies: the distress of the world's second largest oil company was its rivals' opportunity. The Admiralty doted on the still unproved Anglo-Persian and, perhaps deliberately, perhaps accidentally, helped Standard Oil increase sales at the combine's expense. Its marketing firm was stripped from it, and Mesopotamia, which had come within its grasp in the months just prior to the war, seemed likely to be lost.

The leading personalities at Royal Dutch–Shell were not prone to hysteria, but certainly these conditions would have made the most hardy souls uneasy. New oil supplies had to be found, access to shipping guaranteed, and markets protected from further encroachment by rival firms, especially Standard Oil and Anglo-Persian. Royal Dutch–Shell launched a political offensive to counter the influence of Anglo-Persian in Britain and to correct the gross imbalance in critical shipping resources that benefited Standard Oil. To remedy lost production, the company had to get a share of the spoils of war. The company's strategy had worldwide consequences.[18] France would soon become the direct instrument of Royal Dutch–Shell's response to the war.

The Feigned "Oil Crisis" of 1917

The short-term objective of French wartime oil policy was to meet the needs of the civilian population (lamp oil) as well as the rising

[16] Ferrier 1982, pp. 205–206, 217–219.

[17] Ferrier 1982, p. 218, cites one government official as admitting that commercial operations were not "what was intended when the Government went into Persian Oil. The object was to secure navy supplies." Funding the company's commercial sales is hard to justify from any strategic imperative but is consistent with the evolution of British oil policy.

[18] Mexico, for example, was also directly affected; L'Espagnol de la Trameryе 1921, pp. 111–124, describes oil companies backing rival revolutionary factions. Royal Dutch–Shell

demand for military operations. The long-term objectives included the postwar division of spoils. Both short-term measures for domestic procurement and long-term diplomatic goals were decisively influenced by the Royal Dutch–Shell group, which enlisted the French government as an ally. Because French policy answered the needs of Royal Dutch–Shell more than any real strategic concerns, no serious intervention occurred in the oil market until about six months before the German capitulation, and the state's most dramatic regulatory moves were made not during the war but after it.

At first the French government bought its war needs from the major domestic importers, the Cartel des Dix that had dominated the French oil market for decades. These contracts remained in force until March 1918. The importers depended primarily on American oil. A ministry of commerce official exaggerated when he said that "currently it is Standard Oil alone that supplies France," but the reality was not far off.[19] Russian and Romanian sources constituted 44 percent of the prewar French market; by 1917 the United States supplied 82.6% of total needs.[20] In 1917 Standard Oil of New Jersey alone supplied 47 percent of all the oil sold to France.[21] To this one could add the other Standard Oil companies, especially Socony; they unquestionably provided the most oil to France. Oil from other important companies, such as Gulf, often passed through Standard Oil's ships, refineries, and pipelines.

What Standard Oil could not supply directly it could arrange, even with huge increases in allied consumption. Its near monopoly on world export markets and near monopsony on American oil made a government-sponsored purchasing consortium unnecessary. The world monopoly gave French importers the luxury of following an unsupervised "*politique libérale,*" which in fact continued their prewar cartel. Contracts between French cartel members and the state in 1914 committed the refiners to maintaining reserves of 22,000 tons and also fixed prices, subject to periodic revision in accordance with changes in the price of American oil.[22] By 1918 the United States' share of France's oil

acquired Pearson's Mexican Eagle oil holdings in 1919. See also Meyer 1977, Nash 1978, and especially Katz 1981. Bates 1963, p. 97, cites Senator Borah: "To promote their vast designs these oil magnates are capable of starting revolutions in Mexico, instigating civil wars in Asia, of setting fire to Europe and the world to crush a competitor."

[19] Arch.Nat. F[23]81, file "Pétroles et essences 1914," "Note" dated 19 December 1914.

[20] Calculated from figure 2; see also Nayberg 1983, p. 35.

[21] Gibb and Knowlton 1957, p. 224. Data for a full statistical breakdown of market share for each corporation are not available; such figures as occur in both secondary and archival sources are incomplete.

[22] Nayberg 1983, pp. 33–34.

requirements had risen to 89.4 percent.[23] The United States could easily have furnished the remaining 10.6 percent, but had it not been able to do so, no change in the French 1914 purchasing arrangements was really needed. Yet in March 1918 the French government established a state import monopoly that purchased oil directly abroad and then resold it to the domestic companies. Why did this interventionist move occur at this late date?

The specialized histories argue that the German submarine campaign, together with rising wartime demand, caused a shortage of tankers and forced the French government to take bold steps to ensure supplies, narrowly averting a critical oil shortage. The French state "became aware" of the strategic value of oil as a result of war and, faced with undependable market suppliers, had to take action. As Melby writes, the "French State, because of wartime needs, took complete responsibility for procuring oil supplies."[24] Strategic need thus galvanized the French state's drive to institute a comprehensive French policy over the next ten years.[25]

The state-centric view is not entirely without foundation. Demand had gone up, and after the start of the January 1917 German submarine offensive, tankers were going down. In 1917 the Germans sank a number of tankers: four British, one American, and one French. Six British and two French ships had to undergo repairs.[26] The convoy system reduced the submarine threat but also cut effective fleet capacity by one-sixth.

After the first battle of the Marne in 1914, when the Paris taxi fleet moved troops rapidly in response to German advances, the importance of mechanized transport grew apace. France consumed 476,000 tons of oil in 1914, rising to one million tons in 1918.[27] Not all of the increase went for combat, although anything that affected French economic performance had an indirect strategic character. Provoked by the labor shortages resulting from mass mobilization, agriculture mechanized more rapidly.[28]

In reality, the struggle for control began slowly in response to the Shell combine's commercial needs and intensified as the war progressed. Military factors were important, but the German submarine campaign of 1917 weighed little next to the failure of Churchill's expe-

[23] Nayberg 1983, p. 401.

[24] Melby 1981, p. 11.

[25] Similar interpretations, with varying degrees of detail, are found in Murat 1969, pp. 13–14; Touret 1968, p. 23; Thomas 1934, p. 45; and Nayberg 1983, pp. 118–128. Neither Shonfield 1965, p. 82, nor Krasner 1978, pp. 58–61, gives oil history sources for his state-centric interpretation; it has become received wisdom.

[26] Nayberg 1983, pp. 72–73, 76.

[27] L'Espagnol de la Trameryc 1921, p. 178.

[28] Chauveau 1916, pp. 117–232. See also Augé-Laribé 1950, pp. 356–366.

dition at Gallipoli. This major "realist" event determined access to supplies of both commercial and strategic value, which the "realist-mercantilist" oil histories have missed.

The Gallipoli expedition began in February 1915 and failed a year later. The campaign aimed to knock Turkey out of the war early and promised to liberate the oil of the Caucasus and Romania. The oil supply lines to Britain and France would be shorter. Opening the Dardanelles would have liberated more tanker tonnage for Shell than ever was claimed by German submarines. Hoping for military success, the Shell group did not claim more shipping from allied governments until the Gallipoli campaign had failed.

After that failure, the combine could no longer delay seeking more tankers to help with its extended supply routes. This caused the French government's first intervention on the company's behalf. France had fourteen tankers of its own, but these were sailing under the British flag to avoid French taxation and regulation. The British government requisitioned the French vessels and used them to bring oil to Britain. The Admiralty benefited from shrewd requisitioning and the windfall terms of Royal Dutch–Shell leasing. It had ninety-one tankers at its disposal and wanted to control all oil deliveries to the western front. When France's refiners wanted to put four or five of the French-owned ships back into French service, the Admiralty advised them to lease neutral tonnage or buy new tankers.[29] The French refiners appealed to their own government's Ministère de la Guerre for help, and in January 1916 four ships were transferred to French use.

The French government immediately turned the tankers over to the British shipping firm of Lane and McAndrew, which already operated four other French oil ships.[30] Frederick Lane was a long-standing business associate of Marcus Samuel, the Rothschilds, and Calouste Gulbenkian; he also sat on the board of directors of Royal Dutch–Shell.[31] Juridically distinct from Royal Dutch–Shell, Lane and McAndrew's interests were indissociable from it. As Churchill's expedition withdrew from Gallipoli from November 1915 through February 1916, the Shell combine began its quest to reacquire the tankers it had lost to the Admiralty and indirectly to Standard Oil.

This attempt was the first in a series of French initiatives to shield the Royal Dutch–Shell group from the hostility of the Admiralty, from Standard Oil, and from the effects of the tanker shortage. The shipping struggle escalated in spring 1917. This period corresponds with

[29] Nayberg 1983, pp. 36, 50–52.
[30] Nayberg 1983, pp. 53–54.
[31] Henriques 1966, pp. 130–131.

the German submarine offensive and the belligerent status of American shipping with the United States' April entry into the war.[32] French oil policy's development would be tailored to the needs of Royal Dutch–Shell under the guidance of Calouste Gulbenkian, who sought to influence not just shipping, but also market division and supply allocation. Gulbenkian wanted to engineer an oil purchasing monopoly to bypass Standard Oil and the "ring of French refiners" that depended on it. It was, he said, "a tough problem but a most attractive and interesting one."[33]

Gulbenkian also wanted France to rescue Royal Dutch–Shell's Mesopotamian rights from threats in Great Britain. In November 1915 the Foreign Office told Anglo-Persian that it no longer considered the Turkish Petroleum Company's contract, ratified by the Foreign Office in March 1914, to be valid. The nullification held out the enticing prospect of awarding all the Mesopotamian oil fields to Anglo-Persian.[34] Anglo-Persian's director, Sir Charles Greenway, refused to continue paying his company's share of the Turkish Petroleum Company's routine operating costs.[35] Gulbenkian reported as much to Henry Deterding, "who immediately jumped to the conclusion that the Anglo-Persian was again up to some 'tricks' to see us all out [of Mesopotamia]. Mr. Deterding thought that they were labouring with the belief that once the war was over the whole of Mesopotamia would become British, and then there would be no room for any foreign Companies [Royal Dutch–Shell]. He became, as usual, very vehement and he said he would see them d[amned] first."[36]

Royal Dutch–Shell voiced no opposition to Anglo-Persian's claims in Mesopotamia, however, until the British militarily controlled the oil.[37] With the British recapture of Baghdad in March 1917, Mesopotamian oil and French regulation became linked for the rest of the pre–World

[32] Art 1973.

[33] Gulbenkian 1945, p. 17.

[34] Kent 1976, p. 113. See also Gulbenkian 1945, p. 21, and Hewins 1957, p. 108.

[35] Gulbenkian 1945, p. 16. Gulbenkian, writing from memory, dates the time as "1915 or 1916." He says he met Bérenger early in 1916, but there is no evidence that Bérenger had anything to do with oil policy then. Bérenger's actions, as described by Gulbenkian, correspond to early 1917. My interpretation of Gulbenkian's dating is that he met with Greenway shortly after the Foreign Office invalidated the Turkish Petroleum Company (late 1915 or early 1916); but Royal Dutch–Shell took no action and opted to fund the company itself; the military situation had rendered Anglo-Persian's pretensions within the Ottoman Empire meaningless. The Dardanelles expedition withdrew from November 1915 to February 1916. In April 1916 the British army in Iraq was captured by the Turks. The Arab uprising did not begin until June 1916. Between the April 1916 defeat and the British recapture of Baghdad in March 1917, Gulbenkian and Deterding did not bother seriously fighting Anglo-Persian for oil over which it had no control.

[36] Gulbenkian 1945, p. 17.

[37] See note 35 above on Gulbenkian's dating.

War II period. Gulbenkian had to obtain French government assistance to suport Royal Dutch–Shell's claims in Mesopotamia; without an Allied government's backing, those claims would evaporate.

Gulbenkian exerted parallel efforts to protect and to increase the market share of Royal Dutch–Shell in France. These two goals were interconnected:

> I was nominated Foreign Delegate of the Royal Dutch combine, my mission being to place myself at the disposal of the French Government to facilitate their oil supplies; also, at the same time, to endeavour to break the ring of French refiners and increase our trade and influence on the Continent against the American interests. It was a tough problem but a most attractive and interesting one. . . . Although from a material point of view the prize attached to success was very handsome,[38] what I had at heart was to render a great service to France in the matter of creating a sound oil policy and economics. The French oil groups . . . were nothing else than a monopolistic association of grocers.[39] I mean to say that they had no industry and France, in the matter of petroleum, was in a pitiful condition in spite of the fact that French refiners had accumulated enormous fortunes by price rigging and dubious methods such as bribing the press, etc., (to which I need not refer).[40, 41]

Gulbenkian may have had some warm feelings of personal attachment to France, but he held passports from Turkey, Great Britain, France, and later Persia, and he died in Portugal. He chiefly followed his own interests. He observed that "in peace-time, the realization of [my] plans would have been almost impossible. The grip of the French refiners (political, press, Chamber of Deputies and outside) was such that they could unscrupulously checkmate the activities of anyone working honestly in the interests of the country. . . . Fortunately for my schemes. . . . [d]isorder and confusion were such that the Government was obliged to take special steps to reorganize the petroleum

[38] His eventual capture of 5 percent of the Iraq Petroleum Company.

[39] "Grocers": the French national refining companies, unlike the multinationals, were not as vertically integrated and sold many of their products in cans through retail *épiceries.*

[40] Gulbenkian's bashfulness illustrates how difficult it is, even in a very frank document, to document rent seeking, bribery, etc. Gulbenkian is described by his biographer as having "mastered the mysteries of baksheesh" (Hewins, 1957, photo caption between pp. 140 and 141; see also p. 71). Cf. Feis 1964, p. 133: "The financial press was unreliable and venal," and pp. 157–158: "The press was bribed—sometimes with the knowledge and advice of the French Ministers who also had to reckon with outside opinion; the way was opened for blackmail by the press."

[41] Gulbenkian 1945, p. 17. Gulbenkian dates his mission at "the end of 1915 or at the beginning of 1916," but as mentioned above (note 35), his meetings with Bérenger could not have started until 1917. Capitalization in all quotations follows original.

and fuel administration, I advised the competent authorities, through influential channels, to create a special Commissariat."[42]

At the beginning of 1917 Gulbenkian met with French senator Henry Bérenger, whom he called "an astute parliamentarian, fully acquainted with all the political intricacies, and to my great satisfaction he immediately saw the advantages to be derived from my schemes, not only from his personal political standpoint, but also from the national interest point of view. Mr. Henri Bérenger was a very well-known political figure and being connected with all the Governmental Departments, I soon got in touch with all the elements that would facilitate my task."[43] Bérenger's "personal political standpoint" was the advantage of helping the powerful Rothschild banking interests in oil matters. Whatever his personal motives, he saw in Royal Dutch–Shell's struggle with Anglo-Persian an opportunity for France to increase its economic power. The combine's "Frenchness" was limited to the Rothschilds' shares, but this was more than any French citizen had in Standard Oil or any other major world oil company. For Gulbenkian, talk of the "national interest point of view" was bait to dangle before the French government; for Bérenger, it was a chance to unite service to country and career.

Royal Dutch–Shell was not France's only chance for Mesopotamian oil: the Sykes-Picot agreement had already been negotiated, and this agreement may have been part of a larger effort by other French banking interests to outmaneuver the Rothschild/Royal Dutch–Shell coalition. Bérenger's rise was part of an interbank rivalry in France over Mesopotamian oil. Bérenger ascended rapidly in oil policy because of his association with Gulbenkian and, through him, the Rothschild bank. Before then Bérenger had no presence in French oil policy as, for example, in the 1914 monopoly proposal. He learned rapidly, however: "I imparted all the knowledge I had at my disposal to the new Commissaire [Bérenger] for the improvement of the French oil economy. This Department was thus able to gain great influence in governmental circles and it created for me personally a genuine atmosphere of cordiality and influence."[44]

[42] Gulbenkian 1945, pp. 18–19.

[43] Gulbenkian 1945, p. 19. Again, Gulbenkian's memoir mistakenly places these events in 1916. Bérenger's first and last names, here and elsewhere, were subject to numerous misspellings, the most common being "Henri" instead of "Henry" and "Béranger" instead of "Bérenger." Bérenger himself would sometimes sign "Henri," sometimes "Henry"; compare his letter dated 4 December 1917 with another dated 21 September 1917, in Arch.Nat. $F^{12}7716$.

[44] Gulbenkian 1945, p. 19. Gulbenkian exhibits a tendency to exaggerate his claims to anything "positive" in the course of his career. He could be considered vainglorious and boastful, at least in this memoir, but in his life he scorned publicity and honor, refusing,

Gulbenkian's influence stemmed from his representing the Rothschild oil interests. His claims, which show little personal modesty, do not contradict other elements of the historical record. Bérenger's rise to preeminence in French oil coincides with the beginning of Gulbenkian's efforts in France, and Bérenger's early policy recommendations are those of a convert to a newfound cause. For him the transitory tanker crisis was but a prelude to developing a policy to replace coal's traditional uses with oil and to rationalize the development of electricity.[45]

American entry into the war also brought economic mobilization: Standard Oil captured the commanding heights of the American economy. Walter Teagle, president of the company, became president of the United States Shipping Board. A. C. Bedford, chairman of the company's board of directors, assumed the presidency of the Petroleum Committee, formed two weeks before the formal declaration of war.[46] The expertise of these two men undoubtedly helped them coordinate shipping needs for the conduct of a world war, and on the United States Shipping Board Teagle was surely sagacious about the host of commodities whose transportation lay under its jurisdiction. The two nonetheless were well placed to defend their interests in oil policy.

The French formed the Comité Général des Pétroles on the heels of regulation in the United States: tankers, markets, and supplies were at stake. The Petroleum Committee in the United States was really a government-sponsored cartel of producers. In France the government divided the market among importers. Short on crude, Royal Dutch–Shell was at a serious disadvantage. But political power, like military power, is only partially related to size. Though holding one-fifth the market of the American companies, Royal Dutch–Shell had ties to a major bank and greater experience in European politics; Standard Oil sold to French companies but was far less well connected politically. It did not deal directly with the French government but was represented through its client companies in France. These were not subsidiaries,

---

for example, a British knighthood. When Gulbenkian made exaggerated claims, it was usually with the intent of protecting some benefit or acquired position. One might be tempted to dismiss him as a fop were it not that he parlayed his connections into one of the greatest fortunes in the world. The operating assumption in referring to this source is that he could not have gotten so wealthy, nor negotiated toe-to-toe with the world's great powers from 1907 to 1914 and again in the 1920s, if there were not some truth to his account. The alternative is to believe that he stumbled into his fortune, which seems unlikely.

[45] Doc.Sén., 14 June 1917, annexe 201, pp. 316–320. See also Nayberg 1983, pp. 107–108.

[46] Gibb and Knowlton 1957, pp. 237–241; Nayberg 1983, p. 145; Bates 1963, pp. 99–100. The Standard Oil historians unabashedly describe the Petroleum Committee as "Mr. Bedford's Committee."

and they only partially expressed the interests of Standard Oil. As semi-independents, they wanted some freedom of maneuver from the American oil behemoth.[47]

Bérenger jumped into the melee of regulatory politics following a May 1917 report that oil supplies were inadequate. Denouncing competition among the companies (there was very little) as injurious to the national interest, he pushed for the creation of the Comité Général des Pétroles, created 13 July 1917, and was nominated by Premier Alexandre Ribot to be its head. The Comité Général des Pétroles, the official name of what Gulbenkian calls a "commissariat," had a purely consultative role, however. Administrative power was in the hands of the Ministère de l'Agriculture et du Ravitaillement (ministry of agriculture and supplies). The Comité Général spent months arguing over whether oil supplies were in fact inadequate, as stated in the May 1917 memorandum. Maurice Violette, head of the Ministère du Ravitaillement, thought Bérenger's claims of an oil shortage were exaggerated.[48] Bérenger was adamant. Members of the oil industry sat on the consultative committee, and Standard Oil, which was excluded from the Comité Général des Pétroles, learned of the debate through channels. Violette was sympathetic to Fenaille et Despeaux, which maintained good terms with Standard Oil; and in America the French oil representative Edmond Paix was so flagrant in passing information among the oil companies that he was reprimanded.[49] Through Teagle and Bedford, Standard Oil could directly monitor over 80 percent of the oil supplies going to France, and they no doubt had additional information about the remaining 20 percent coming from Mexico and Indonesia.

In short, France's oil situation was disputed among its top officials;

[47] For example, in April 1917 Desmarais Frères, Deutsch de la Meurthe, and Fenaille et Despeaux, operating as a private consortium, requested five tankers, to be contracted through December, for oil deliveries to France. Desmarais Frères was historically more "independent," having had investments in Romania before the war. Deutsch de la Meurthe overtly worked with Royal Dutch–Shell, while Fenaille et Despeaux worked with Standard Oil. All wanted oil from Standard, but that says nothing about their political intentions. France's 1916 tanker requisitions from Britain all ended up in the hands of the Royal Dutch–Shell group, so Standard Oil's skepticism about the French consortium is understandable. Standard was furnishing oil to its principal rivals in the French market, and the appearance of the Fenaille et Despeaux in their consortium, far from being reassuring, could only call into question Fenaille et Despeaux's long-range plans. Standard Oil at first denied the request but agreed to furnish three tankers after the three importers called on the French government for help. See Nayberg 1983, pp. 77–81, 87.

[48] Nayberg 1983 claims that Violette was too optimistic (p. 92) but that Bérenger's forecasts were "alarmist" and "a good deal greater than the figures warranted" (p. 116).

[49] Nayberg 1983, p. 285. Violette thought that important matters should not be up for "public discussion" in the Comité Général: what was said there did not remain confidential (Arch.Nat. F[12]7715, Comité Général des Pétroles, letter dated 7 September 1917).

Standard Oil knew that and had its own information that France was getting what it needed. The company had commercial information concerning sales, plus information from its cooperating client companies in France about the stocks on hand. Later the company agreed to build up French oil reserves, but the extra four or five tanker loads required to do this hardly merited, from Standard Oil's viewpoint, the dramatic reorganization of delivery procedures that Bérenger wanted.

The only person complaining about an oil shortage was one French senator, newly entered into oil matters and working closely with top representatives of Standard Oil's number one rival in world markets. In fall 1917 the American company could assume that Bérenger's blusterings about oil deliveries were designed to get Standard Oil to allocate more ships than were necessary to supply the French market. More ships in the European trade was a ploy to limit Standard Oil's expansion into the Shell combine's Asian markets.

German submarines sank some of Standard Oil's fleet, but the company prospered through the crisis. Eight tankers were lost in 1917 and ten new ones were launched, and in 1918 nine ships were lost and nine new ones launched. The newer ships were bigger, and though some of the new ships were lost, the stable number of ships in the company fleet (about forty) masked an overall increase in carrying capacity. "With the high risks went high profits for those companies fortunate enough to own tanker fleets," write the company's historians, and "if consideration is also given to high war profits earned, Jersey's actual monetary losses on the sunken vessels were probably far from severe."[50] The company was not risk averse about supplying France, in spite of the submarine peril, and it found more than enough financial compensation for the dangers involved. The company could cover French requirements and also expand into Asian markets.

Yet Standard Oil's actions in late 1917 would later be cited in the French parliament as an example of the company's perfidy, a perfidy sufficient to justify state intervention in the market. Little malevolence is evident beyond a desire to beat Royal Dutch–Shell. In the summer of 1917 the American government requisitioned the Royal Dutch–Shell tanker *Goldshell*, ostensibly because its freight rates were set too high.[51] The tanker had been scheduled to carry a load of kerosene for Deutsch de la Meurthe, the French firm closest to Royal Dutch–Shell. Advised of the problem by Deutsch de la Meurthe, the French high commissioner in charge of French procurement in the United States, André

[50] Gibb and Knowlton 1957, pp. 224, 228–229.

[51] Arch.Nat. F^{12}7715, Comité Général des Pétroles, letter dated 25 August 1917; also letter dated 31 August 1917 in same file.

Tardieu, tried to get the *Muskogee,* a Standard Oil ship of superior tonnage, assigned to carry the cargo. Standard Oil, however, wanted the ship to carry its own oil to France, to replace another of its fleet in French service that had recently been sunk (the *Archbold*). In Tardieu's words, the "American government, not wanting to favor the Shell Company, wanted it [Shell] to replace the *Goldshell* by its own means . . . the impression here is that the requisitioning of the *Muskogee* to replace the *Goldshell* displeased Standard Oil, which profited from its situation in the Petroleum Committee to checkmate the scheme, because of its hostility to Shell, while at the same time getting a more advantageous freight rate by assigning the *Muskogee* to replace the *Archbold.*" The high commissioner admitted that France had two procurement systems. The French refiners operated under the terms of their 1914 contracts, which made them responsible for finding freight. Then there was direct procurement by the French government, which began with André Tardieu's arrival in the United States in the summer of 1917. With France represented by two sets of authorized personnel, Standard Oil found it "commercial and legitimate" to choose the higher offer. The American company of course knew that the "official" French government representative advocated the procurement policy most favorable to Royal Dutch–Shell. Standard Oil's *Muskogee* was given a permanent assignment to replace the *Archbold,* keeping it within the control of Standard Oil and its most closely allied French firms. Disappointed, Tardieu went to discuss with A. C. Bedford some other replacement for the *Goldshell.* As president of the Petroleum Committee and the Standard Oil Company, Bedford promised "to study very closely the possibility of giving satisfaction" to the high commissioner, but "he did not want to make any promises."[52]

Tardieu would later cite this incident as proof of collusion between French refiners and Standard Oil (as indeed it was) to the detriment of France, which is less certain.[53] Standard Oil filled the *Muskogee* with oil and sent it to France. The French government may have gotten a bad deal on the price, but even this is unclear, since the Shell combine may have been charging less for the oil and more for the freight. Tardieu's own competence may be questioned, for he perhaps would have had more luck negotiating a lower price than trying to get Standard Oil's tanker assigned to the Shell combine. The high commissioner did not state these nuances before parliament. His scapegoat version of the uncooperative American oil giant became part of the mythology

[52] Arch.Nat. F[12]7715, Comité Général des Pétroles, letter dated 4 September 1917.

[53] See Tardieu's testimony in Déb.Ch., 26 April 1920, p. 1382; also Nowell 1983, p. 236, which incorrectly accepts Tardieu's version.

of crisis used to justify French oil regulation a decade later and in the decades to come.

Frustrated by Standard Oil and its ally Fenaille et Despeaux, Bérenger tried to take procurement policies into his own hands. He sought to expand the authority of the Comité Général, which he headed, and wrote to Etienne Clémentel, then minister of commerce and industry.[54] Clémentel approved Bérenger's proposals. He felt that if France were to become a major exporting country after the war, it would have to be able to get "raw materials under advantageous conditions" and that "to have these at the lowest price, we will have to suppress, at least amongst ourselves, ruinous competition."[55] His reasoning is mysterious. "Ruinous competition" usually means prices that fall so low as to cause financial difficulty among marginal firms; suppressing that competition could not ensure the "lowest price." On the contrary, it would ensure the survival of less efficient firms. Clémentel meant that government controls on Standard Oil's market share would reduce "ruinous competition," from Royal Dutch–Shell's viewpoint, and if in exchange Royal Dutch–Shell would refrain from price gouging, Standard Oil's share of the market could be reduced. Still, plans at this time were not to diversify the market but to turn it over to the combine: one monopoly would be exchanged for another. Bérenger urged Tardieu to circumvent Standard Oil's control over shipping and oil by appealing to President Wilson while Clémentel and he worked on a scheme whose goal would be to put France's oil trade in the hands of the Shell combine.[56]

The Shell group pursued other avenues to get more tankers. In 1917 the Admiralty requisitioned Socony's entire Hong Kong oil fleet of nineteen ships under the British flag. Socony interpreted this strong, state-centric move in the light not of strategy but of regulatory politics, which the company's officials believed originated with Asiatic, Royal Dutch–Shell's marketing concern in Far East markets.[57] A. C. Bedford concurred with Socony's view that bringing oil to Europe from the Dutch East Indies, particularly on any Standard Oil Company ships,

[54] Arch.Nat. F[12]7715, Comité Général des Pétroles, letter dated 21 September 1917.

[55] Arch.Nat. F[7]7662, Commission du Budget, 17 May 1918, testimony of Etienne Clémentel.

[56] Arch.Nat. F[12]7716, letter to Tardieu dated 21 September 1917. The language of this letter is in places nearly identical to Premier Clemenceau's 15 December telegram to President Wilson; Nayberg 1983, p. 181, affirms that the telegram was in fact written by Bérenger.

[57] Gibb and Knowlton 1957, pp. 241–242. Though technically independent of Standard Oil of New Jersey, the Standard Oil companies were not entirely autonomous.

was a waste of tonnage. The transit route from American suppliers was significantly shorter.[58]

Why was the Admiralty suddenly favoring the company to which it had dealt such serious blows? Its antagonism toward the combine grew out of the desire to favor the interests of the Burmah Oil Company and Anglo-Persian; this strategy meant that the combine should be kept down, but not out. Anglo-Persian's principal marketing contracts were with Royal Dutch–Shell. The combine also provided technically essential services, saving the British Empire's oil champion from difficulties in building refining and marketing facilities. In the Asian market, cooperation between the Burmah Oil Company and Royal Dutch–Shell was essential to resist the onslaught of the Standard Oil corporations.[59] The Admiralty policy was to favor Anglo-Persian in Mesopotamia and Britain, but to help the Shell combine maintain its position in the Far East. This solicitude for commercial considerations is not what we would expect from a military bureaucracy deep in what was then the greatest war in human history, but it was of the same ilk as other decisions such as helping Anglo-Persian to develop commercial marketing operations in 1917.[60]

Royal Dutch–Shell had another card to play: its Borneo crude contained a small fraction of toluol, used for making explosives (TNT). The extraction process used only 2 percent of the crude. If the other 98 percent was not to be wasted, it had to be marketed as ordinary oil products. Toluol could also be extracted from by-products of industrial coal, but the demand was large, and both the French and the British were ready to buy all they could.[61]

Out of this 2 percent fraction of oil grew the strategic justification for preserving the entire Asian market of Royal Dutch–Shell. To produce toluol economically, the other 98 percent of the oil would also have to be sold. To sell it the company needed ships. The Admiralty would not relinquish ships it had leased from the combine at the beginning of the war, but it knew that additional ships would supply some toluol *and* protect the collective interests of Royal Dutch–Shell, Burmah Oil, and Anglo-Persian. Hence Socony's nineteen ships were commandeered, not just to bring the 2 percent toluol fraction of Royal Dutch–Shell's Borneo production to Europe, but to maintain that company's sales in the Far East markets Socony coveted. Socony interpreted the move as hostile. Something less drastic, such as having Socony buy oil

[58] Gibb and Knowlton 1957, p. 242.
[59] Kent 1976, pp. 127–136.
[60] Ferrier 1982, pp. 217–219.
[61] Henriques 1960, pp. 597–605.

for Asian distribution from Royal Dutch while the fraction of toluol went to Europe, would have been more to its liking.

The Admiralty's growing willingness to help Shell was also evident when Robert Waley Cohen, the Shell group's second-in-command after Henry Deterding, was asked by Sir Frederick Black of the Admiralty to go as part of a British war mission to the United States, where Cohen would have the job of coordinating Allied supplies of oil.[62] Cohen's arrival in New York in August 1917 placed him squarely in the ongoing shipping struggle between Standard Oil, Tardieu, Deutsch de la Meurthe, and Fenaille et Despeaux in the *Muskogee-Goldshell* incident.[63] Henriques remarks, "If this commercial war [between Standard Oil and the Shell combine] was now to be halted in the interests of Allied economy, it seemed unlikely that [Cohen], the foremost natural enemy (after Deterding) of the Standard, was a suitable leader of the British group . . . for the pooling of all British and American Shipping."[64] The point, however, was not to end the commercial war but to wage it on terms more favorable to British business. Waley Cohen was the best man to stabilize the Far East markets not in spite of, but because of, the fact that he was Standard Oil's "foremost natural enemy."

This was the world commercial context in which France traversed its first "oil crisis." André Tardieu, in the United States to take care of supply problems, in late October knew nothing about any oil supply crisis and confessed a "complete lack of knowledge concerning France's exact needs."[65] He was told by telegram, on 20 October 1917, that forecasts showed no stocks of lamp oil in France by 1 January 1918.[66] Forecasts for other oil supplies were similar. Socony smelled a rat and was convinced that Royal Dutch–Shell was feeding misleading figures to governments to make the Standard Oil companies look bad.[67] The British exaggerated their needs during this period in order to build up their stocks, and the French did the same. In December 1917 a new set of French forecasts estimated that there would be about 30,000 tons of gasoline and 12,000 tons of oil in January, substantially higher than the catastrophic forecasts of October. France would now run out of oil in March, when the Germans were expected to launch their final offensive to win the war before the American expeditionary forces arrived

[62] Henriques 1966, p. 230.

[63] Cohen is mentioned in the French documents, but his role in the affair is obscure. See for example Arch.Nat. F[12]7715, Comité Général des Pétroles, letters dated 31 August 1917 and 8 September 1917.

[64] Henriques 1966, p. 230.

[65] Arch.Nat. F[12]7716, letter dated 20 October 1917.

[66] Nayberg 1983, p. 153.

[67] Gibb and Knowlton 1957, p. 241.

in great numbers.[68] With this new prediction, Bérenger persuaded Premier Clemenceau to intervene personally to get more tonnage attributed to the French, and Clemenceau telegraphed as follows on 15 December 1917, requesting that gasoline stocks be increased:

> These measures can and must be taken without one day's delay for the common safety of the Allies, which can only be done if President Wilson requires from the American companies the supplementary tankers of 100,000 tons. This is necessary to the armies and people of France. These tankers exist. They are sailing at this moment in the Pacific Ocean, and not in the Atlantic. An additional part could be obtained from new tankers under construction in the United States.
>
> Premier Clemenceau personally asks President Wilson to use the authority of his government to direct these tankers immediately toward French ports.
>
> This is a question of interallied safety. If the Allies do not want to lose the war, France must have, in its hour of supreme struggle with the Germans, the gasoline which is as necessary as blood in tomorrow's battles.[69]

Thus France's alleged oil problems were due to American tankers plying Pacific waters. Given the months it takes to get a tanker from Asia to Europe, had the situation truly been dire, the time to have sent the telegram, which speaks of measures to be taken "without one day's delay," was not in December 1917 but in October, when it was first forecast that France would be out of oil in January. Clemenceau then told the American ambassador Sharp that he might sign a separate peace with the Germans if France could not get enough oil.[70]

In fact the oil situation was not bleak at all. By January France had on hand not zero oil reserves, as forecast in October, nor 30,000 tons, as forecast in December, but 50,000 tons, and the revised figures for March now stood at 36,000 tons of oil on hand rather than nothing.[71] This was without any change in tanker allocation procedures whatever. Within ten days of Clemenceau's telegram, four new tankers were allocated to France, and the fictitious strategic "crisis" was over.[72] The American Pacific fleet, which Bérenger and Clemenceau were so eager to pull out of Asian waters, had absolutely nothing to do with augmenting French oil supplies, for four ships were found and allocated

[68] Nayberg 1983, pp. 192, 385.

[69] See Nowell 1983, p. 236, for complete text; L'Espagnol de la Tramereye 1921, pp. 104–105, for the French.

[70] Nayberg 1983, p. 183.

[71] Nayberg 1983, pp. 205–206.

[72] Nayberg 1983, pp. 201–202. Nayberg interprets the data differently, but he agrees that Bérenger overstated French needs.

for French needs well before any vessel in Asian waters could have been positioned to service the European market. The references to the Pacific trade were Bérenger's efforts to protect the Asian markets of Royal Dutch–Shell.

If the French faced any crisis at all, it was that they wanted to build up their reserve stock to about 100,000 tons to face surges in demand associated with military offensives. Even the skeptical officers of Standard Oil, Bedford and Teagle, could agree to this.[73] The ordinary, predictable needs of France, including the military's needs when not engaged in a full offensive, were met by ongoing arrivals of oil from the United States.[74] By the end of February 1918 stocks on hand had climbed to 78,000 tons, and ongoing deliveries kept them at that level during Germany's spring offensive.[75]

In fact, oil flowed into France faster than the country could absorb it, in spite of the offensive. By June 1918 France was so engorged with oil that it could take no more deliveries; Bérenger had two arriving tankers of oil passed on to Italy with a mendacious explanation to the Americans that their help had "allowed us to get through the battle and minimally augment our diminished stocks." At the same time he informed Tardieu, "for your strictly personal information our coastal reservoirs on the Channel and on the Atlantic are full and domestic distribution is hindered by the lack of railroad cars."[76] Not to be fooled any longer, the Standard Oil Company pulled eight tankers out of French service and reassigned them to other markets.[77]

The strategic objective of increasing reserves to 100,000 tons was used to try to force Standard Oil's fleet out of the Pacific. Standard Oil increased the strategic reserve in its own way; blustering and threatening a separate peace, Bérenger only proved to the world's foremost oil supplier that French policy was set on a course directly antagonistic to its interests. As a result of this ill-considered favoritism, the French government lost all credibility with Standard Oil.

The French "oil crisis" of 1917 combined a request to increase reserves with a ploy to hamper Standard Oil in the Far East. The effort also aimed to discredit Standard Oil in France and make Royal Dutch–Shell look pro-French. The "crisis" of 1917 was not an oil shortfall

[73] Nayberg 1983, p. 145.

[74] Clémentel testified about low French reserves and simultaneously argued that in October and November 1917 expenditures on oil were double those of the year before. Even taking inflation into account, his figures argue that somewhere, someone in France was buying oil. See his letter to Bérenger dated 29 December 1917 (Arch.Nat. F[12]7715, folder "Lettre du Ministre du Commerce au Ministre de l'Agriculture").

[75] Nayberg 1983, p. 232.

[76] Arch.Nat. F[12]7715, letter to Tardieu dated 3 June 1918.

[77] Nayberg 1983, p. 338.

imputable to American nefariousness. The French desire to increase reserves was met promptly with four additional tankers ten days after diplomatic intervention. Six months later, France was glutted with oil. The ad hoc procedures in place were more than adequate to deal with France's oil supplies, even at the height of submarine warfare. Yet by the end of December 1917 Clémentel and Bérenger were forging plans to purchase France's oil through a government consortium. Royal Dutch–Shell had, as Gulbenkian writes, made astute political connections via Bérenger with key persons in the French government, to whom it fed information critical to the formulation of national policy. During the crucial months toward the end of 1917 Bérenger maintained close contact with the Shell group, on which he depended for technical assistance and information. In a letter to Gulbenkian, he asked that his gratitude to Deterding be passed along, for having put "at the disposition of our committee [Comité Général des Pétroles] the vast experience of his various technical services."[78]

Mesopotamian oil was also a powerful lure for practitioners of realpolitik, and the alliance with Royal Dutch–Shell fit Bérenger's and Clemenceau's desiderata on this score. The need to use the Shell combine to get France into Mesopotamian oil was hardly a consensus matter among financial and political elites. Even if we concede the point, to say this is quite different from saying that Standard Oil failed France in a dark hour and that this is why the government regulated the market. The shenanigans over oil tankers in 1917 were a sideshow used to position the Shell combine to represent France's oil interests in Mesopotamia.

THE CONSORTIUM AND THE INTERALLIED PETROLEUM CONFERENCE

The Shell combine succeeded in getting Standard Oil to assign four more tankers to France. Now Standard Oil and its French affiliates had to be kept from gaining market share in France when those tankers delivered. Under the reserve system inaugurated with the war, importing companies had to maintain reserve stocks; if the increased oil arriving from the United States was, under that system, stored by Standard Oil's allied French companies, they would have a tremendous advantage. To prevent this, a regulatory means of allocating market share was needed, which in fact was part and parcel of the December 1917 "crisis."

[78] Arch.Nat. F[12]7716, letter dated 4 December 1917. This is not Gulbenkian writing about Gulbenkian, and it helps corroborate the Armenian's memoirs.

A simple import quota system would not have worked because too much of the world's oil was in Standard Oil's hands. Regulators would need to procure oil from all available sources, including Standard Oil, and then reallocate it so that Standard Oil's French client firms did not totally dominate France. In a word, Standard Oil had to be forced to supply its competitors, foremost of which were Royal Dutch–Shell and its client firms.

Therefore the recklessly preferential treatment for the Shell group increased at the beginning of 1918. Clémentel asked for a state import monopoly in a letter sent to Bérenger on 29 December 1917.[79] The minister of commerce claimed that increased payments to cover the rising oil imports required an import consortium; the state should purchase all oil and divide it among companies. Clémentel mentioned the cost of paying for the cargoes on the four tankers allocated by Standard Oil to build up reserves. Bérenger wanted a monopoly because Standard Oil allegedly was not delivering, while Clémentel wanted it because the company *was* delivering: the two could not get their story straight.

Oil expenditures had doubled from the year before, reaching in October and November of 1917 about 20 percent of France's total import costs.[80] Independent French oil concerns had bid against the authorized government representative, as in the *Muskogee-Goldshell* incident, and this may have raised costs. If that were true, the French domestic companies ought to have supported the import monopoly; in fact, they split along pro- and anti-Standard lines.

French policy wholeheartedly favored the Shell group. Just as Clemenceau's tanker request had mixed favoritism for the Shell group in Asia with the strategic objective of increasing reserves, so the financial need for an import monopoly became embroiled in more favoritism. Standard Oil interpreted the state monopoly as an attack on its interests: its Bedford Petroleum Company[81] wrote to Clémentel on 30 January 1918. Surely, the company reminded Clémentel, the world's largest oil exporter would be consulted if such a monopoly were indeed contemplated. It observed that "unless there are new facts that we do not know about, such a measure could only add to the current difficulties of the country, and be prejudicial to the interests of the French and

[79] Arch.Nat. F^{12}7715, folder "Lettre du Ministre du Commerce au Ministre de l'Agriculture," letter dated 29 December 1917.

[80] See previous citation and also Clémentel's deposition before parliament (Arch.Nat. F^{7}7662, Commission du Budget, 1918).

[81] Standard Oil's only subsidiary in France specialized in lubrication products and was kept small because of pre–World War I agreements with the French refiners' cartel.

the allies." The letter concluded, "We respectfully but energetically protest such a plan."[82]

Standard Oil gathered additional support. Another letter addressed to Clémentel was sent by the Chambre Syndicale des Huiles et Graisses Industrielles, which represented all American companies exporting to France. Among these were Bedford Petroleum Company (Standard Oil of New Jersey), Vacuum Oil (Standard Oil of New York), the Union Petroleum Company, the Pure Oil Refining Company (another Standard Oil of New Jersey firm), Gulf Refining, and the Atlantic Refining Company (separated from Standard Oil in the breakup of 1911). The companies pointedly reminded Clémentel that they brought in 75 to 80 percent of France's oil. "You cannot keep separated from the consortium, Monsieur le Ministre, the majority of French importers in favor of one group representing the minority [of the oil market], which allows the market to be grabbed up by certain privileged people."[83] The companies asked to be part of Bérenger's Comité Général des Pétroles, whose scare tactics were causing them so much difficulty, "so that we can defend our interests at the heart of this commission." Fenaille et Despeaux, Standard Oil's closest French ally, also opposed the state monopoly and distribution consortium.[84] Bérenger, backed by Clémentel, ignored the protests, and the government adopted the monopoly project on 29 March 1918.

In the first six months of 1918, over American protests, the British and French maneuvered to control American oil shipping. The French also sought to control the American presence in their market. The state import monopoly led to a reorganization of the Comité Général des Pétroles, creating an agency with full administrative powers over the French oil market. The shared British and French concerns about shipping led to the Interallied Petroleum Conference, which united the interests of disadvantaged participants in the world oil market against Standard Oil. Only months after entering oil affairs, in June 1917, Bérenger had proposed such an international authority, echoing Royal

[82] Arch.Nat. F^{12}7716, letter dated 28 January 1918.

[83] Arch.Nat. F^{12}7716, letter dated 30 January 1918.

[84] Nayberg 1983, p. 284. Some historians consider Clemenceau's 15 December 1917 telegram to Wilson, and later consortium, to have been provoked by a 5 December 1917 letter from the Chambre Syndicale de l'Industrie de Pétrole. The letter was signed by G. Despeaux, the president of the group and also a partner in Fenaille and Despeaux. He warned that stocks would be at the "zero level" by March. (See Murat 1969, p. 14; Touret 1968, p. 23; Nayberg 1983, p. 171.) Given Despeaux's opposition to the consortium two months later and the unanimous opposition of American firms, as well as the inaccurate data in Despeaux's letter, it seems the firm was uncertain how its long-term interests could best be served and therefore vacillated between allegiance to Standard Oil and to the Shell combine.

Dutch–Shell's concerns about the "waste" that competition created.[85] His plans were stated in somewhat greater detail in an October letter to André Tardieu.[86] By February 1918 the British wanted to allocate oil tonnage, because the ships Standard Oil gave to France in December 1917 had originally been destined for Great Britain.[87] Britain's ruffled feathers should be kept in perspective: it had requisitioned nineteen ships of Socony's Hong Kong fleet in November, and on balance it was still ahead.

Etienne Clémentel, Sir Walter Long, and Sir John Cadman held preliminary meetings of the Interallied Petroleum Conference in February 1918; the American delegation, skeptical of British and French motives, did not join until April. The British tried to force France into backing their petroleum policy by demanding the return of French-owned oil tankers under British registry, almost the whole of the French oil fleet.[88] The French found this overbearing and, in spite of their support for tanker rationing, refused to cooperate until the Americans joined the conference.

In this French policy was not so much pro-American as realistic. The Americans controlled the vast majority of oil and shipping; an Interallied Petroleum Conference without the Americans would have surrendered French policy to the British. The same dominance of American resources made the Interallied Petroleum Conference largely symbolic: the real world powers in shipping and oil distribution, guided by Standard Oil, were the Shipping Board and the Petroleum Committee, both in the United States.

The British wanted to divide European markets and allocate tanker tonnage so as to justify the long haul of imports from the Far East. The French supported them. Because Transcaucasian and Romanian oil sources were shut down, 40 percent of the French market had opened up to other companies. Royal Dutch–Shell's direct imports into France grew from 1.7 percent in 1913 to 14 percent in 1917; American companies, chiefly of the Standard group, took the rest. The Americans conceded that Borneo imports for toluol were justifiable, but the total tanker tonnage needed for this purpose was small.[89] Bérenger made extravagant claims to allocate tankers to the Shell group, maintaining that the net turnaround disadvantage of Dutch East Indies

[85] Nayberg 1983, p. 103.

[86] Arch.Nat. F[12]7716, letter dated 20 October 1917.

[87] Nayberg 1983, p. 295.

[88] Nayberg 1983, pp. 291, 296–311.

[89] The British wanted 132,000 tons of oil annually from Borneo, of which 30,000 was for explosives. The Americans wanted to limit deliveries to the 30,000 tons for explosives. Arch.Nat. F[12]7715, letter dated 3 June 1918.

imports, compared with American imports, was only 1.5 percent.[90] He also argued that Borneo oil had political advantages because of diplomatic relations with Holland, one of the host countries of Royal Dutch–Shell, and the alleged "tranquillity" of the Asian populations of the French empire. His intricate sophistry argued that the Asians would revolt if their lamp oil prices rose precipitously. This claim assumed a direct correlation between lamp oil prices and political restiveness and also assumed that if the Royal Dutch–Shell group did not furnish oil to these areas no one would. This was false: the whole struggle centered precisely on keeping alternative suppliers and their oil out of Asia.

Bérenger admitted that his favoritism of Royal Dutch–Shell took the risk of "indisposing Standard Oil. I think we can go ahead anyhow because Standard already has a serious advantage and cannot hope to deny Royal Dutch all business."[91] Standard Oil's delegate at the Interallied Petroleum Conference tactfully acknowledged that "the discussed political considerations had their value." French supply contract negotiations with the Shell group were under way before the Interallied Petroleum Conference, suggesting plans to increase that company's presence on the national market regardless of Standard Oil's approval. The Shell group was to supply all oil to the Mediterranean French ports, eliminating the Americans from that market.[92]

Standard Oil realized it could not use wartime circumstances to eliminate Royal Dutch–Shell, and it bowed to the pro-Shell policies inspired by the French and British governments' desire for tankers to bring in Dutch East Indies oil. Bérenger, meanwhile, sought to increase his own influence in French oil affairs and, helped by Clémentel in the commerce ministry, wrested control over policy from the ministry of agriculture and supplies. Any move that increased Bérenger's personal power also benefited Royal Dutch–Shell. Though still under the ministry of agriculture and supplies, the new Commissariat General aux Essences et Combustibles was formed 21 August 1918, with Bérenger

[90] Arch. Nat. F[12]7715, letter dated 3 June 1918. Assuming tanker speeds of eleven knots an hour, a ten-day layover on each side, departures from Texas for American oil and from Borneo for Dutch East Indies oil, with delivery at Marseilles, which biases the calculations in favor of Far East oil, round-trip time for American oil comes out to sixty-four days versus seventy-four days for Indonesian, an increase of 15 percent. Since much American oil was delivered from New Jersey, that tanker turnaround time would be even more advantageous for the Americans. Bérenger also argued that Dutch East Indies oil was better priced, but given the liberty he takes with geography, his price data may also be questionable. See also discussion of toluol above.

[91] Arch.Nat F[12]7715, Bérenger to Tardieu, letter dated 3 June 1918.

[92] Arch.Nat. F[12]7715, letter dated 3 June 1918. Nayberg 1983, pp. 323–324, puts the contract negotiations in February 1918.

at its head: unlike the Comité Générale des Pétroles, this new agency had completely centralized control over oil policy.[93]

Royal Dutch-Shell and the British-French Oil Alliance

France's strong tilt toward the Royal Dutch–Shell group helped it and the Burmah Oil Company meet the challenge of American penetration in Asian oil markets. The Interallied Petroleum Conference marked the beginning of an anti-American French and British alliance in oil policy.

The spectacular growth of wartime oil use is said to have forced the French government to "realize" that this policy area was too important to leave to the private sector. But state regulatory measures regarding market shares (the import monopoly) and tanker allocations (the almost exclusive concern of the Interallied Petroleum Conference) were not necessary to ensure the delivery of oil: they were necessary to control the growth of Standard Oil. Strategic motivations for this favoritism lay more in Royal Dutch–Shell's large Rothschild stockholdings than in any inherent differences between the two companies as suppliers. Still, the Shell combine dangled Mesopotamia as bait; the American firm could not. Royal Dutch–Shell selected itself to be France's chosen instrument, but that made it no less appealing to Clemenceau and Bérenger.

French interventionism in the pre–World War I period, except during 1914, was limited to tariff policies and balancing the lamp oil market between oil and the rival agricultural alcohol lobby. Mesopotamia was off limits to government policy because the Rothschild banking interests and the Shell group had reached a satisfactory deal with the Deutsche Bank in the months before the war. French intervention in oil policy abroad was neither welcome nor needed; at home the strategic dimension of oil policy was deliberately ignored as the coal sector kept France out of the naval oil conversion programs that Germany, Italy, Russia, and Great Britain had embraced.[94] The most interventionist policy was sponsored by the alcohol lobby, which—as an extension of the powerful French wine growers—attempted to strangle the domestic lamp oil market. In spite of repeated interventions in tariff and taxation structure as well as proposed fuel monopolies, France had no coherent oil policy encompassing international objectives.

World War I changed this: French oil policy in 1918 and 1919 was

[93] Nayberg 1983, pp. 335–336.
[94] See previous chapter.

coherently organized around precise objectives. Neither strategic factors nor the state's desire to maximize power, but rather war-related shifts in the national and international political economy, caused this new concern. Shell worried that if the Turkish Petroleum Company dissolved, Mesopotamia would go over to the Anglo-Persian and Burmah Oil interests. Increased oil fuel needs and the loss of production in Romania and Russia destroyed the pan-European coalition of interests that had hitherto kept France from having an oil policy with an international perspective.

In France, Royal Dutch–Shell seized the political high ground. Its objectives were to stop Standard Oil in Far East markets and to stabilize its market share by securing the necessary tankers and regulated markets. The Shell combine's desire to maintain its hold on Mesopotamia caused it to recruit France for an aggressive oil policy in that region. As Gulbenkian wrote, "Without trespassing on the limits of modesty, had I not directed Mr. Bérenger, the French Government would never have entered the Iraq Petroleum Company. . . . Our French friends now pride themselves on their rights under the Treaty of San Remo but—and this is probably due to the fear of feeling themselves under obligation—they always avoid remembering that it was due to my conception."[95]

NEGOTIATIONS OVER MOSUL

Mesopotamian oil was integral to the division of the Ottoman Empire. Negotiations begun in 1915–1916 led to the Sykes-Picot agreement signed in mid-1916. France gained title to Mosul, which lay in the heart of the oil-bearing regions of modern Iraq. Given the oil wealth of this prize, Britain's willingness to cede the area to France in 1916, and France's willingness to give it back in 1918, are a puzzle.

Three reasons are advanced for Britain's ceding Mosul to the French in 1916 and its later successful attempt to retrieve it. First, Britain wanted France to maintain nominal control over the area as a buffer zone between itself and the Russian empire, which in 1916 still existed;[96] after the Russian Revolution, Britain lost interest in a buffer zone.[97] Second, after the Turkish capture of General Townshend's

[95] Gulbenkian 1945, p. 32.

[96] See map of the division of spoils, figure 3.

[97] A widely held view; but see, for example, R. de Gontaut Biron's published report, *La France et la question de Mosoul* in MAE Levant, Turquie, 1918–40, vol. 430, p. 5 (of the report); also MAE E-Levant, Irak-Mésopotamie, vol. 32, letter dated 11 March 1920: "Regarding Mosul, it is the English who pushed us with all their might to take this area,

forces on 28 April 1916, the British wanted French participation in the Middle East campaign and ceded Mosul to give an incentive: but an overextended France committed only a small number of troops, and the British alone wrested control of Iraq back from the Turks.[98] Third, the cession of Mosul was a British blunder that was quickly perceived as such and corrected. These explanations are not mutually exclusive, and there is no need to choose between them here; of more interest is why the French, once handed the prize, gave it back.[99]

According to Kent, the French request for Mosul was, in Sykes's opinion, motivated by "Franco-Levantine financiers" interested in expanding their railroad network.[100] The division of the Ottoman Empire among the Entente (see map in fig. 3), compared with the structure of railroad ownership and development in the area (see map in fig. 4), does explain French commercial interest in Syria and the possibilities Mosul represented for expansion. No financier in 1916 could have been interested in Mosul and its railroad without being aware of the petroleum mineral rights that went with the railroad.[101] Yet the award of Mosul to the French on 5 January 1916 specifically excluded rights over the oil-bearing territories: the British protected their Turkish Petroleum Company concession and therefore Anglo-Persian's interests.[102] Senator Henry Bérenger himself would deprecate the 1916 accord, writing that "we had influence over Mosul, but none whatever over the oils of Mosul!"[103]

The French "Mosul without oil" policy has two explanations. The first is that the French were unable to get the British to give up more and took what they could get. Second, interbank rivalry in France played a part. The secretary general of the French foreign affairs office, Jules Cambon, was intimately linked to the Banque de Paris et des Pays Bas, serving on its board of directors and other associated interests

because at that time Russia still existed and they were practitioners of the theory of buffer states."

[98] See Neumann 1935, p. 145.

[99] The "mistake thesis" about Mosul is advanced by Kent 1976, pp. 122–123, who cites evidence for the "buffer zone" explanation and follows with the members of government opposed (pp. 124–125) to giving up the area.

[100] Kent 1976, p. 122.

[101] Curiously, an "oil expert" sent to help Paul Cambon negotiate with the British insisted that the greatest economic interest in Mesopotamia lay in its vast potential for wheat production. Cambon considered his helper a worthless eccentric (MAE E-Levant, Mésopotamie-Irak, vol. 32, letter dated 11 March 1920).

[102] Melby 1981, p. 23.

[103] MAE Papiers d'Agents, André Tardieu, vol. 56, B: Comm. Gén. aux Essences. "Note Resumée sur les Pétroles de Mésopotamie et les Accords Franco-Britanniques," by Henry Bérenger, dated 10 May 1920. Exclamation point in original.

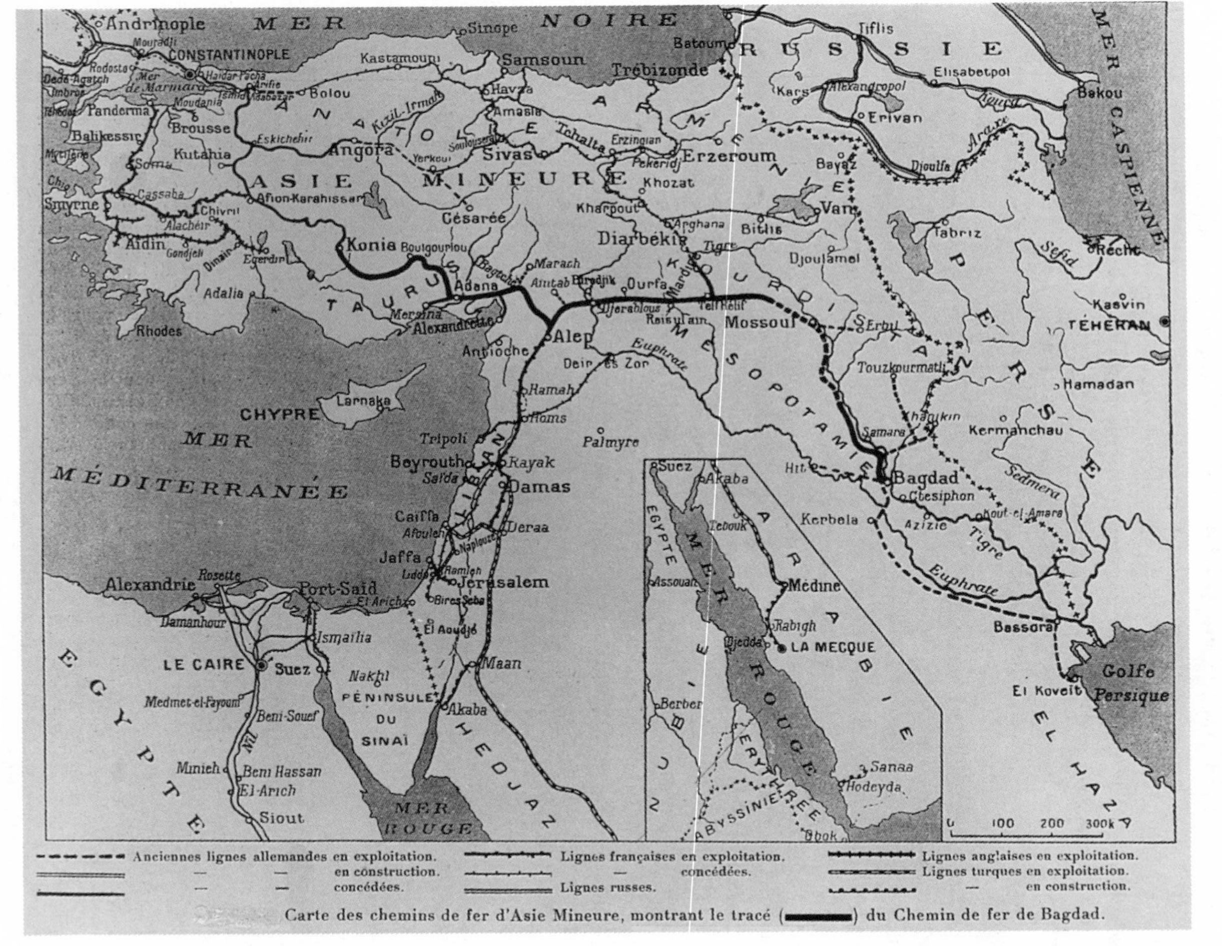

Figure 4. Railroad ownership in the Middle East. *Source:* Calfas 1921, p. 26. Phot. Bibl. Nat. Paris.

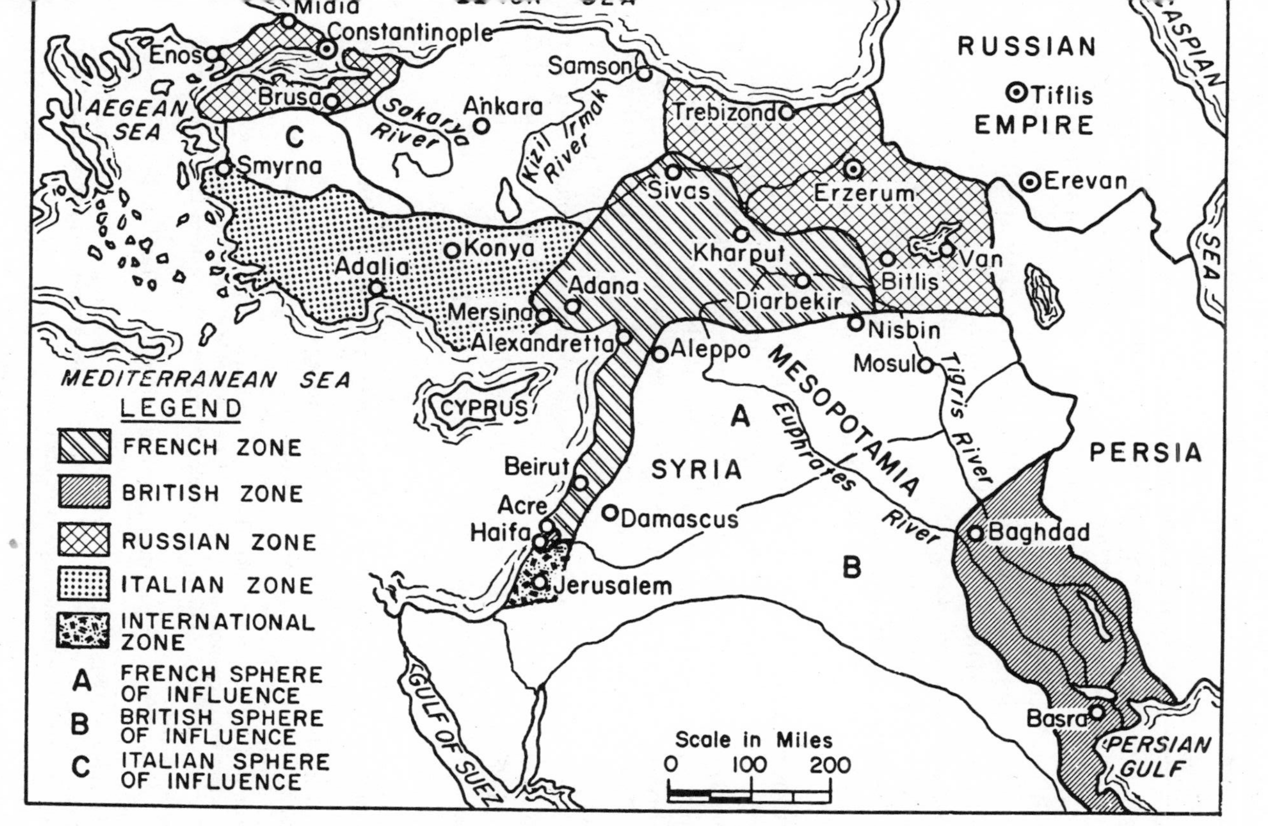

Figure 3. The planned partition of the Ottoman Empire: the Sykes-Picot agreement, 1916. *Source:* Richard Hovannisian, *Armenia on the Road to Independence, 1918,* p. 61. Copyright © 1967 The Regents of the University of California.

after his retirement in 1922.[104] In 1920 Jules Cambon would become a member of the board of directors of Standard Oil's new subsidiary in France, Standard Franco-Américaine.[105] His brother, Paul Cambon, was ambassador to London, and on 15–16 May 1916 he exchanged letters with Sir Edward Grey on the subject of preserving Britain's previously acquired economic rights in the area to be ceded to France.[106] Aristide Briand, also on the French side of the negotiations, was sympathetic to the same banking group.[107] Fifteen years later Aristide Briand, as foreign affairs minister, testified before a closed parliamentary committee that in 1916, "We made the greatest effort to have [in our mandate] the Vilayet of Mosul. *I won't hide the fact that what made us put our heart into the negotiations on this point was our certainty that important oil reserves were to be found there.*"[108] The Banque de Paris et des Pays Bas owned the largest French firm operating in Romania before World War I and was naturally interested in Mesopotamian oil.[109]

France settled for Mosul without oil rights, even though everyone at the conference knew what he was bargaining for. The arrangement was most likely an effort by the Banque de Paris et des Pays Bas to stop the ambitious expansion of the rival Rothschild banking interests. Before the war, the Banque de Paris et des Pays Bas, knowing all about the Baghdad railroad and Mesopotamian oil project, nonetheless favored the division of Syria and Turkey into spheres of interest; Syria became "French," the Baghdad railway became "German." But the Banque de Paris et des Pays Bas got part of the deal by floating the German securities in Paris, one way for the bank to profit from the plan.[110] The Rothschild bank, as shareholder in Royal Dutch–Shell,

[104] See biographical details in the *Dictionnaire de biographie française*, 7:962–963. Cambon's influence diminished with Clemenceau's rise to power in 1915, but he was named to the Peace Conference in 1919. The Banque de Paris et des Pays Bas became the official Standard Oil bank in 1920, placing it firmly opposite the Rothschild alliance with Royal Dutch–Shell. Gibb and Knowlton 1956, p. 282, follow Gulbenkian's own dating, which I contend is faulty, and therefore suggest that his actions helped lead to the awarding of Mosul to the French. Their interpretation would simplify matters greatly, but it does not correspond to events as I have been able to establish them.

[105] MAE Etats-Unis, 1918–1929, vol. 250, Article "La Mercante et le pétrole," dated 24 October 1920.

[106] As related by Bérenger (MAE Papiers d'Agents, André Tardieu, vol. 56, A, letter dated 15 March 1920).

[107] Jeanneney 1981, pp. 131–190, chronicles the intricate connections between Briand and Finaly of the Banque de Paris et des Pays Bas during a later financial scandal. Finaly boasted (p. 173), "I hold the majority of the Chambre."

[108] Ass.Nat., Procès Verbaux de la Commission des Affaires Etrangères, audition de M. Aristide Briand, 9 February 1931. Emphasis added.

[109] On the Omnium, the Banque de Paris et des Pays Bas, and Romania, see below.

[110] Feis 1964, pp. 197, 357.

had only two years earlier verged on stupendous success in the Middle East, through the concession signed for the Turkish Petroleum Company. Those rights were in theory preserved by the 1916 agreement.

In 1916, however, the extension of French sovereignty over Mosul would have greatly complicated the 1914 arrangements. The new authority of the French government provided an opening for other French interest groups to make demands to modify the Turkish Petroleum Company. Mosul "without the oil" was a logical first move to break the Rothschilds' exclusive participation in the Mesopotamian oil bonanza. That is exactly what occurred in the end: the Banque de Paris et des Pays Bas became the dominant French bank in Mesopotamian oil, through the Compagnie Française des Pétroles. But in 1916 the Banque de Paris et des Pays Bas envisioned the glorious international prize going to its Romanian oil subsidiary; shuttling sovereignty over Mosul was the first in a series of engagements with the Shell–Rothschild interests.

The Shell combine's claim to Mesopotamia in 1916 was under a double attack: from Anglo-Persian and the British government, and from the Banque de Paris et des Pays Bas. The British had fielded an army in Mesopotamia, and it was they who principally threatened the combine's interests. To get diplomatic leverage against Britain, Gulbenkian proposed that the French participate in the Mesopotamian oil concession as elaborated in the 1914 oil agreement:

> It occurred to me that I would be rendering a valuable service to the French Government, and that it would enhance also the popularity of the interests I was representing, if I could assure to the French a direct source of oil supply. At that the time, round about 1915/1916, the oil participation of the Deutsche Bank, as explained elsewhere, had been taken over by the British Treasury and the Custodian as enemy property. . . . I felt the opportunity was excellent if I could arrange for this German participation to be handed over to the French State, after the war. . . . I took great pains in convincing Sir Henry Deterding, who finally agreed, subject to my arranging a French Company in which the French and the Royal Dutch would be jointly interested. . . . Sir Henry Deterding . . . had at first been reluctant to agree to the French participation, then thought, at my instigation, that the joining in of the French government as a participant in our Company, would be another trump to deter the designs of the Anglo-Persian Company, who only came in definitely to take up their participation . . . after the San Remo Treaty.[111]

[111] Gulbenkian 1945, p. 20. Ferrier 1982, p. 258, corroborates Gulbenkian's role, calling him "one of the principal intermediaries in negotiations over the Mesopotamian Concession on their [Royal Dutch's] behalf."

The French government could pursue Middle East oil with the Sykes-Picot accord, which potentially favored the Banque de Paris et des Pays Bas, or it could protect the 1914 agreement as desired by Royal Dutch–Shell. The Foreign Office, contending the 1914 agreement was no longer valid, had backed Anglo-Persian's aggressive claim to sole oil rights in Mesopotamia. Of the Anglo-Persian's strategy, Gulbenkian records, "Sir Henry Deterding was of the opinion that [it] had been engineered by Sir Charles Greenway who had a great pull at the Foreign Office, so as to justify the stand he had taken in not carrying out the stipulations of the March 1914 Agreement."[112]

In short, with the British recapture of Mesopotamia in 1917, Anglo-Persian envisioned annexation by conquest. Royal Dutch–Shell could hang on only by maintaining the validity of the Turkish Petroleum Company's contract in what was now an enemy country. By bringing France in to claim the one-quarter of the concession held originally by the Deutsche Bank, the Shell combine breathed new life into a moribund document. The Turkish Petroleum Company shareholdings became the cornerstone of French oil policy in the interwar period, but it originated in the maneuvering of Royal Dutch–Shell, which needed France as "another trump" in the complex relationships with Anglo-Persian and the British government. Without Royal Dutch–Shell, France's oil concerns might have found another avenue of expression, but the company's decision to ally with France propped up the 1914 Turkish Petroleum Company agreement and had a durable effect on Middle East oil policy.

The key to this masterly diplomatic strategy was Deterding's proviso that the "French" share of Mesopotamian oil be in fact held by a joint company in which the Shell group would be the principal member. France would help the Shell group, but Shell, though it would reward France with a share, would control all operations. The Shell group would have control of its 25 percent, plus the French 25 percent of the concession, and become the dominant partner. This was offensive enough to the British. But the greatest injury, on the French side, was that preserving the 1914 agreement left out the rival Banque de Paris et des Pays Bas. The exclusion of this bank, which was allied with the Standard Oil Company, guaranteed the later collapse of Gulbenkian's and Deterding's schemes to control Mesopotamia and the French oil market.

The Shell combine's alliance with France also thwarted an attempt to make Royal Dutch–Shell "British." Between November 1917 and May 1919 proposals were discussed in Britain to amalgamate the Royal Dutch–Shell and Anglo-Persian interests, including their claims to the

[112] Gulbenkian 1945, p. 21.

Mesopotamian concession. A fight erupted over which company would swallow which. The British government wanted the Dutch half of Royal Dutch–Shell to be under British control, or the "British" (Shell) half of the company to be controlled by the British government. One variant of these plans even included a shareholding for France and was approved on 5 May 1919 by the British cabinet.[113] The plan was a preposterous abdication of power for Deterding, for which even Mesopotamia could not compensate. The accord, which was initialed but not signed, had a key clause: even if approved by the British government, it also had to be ratified by the Dutch directors of the company. Deterding could count on them to veto the plan. Plainly put, Deterding was stalling: he could not possibly have met the demands of this plan and delivered on what he was negotiating with the French government at the same time.

The Shell combine's dilatory negotiations with the British over "anglicization" allowed the company to consolidate its alliance with France. While the company proposed British control over its activities in London, across the Channel it sought to become the official oil company of France. The combine could not be under the ultimate jurisdiction of the British and also pledge unconditionally to give priority to French needs. Collectively, the proposals were absurd: one might have had "Shell" ships under British national control, with no authority over the oil fields owned by the "Dutch" half of the company; and the corporation as a whole obliged to put service to France as its highest priority.

Because the "French" 25 percent of Mesopotamia was to be managed by Royal Dutch–Shell, the combine gained an additional 25 percent that would nominally be French and evaded concessions to the British. On the English side there was unending competition and departmental infighting with Anglo-Persian.[114] Britain's bait to gain control over the Royal Dutch–Shell combine was participation in Mesopotamia. The combine's ability to get this booty with French help destroyed the basis for making the group British. Deterding and the Dutch half of the company remained in control of the combine and effectively free of any kind of government control anywhere.

Britain's failure to anglicize the Shell combine was due neither, as

[113] Ferrier's history (1982, pp. 245–261) of the plans to make the Shell group British is detailed, but I disagree with some of his analysis (e.g., Deterding had no reason to be disappointed with Paris in January 1919, as Ferrier says on p. 259) as well as the reasons he gives for the plan's failure. Kent 1976, pp. 178–182, reproduces the text of the plan.

[114] In spite of what Gulbenkian called Greenway's (of Anglo-Persian) "pull" at the Foreign Office, he and the other officers of Royal Dutch–Shell did not lack influence. Gulbenkian 1945, p. 20, claims to have cleared his plans for French participation with Sir William Tyrell, undersecretary of state, and Walter Long of the Foreign Office before approaching the French government with his plans; but within the Foreign Office much opposition remained.

Ferrier contends, to "lack of political will" nor to an easing of competitive pressures worldwide.[115] Royal Dutch–Shell kept anglicization as a backup plan in case the primary strategy, using France to preserve the March 1914 Turkish Petroleum Company agreement, did not work. Since it did work, Gulbenkian did not even find it necessary to discuss anglicization of the Shell combine in his memoirs.[116]

When Germany surrendered, France lost the fig-leaf "strategic" need to control market shares and shipping. Yet precisely at this time French state intervention grew, closely coordinated with the Royal Dutch–Shell group. In early 1918 the import monopoly merely preserved the Shell group's market share. By late 1918 the intention was to turn the French market entirely over to Royal Dutch–Shell, which would need help to have enough oil to supply it. Favoritism to the combine was explicit by the end of 1918. France's Premier Clemenceau wrote to his London ambassador: "It is Royal Dutch–Shell that will serve us as intermediary in this combination [partnership in Mesopotamia]. Senator Bérenger tells of the very favorable disposition of the president of this association, who, desiring the support in the East and the Far East of a great occidental power and convinced of the benefit of a strong collaboration with our country in particular, would be willing to offer us his support to help us bring about our oil policy."[117] Clemenceau's text is nearly identical to a letter sent to him two weeks earlier by Bérenger; he is approving policies put forward by the senator.[118] Clemenceau approved of the Shell combine's offer to create a French company that would take charge ("prendrait à son compte") of all shares France might procure in all producing countries, not just Mesopotamia.[119]

The alignment with the Royal Dutch–Shell group also had the approval of Clémentel in the ministry of commerce, but other officials worried. A. Guiselin in the finance ministry argued that the plan came "up against a number of all-powerful private interests," and that "the question of oil touches too many interests, who are formidably organized, to envision a policy, as does M. Bérenger, with the sole support

[115] Ferrier 1982, p. 260.

[116] As for Cadman's (of Anglo-Persian) assertions, cited by Ferrier 1982, p. 260, that the Standard Oil Company had offered to work the Mesopotamian oil concession on behalf of the French, there is little basis to see a pro–Standard Oil tilt in French policy in 1919. There were two meetings between high officials of the French government and Standard Oil's chairman in the spring of that year (see below), about which Cadman may have been misinformed (or trying to mislead others), but French policy was anti-Standard to an extreme degree.

[117] Arch.Nat. F^{12}7716, Clemenceau to Cambon, 17 December 1918.

[118] Bérenger to Clemenceau, in MAE E-Levant: Irak-Mésopotamie, vol. 32, letter dated 29 November 1918.

[119] Arch.Nat. F^{12}7716, Clemenceau to Cambon, 17 December 1918.

of a single company, no matter how powerful it might be, even Royal Dutch."[120] This stellar perspicacity was ignored. An exclusive concession for "the sale and refining of Algerian oil, as well as the importation and sale of imported refined oils" and also "all other minerals" was given in 1919 to British and French investors closely linked with the Shell group.[121] The Algerian concession was a lagniappe in the world oil market, but the collaboration with Britain, an outgrowth of using Royal Dutch–Shell as "national champion" of France in Mesopotamia, is clearly seen in the disdainful comment, "It is indeed Algerian oil to the English, just as Mr. H. B. desires."[122] But "Mr. H. B.," Henry Bérenger, enjoyed Clemenceau's support. The president instructed the ambassador in London that "we [should] let the British acquire Algeria, as well as our assent to an oil pipeline from Persia (and Mesopotamia) to a Mediterranean port," so as to have "good negotiating points" for French participation in Mesopotamia and Persia.[123] It does not appear that the French government was ever seriously committed to gaining access to Persia, but Royal Dutch–Shell certainly was, and it formed a company for the purpose in 1920.[124] The reference to Persia reflects the Shell combine's agenda, not the feasible diplomatic objectives of the French government. Favoritism to the combine's interests in Algeria was all of a piece with French Mesopotamian diplomacy: Royal Dutch–Shell carried the day.

With Clemenceau's blessing, Bérenger negotiated with the British

[120] Arch.Nat. F^{12}7716, letter dated 20 February 1919. The document is signed by A. Guiselin, who was "directeur du mouvement des fonds."

[121] The company was the Société d'Études, de Recherches, et d' Exploitation des Pétroles en Algérie. A. Guiselin (Arch.Nat. F^{12}7716, letter dated 19 March 1919) analyzed the ownership structure in detail and concluded: "This is a disguised English company." Even that is questionable, since in these elevated financial circles nationality was more a matter of form than of content. The company's two principal stockholders, Basil Zaharoff and S. Pearson and Sons, were intimately linked to the Shell group; the British Pearson group was selling its Mexican oil to Shell. Zaharoff, a fabulously wealthy Greek arms merchant and financier, gained French nationality in 1918 through questionable procedures (debated in parliament 26 July 1918, according to Neumann 1935, p. 150) and was decorated by Clemenceau with the Légion d'Honneur in 1919. Zaharoff worked with the Banque de l'Union Parisienne, the Shell group's business bank in France; the Algerian company was dominated by the Pearson group, Zaharoff's personal shareholdings, and those of the bank he owned, the Banque de la Seine. Zaharoff's biographer gives details of these connections (Neumann 1935, pp. 153–155). Clemenceau called the Algerian group "British," meaning the Pearson interests, but this enterprise flew the flag of Royal Dutch–Shell.

[122] Arch.Nat. F^{12}7716, letter dated 19 March 1919. The words are A. Guiselin's. Whether the oil wealth of Algeria was deliberately not developed before World War II to avoid overproduction is an open point. Oil journals from the period show suspected deposits along the Algerian littoral, not deep inland where they were later found. The theory of oil suppression can be found in Fontaine 1961, pp. 137–143.

[123] Arch.Nat. F^{12}7716, letter dated 17 December 1918.

[124] See p. 143.

on almost all world oil policy issues. In every case he helped Royal Dutch–Shell: in Mesopotamia, to preserve the Turkish Petroleum Company agreements of March 1914; in Romania, to compensate for the confiscation of Royal Dutch–Shell holdings, deemed "German" by virtue of participation with the Deutsche Bank in Epu during the war;[125] in the Transcaucasus, to reclaim the extensive Rothschild holdings that had been sold to the Shell group in 1911. These were three of the world's top oil-producing areas. Bérenger, always explicit about using Royal Dutch–Shell, brought the matter up with Walter Long of the Foreign Office while in Britain for a meeting of the Interallied Petroleum Conference from 16–23 November 1918. If the British were trying to anglicize Royal Dutch–Shell until the summer of 1919, part of the objective was to control France's "chosen instrument" and hamstring the French bid for equality in Anglo-French oil relations. The British "confirmed" to Bérenger "the possibility for French interests to take their part in the exploitation of various oil fields, especially concerning Romania, the Caucasus, Asia Minor, Mesopotamia, and Persia." With the help of the Shell combine and British assent, Bérenger thought, France could "procure at all times, by its own means, all quantities of oil products it could need."[126] Bérenger may have selected Royal Dutch–Shell to be France's "national champion," but a more perfect state instrumentality, from Royal Dutch–Shell's viewpoint, could scarcely be found. France's benefits from these intrigues are less certain: favoring Royal Dutch–Shell did not have to mean virulently exclusionary policies vis-à-vis Standard Oil, which were Bérenger's trademark.

Britain and France signed (6 March 1919 in London and 7 April in Paris) an accord covering oil, in Bérenger's words, over "the whole world, especially the East and the Mediterranean basin." A related Anglo-French–Romanian accord on oil was signed by Bérenger and Sir John Cadman, Long's delegate. The Long-Bérenger accord gave France 50 percent of German properties in Romania and 25 percent of the Mesopotamian concession. Bérenger also secured from Royal Dutch–Shell a formal commitment to meeting France's needs. The combine would increase its market share vis-à-vis the Standard Oil Company. The combine assented in a letter on 25 March 1919 to "coop-

[125] France's representative in Romania argued for a Romanian state company with foreign capital participation, "in accordance with the French, the British, and the Shell" (MAE Papiers d'Agents, André Tardieu, vol. 57, letter dated 18 February 1919).

[126] MAE E-Levant, Irak-Mésopotamie, vol. 32, letter dated 29 November 1918.

erate with the [French] government plan concerning the direction and operation of various oil interests that could be reserved for France pursuant to the peace treaty," adding its willingness to give priority to France for a part of its oil production for "civil, military, and naval needs."[127]

The outlines of the Long-Bérenger accord had been drawn as early as November 1918. This indicates why, in December 1918, President Clemenceau gave up his country's claim to direct control of Mosul under the Sykes-Picot agreement of 1916.[128] First, the Anglo-French oil alliance promised France a share of the oil in the area, the chief economic interest of the region. Second, in addition to the future oil-producing areas of Mesopotamia, a general reconciliation of British and French interests promised France an increased presence in other, proven oil-exporting areas, especially Romania and the Transcaucasus. The third reason pertains to suspected oil in Armenia.

Russian oil prospecting in Armenia, carried out behind the Caucasus front during the war, had been promising; the large but imprecisely defined area held five regions with fourteen suspected oil fields.[129] Suspected Armenian oil had also influenced Turkish strategy. The rulers of the disintegrating empire pursued the conflict in the Caucasus at the expense of retaining the Arab provinces. General Liman von Sanders, German military advisor to the Turks, said the "collapse of the Turkish Palestinian front was due to the fact that the Turks, against my orders and advice, sent all their available forces to the Caucasus and Azerbaijan, where they fought the Armenians." American companies became interested in the area for the same reason, contending that "British concessionaires of the Mosul oil fields recognize the Armenian oil as a dangerous competitor because of its closer proximity to the sea."[130] One of these oil fields was even exploited by the Germans and Turks during the war.[131]

Figure 5 shows one French company's requests for oil exploration concessions in mid-1920, based on an earlier preliminary survey. The six circled areas are requested concessions and do not show all the

[127] MAE Papiers d'Agents, André Tardieu, vol. 56, A: dossier Tardieu, letter dated 6 May 1919.

[128] Kent 1976, p. 141; MAE E-Levant, Irak-Mésopotamie, vol. 32, letter dated 11 March 1920.

[129] Délégation de la République Arménienne 1922, p. 33.

[130] American Committee for the Independence of Armenia 1925, pp. 7, 30.

[131] MAE Levant, Syrie-Liban, 1918–1929, no. 344, A: Syrie—Pétroles; letter from Au-blé to Saint-Quentin, dated 14 February 1928.

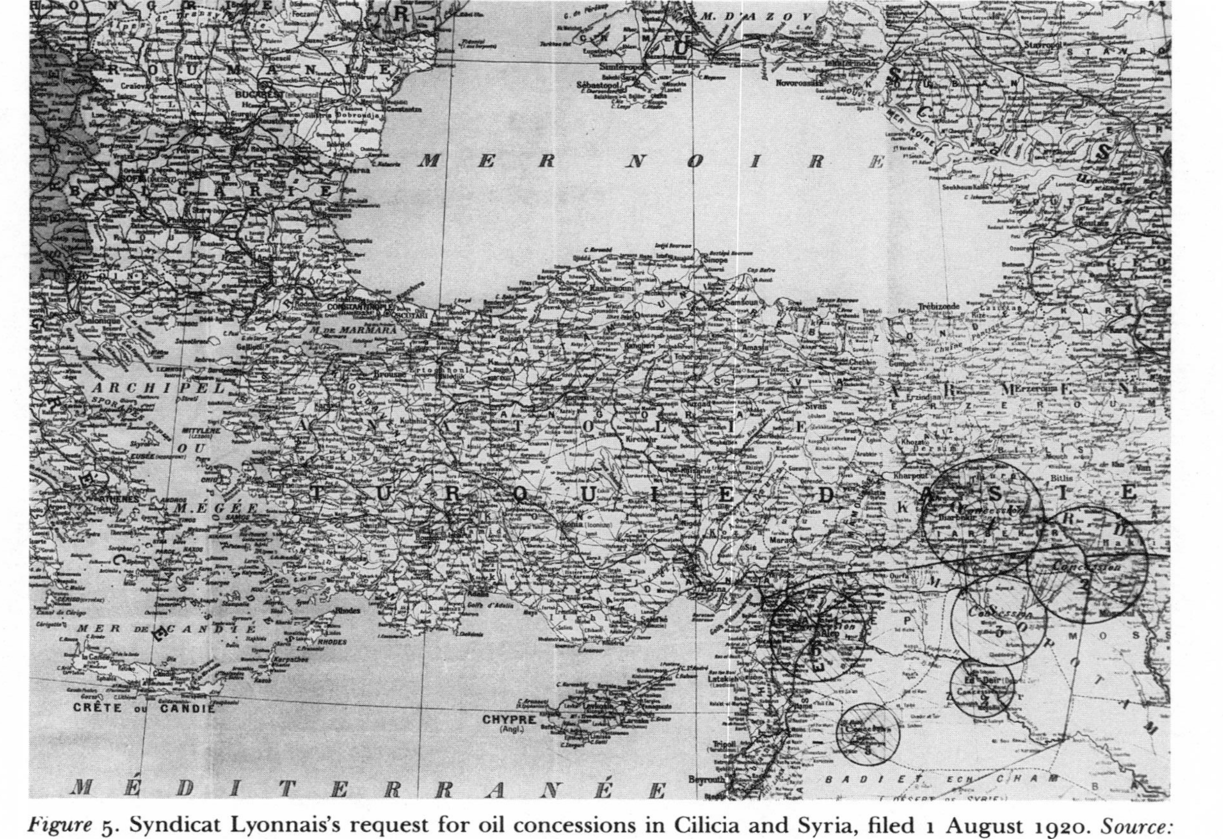

Figure 5. Syndicat Lyonnais's request for oil concessions in Cilicia and Syria, filed 1 August 1920. *Source:* MAE Levant Turquie, 1918–1940, vol. 431. The six circled areas show exploration concession requests of one company, not all suspected oil deposits of the region; the concession requests nonetheless indicate the prevailing "oil fever" regarding the area. Both Syria and Cilicia would later prove to have small workable oil deposits, but these were not comparable to the great finds in Mesopotamia.

suspected oil areas in Armenia and Syria, merely those in which one company was interested. Nonetheless the map is highly suggestive, and we must remember that decisions were based not on where oil is known today to exist, but where it was suspected. Clemenceau traded direct control over Mosul for British support of French claims to Syria and Cilicia (a part of Armenia) not as a major concession but as a quid pro quo. In Ambassador Paul Cambon's words, Clemenceau gave "up all pretensions over Mosul provided that the British cabinet promises to lend us without reserve its support in helping us to realize our just demands concerning Syria and Cilicia."[132] The suspected oil regions were less rich than hoped for, and the nationalist resurgence of Turkey under the Kemalists made holding on to Cilicia impossible.

Armenia played an important part in France's postwar plans for the Middle East: 95 percent of the French military contingent in the Middle East, the Légion d'Orient, was in fact composed of Armenian volunteers.[133] To his credit, Clemenceau did not give up oil rights in Mosul in exchange for oil rights in what is now Turkey. British troops occupied Mosul, and he traded his meaningless paper claim to the city, which had since 1916 excluded oil rights, for a clear shot at Mesopotamian oil and perhaps additional fields in Cilicia and Syria.

The Anglo-French oil alliance culminated in the San Remo agreement of April 1920. The road to this accord was not smooth. The principles of the Long-Bérenger accord were unilaterally revoked by Lloyd George on 21 May 1919. He complained there were "too many companies" on the British side: there were only two, so he meant Royal Dutch–Shell.[134] Bérenger renegotiated the agreement with Sir Hamar Greenwood. The British side of the negotiations was complicated by great internal strife. Efforts to anglicize the Royal Dutch–Shell group were not succeeding, and on this turned the whole question of the Turkish Petroleum Company, and thus the whole of Mesopotamian oil. British foreign policy was torn between the bitter rivalry of the Anglo-Persian and Shell groups and gyrated between opposed positions. France, then solidly behind the Shell combine, had no such contradictions. In December 1919 Sir Hamar Greenwood and Bérenger reaffirmed the basic accord already revoked by Lloyd George in May.[135]

The British and the French wanted to expand into a world market

[132] MAE Papiers d'Agents, Paul Cambon, no. 4, dossier 9, letter dated 29 March 1919.

[133] Hovannisian 1967, p. 66.

[134] MAE Papiers d'Agents, André Tardieu, vol. 56 A: dossier Tardieu, letter from Bérenger to Millerand, dated 15 March 1920.

[135] Kent 1976, pp. 145–157, covers the negotiations from the British side, but with little attention to the rivalry of Anglo-Persian and Shell. Cilicia and the occupation of the Caucasus were also more important than her account suggests.

dominated by Standard Oil while giving up the least amount possible to each other and to other oil firms that, unlike Anglo-Persian and Royal Dutch–Shell, were politically marginal. The British could not be sure how much they needed to give the French to make their own plans possible; the French could not be sure how far they could count on the British.

The uncertainties were part of the geopolitics. Militarily, the British held Mesopotamia and the Caucasus. In the Caucasus the dominant oil interests, after the Nobel group, were the Shell holdings, with the heavy participation of French capital. Saving the Caucasus for capitalism meant saving it for non-British capitalism. Resurgent Russian oil exports meant a strong postwar revival of the Shell group, which Britain was trying to bring to heel. Dividing the Ottoman Empire, in which France had invested heavily, could not ignore the French economic presence. Britain could not simply seize oil assets in Romania, Mesopotamia, and the Caucasus without going against the interests of the Shell group, which was strongly allied with the French and, to complicate matters, headquartered for 40 percent of its worth in Great Britain.

Britain had militarily conquered much of the oil world, but it was an oil world that Royal Dutch–Shell had first conquered commercially. Moreover, the combine was the largest bulwark against the greatest industrial enterprise of all time, Standard Oil. The Shell group was a constant adversary of Anglo-Persian, which it dwarfed; yet Anglo-Persian was twice dwarfed by Standard Oil, which it would face alone, without the cooperation of the Shell group. A modus vivendi between Anglo-Persian and the Shell combine had to be worked out.

Had Royal Dutch–Shell become "British," many imperial oil woes would have vanished. But why should the Shell combine, with its connections to the Dutch government and privileged holdings in the Dutch East Indies (then its greatest source of oil) and other parts of the world, share these in a concocted British government scheme with the diminutive Anglo-Persian? The Shell combine instead became "French," and this was why France had an international oil policy in 1918–1920: the cause is certainly not to be found in the fabricated 1917 "crisis." The Shell combine plus the French government was a difficult mix for the British to digest. Oil concessions were made to the French government that were unwarranted by France's weak strategic presence in the Caucasus and Mesopotamia. But in an oil world dominated by Standard Oil, there was no alternative to collaborating with the Shell group and the French.

Lloyd George and the new French premier Etienne Millerand signed the San Remo accord on 24 April 1920. It harmonized Anglo-French oil policy in Romania, Asia Minor, territories of the former Russian empire in Galicia,[136] and the French and British colonies.[137] The two agreed:

That the French and British governments would help their nationals acquire former enemy oil properties in Romania, especially those formerly belonging to the Deutsche Bank and Disconto Gesellschaft.[138] That new concessions would be sought by Britain and France from the Romanian government.

That all shareholdings acquired from former enemy properties would be divided 50 percent to British interests and 50 percent to French interests. That in acquiring possessions or concessions, the two countries would divide oil benefits equally.

That in the former Russian empire the two governments would both support their nationals in obtaining oil concessions as well as export facilities.

That in Mesopotamia the British would reserve 25 percent of production for French interests, with provisions to keep the price of this oil within reasonable limits.

That the Iraqi local government would have the right to participate in the concession, up to 20 percent.[139]

That the French government would allow an oil pipeline across Syria, en route from Mesopotamia to the Mediterranean. There would be no tax or fee for right-of-way. That France would accord facilities for depots, refineries, etc., for the pipeline.

That the French government would allow Franco-British groups in Africa, with a French participation in excess of 67 percent.[140] That the French government would award a concession in Algeria to groups that had already submitted their project to it for examination, within the limits of French law.[141]

[136] A letter in the MAE files from French firms in Poland wanted to move Poland's eastern boundary into the Ukraine, where recent drilling had found, in territory west of the Bug River, the richest oil deposits in Poland. "Linked to the Baltic Sea and to the port of Danzig by a system of canals that will not fail to be built between the oil fields and the Vistula River, the oil of a Polish Galicia will not fail to augment French supplies and replace American oils" (MAE, Papiers d'Agents, and André Tardieu, Comm. Gén. aux Essences, dossier "Pétroles," letter dated 21 February 1919). Poland got the oil territory, but the producing area never became a major oil source for France.

[137] Kent 1976, pp. 172–178, gives the final text and compares it with previous versions. I have used a French text version in MAE Levant, Turquie, 1918–1940, vol. 429, "Accord sur les pétroles signé par Mm. Millerand et Lloyd George à San Remo, le 24 avril 1920."

[138] Steaua Romana, Concordia, and Vega are cited in the text.

[139] The French saw this as an attempt to reduce their share; it was deleted from later Mesopotamian agreements.

[140] A provision respected only symbolically. See discussion of Pearson and Zaharoff in Algeria above.

[141] Thereby signing the curious Zaharoff-Pearson arrangement in Algeria into international law. See note 121 above.

That Britain, in its empire, would give French nationals the same rights given to British nationals in the French empire.

Standard Oil Destroys the Anglo-French Oil Alliance

American and British interests dominated world oil production for most of this century, until the great nationalizations of the 1970s. France's toehold in Mesopotamia earned it junior partner rank in the Anglo-American alliance. But this was not how matters stood at the end of World War I: the San Remo treaty declared war against the Standard Oil Company and allied groups such as Socony. In later years American firms did not "enlarge" the Anglo-French alliance by "joining it": they destroyed it and forced the creation of a new one more to their liking.

The destruction of the Anglo-French oil alliance was an international conflict with multiple fronts. The abrupt dismantling of the strong regulatory apparatus that had emerged in France in 1918 was just one of the battles won in the course of Standard Oil's war. Standard Oil's campaign against the Anglo-French oil alliance and its two privileged companies, Royal Dutch–Shell and Anglo-Persian, consisted of breaking the French oil import monopoly, halting the division of Mesopotamian spoils until provision was made for American participation, counterattacking against the British in their Persian stronghold, and passing retaliatory United States legislation in order to strike at the Shell combine's American subsidiary with the same discriminatory policies the Shell group pursued in France and elsewhere. Standard Oil's campaign was a commercial "world war."

By 1918 Standard Oil had been consistently outmaneuvered in France by the superior political connections of the Royal Dutch–Shell combine, which included the Rothschild bank and the Banque de l'Union Parisienne. Standard Oil's connections were independent French firms that were not as obedient as subsidiaries. These French importing houses demonstrated neither the power nor the will to protect Standard Oil by stopping projects like the 1918 state import monopoly. Standard Oil needed the support of an *haute banque d'affaires,* the French phrase suggesting a large, politically influential bank.

The Banque de Paris et des Pays Bas was ideal for Standard Oil's purposes: the two had many common goals. Before the war, the Banque de Paris et des Pays Bas had been a coinvestor in Romanian oil with the Deutsche Bank.[142] After the war it wanted to expand oil

[142] Pearton 1971, pp. 45, 57–58.

operations in Romania and throughout the Middle East. The bank's Romanian company, Omnium International des Pétroles, shared Standard Oil's worries about France's favoritism toward the Shell combine.[143] The import monopoly favoring the Shell combine would also keep Omnium's oil out of France as well as Standard's. But Omnium's production, had it been the only issue, could have been given a place in the French market to satisfy the politically weighty bank. The bank was more worried about the Anglo-French alliance. A world oil plan that favored the Royal Dutch–Shell group also favored rival banking interests, the Rothschild bank and the Banque de l'Union Parisienne. The Banque de Paris et des Pays Bas would lose out on the war's most lucrative booty. Allying with Standard Oil gave the bank a powerful industrial ally that countered the constellation of interests around the Shell combine. This was true in Romania and also in Mesopotamia.[144]

Omnium could profit from Standard Oil's strength in international markets even as Standard Oil could profit from the Banque de Paris et des Pays Bas' influence in France, the key diplomatic power employed by the Shell group around the world. Aided by the French banking group, Standard Oil circumvented the national refiner-distributors and entered the French market in 1920 with two new subsidiaries. The first, Economique, was a direct marketing organization; the second, the Compagnie Standard Franco-Américaine, was a holding company controlled 51 percent by the Banque de Paris et des Pays Bas and its French members, who through this instrument held a lucrative 20 percent minority interest in Economique.[145] This alliance did not make for harmony on all details of oil policy, since the interests of

[143] Bérenger cites the Banque de Paris et des Pays Bas as the principal interest behind the Omnium International des Pétroles (MAE Papiers d'Agents, André Tardieu, vol. 56, A: dossier Tardieu, "Rapport sur les pétroles presenté au President du Conseil," dated 6 May 1919, p. 5). The Omnium International des Pétroles was the largest French group in Romania (MAE Papiers d'Agents, André Tardieu, vol. 59. n.d.; prob. 1919, titled "Note").

[144] Standard Oil opposed the Shell group in Romania, where it also produced oil. In January 1919 Henry Deterding, meeting in Paris with Romanian prime minister Ionel Bratianu, won consent to have Astra Romana returned to Royal Dutch–Shell. Shell's ownership of Astra Romana was part of France's alliance with the company, as conceived by Bérenger and Clemenceau. The January 1919 accord in France was a necessary trial run of the accords signed by Long and Bérenger in March 1919. Standard Oil quickly blocked the move: in August 1919 the return of Astra Romana shares to Royal–Dutch Shell had been stalled because of "objections by the Americans" (MAE Levant, Turquie, 1918–1929, vol. 429: letter to René Viviani, Prés. de la Commission des Traités de Paix, dated 8 August 1919).

[145] Gibb and Knowlton 1956, p. 508, called the Standard Franco-Americaine's board of directors "five influential French leaders" (p. 508). I have identified three: Horace Finaly, head of the Banque de Paris et des Pays Bas; General Gassouin; and Jules Cambon, formerly of the foreign affairs office.

Omnium could not entirely mirror those of Standard Oil. But the alliance was durable and led to the rapid rise in French oil politics of Omnium's president, Ernest Mercier, about whom more in the next chapter.

The Royal Dutch–Shell group took the initiative and tried to consolidate its position before Standard Oil and the Banque de Paris et des Pays Bas could stop it. In June 1919, not long after the Long-Bérenger accord, it formed the Société pour l'Exploitation des Pétroles. The new company hoped to control the French portion of Mesopotamia, as stated in the San Remo treaty. The company was created with close consultation between Lucien Villars, on the board of directors of the Banque de l'Union Parisienne and honorary president of same, and the finance ministry. The finance ministry was ostensibly concerned about oil because of the need for credit to cover oil purchases, but it informally helped write the bylaws for this new company, underscoring the influence the Shell group and its banks had with the government.[146]

Simultaneously Alexander Deutsch de la Meurthe, a French oil marketer with long-standing ties to the Rothschild-Shell interests, tried to recruit independent refiners in the oil industry's *chambre syndicale* as stockholders. All the major French firms joined except Fenaille et Despeaux, a Standard Oil ally that boycotted the new combination.[147] Originally established as 60 percent Shell capital and 40 percent French capital, the new company made itself more "French" by changing the capital structure to 49 percent Royal Dutch–Shell and 51 percent "French." The change was cosmetic: the major stockholders, such as Deutsche de la Meurthe and the Rothschild banking interests, were in oil policy indistinguishable from the Shell group.[148]

Among those the Shell group asked to subscribe to the Société pour l'Exploitation des Pétroles were, in addition to the major French refiner-distributors, the Peyerhimoff coal syndicate, the Schneider industrial group, and several other banks, including the Banque de Paris et des Pays Bas.[149] Even so, the Shell group and its allied banks had overwhelming control of the stock structure. As a minor partner, the

[146] MAE E-Levant, Irak-Mésopotamie, 1918–1929, vol. 32, letter from Villars to A. Celier dated 11 June 1919; Celier replied 5 July 1919 (same dossier) that the government could not officially offer an opinion on the new company but observed that "personally it seems to me to conform to the objectives we discussed."

[147] MAE E-Levant, Irak-Mésopotamie, 1918–1929, vol. 32, letter to Celier dated 16 July 1919.

[148] MAE E-Levant, Irak-Mésopotamie, 1918–1929, vol. 32, "Note concernant la Société pour l'exploitation des pétroles," dated 17 October 1921.

[149] Membership structure is in MAE E-Levant, Irak-Mésopotamie, 1918–1929, vol. 32, "Liste des Actionnaires de la Société pour l'exploitation des pétroles," dated 3 February 1922.

Banque de Paris et des Pays Bas bided its time and plotted its revenge, which would come in the name of the Compagnie Française des Pétroles. This "state" company, formed in 1924, would relegate the Royal Dutch–Shell interests to the minority position that Royal Dutch–Shell had reserved for the Banque Paris et des Pays Bas in this first round of the fight for control of French investment in Mesopotamia. The Peyerhimoff group's presence signaled for the first time the coal industry's desire to diversify investment and move into oil, its commercial nemesis. Though the Shell combine and its banks offered something to almost everyone except Standard Oil, even to its commercial enemies in banking and coal, they held on jealously to the largest block of stock in the Société pour l'Exploitation des Pétroles. The strategy failed to neutralize the rivalry of other politically important investors. The Société pour l'Exploitation des Pétroles failed to become the rallying point for French capital. Its organizers also failed to maintain state control of the French oil market and failed to gain control of France's Mesopotamian shares. They did, however, succeed in antagonizing Standard Oil.

The Real Oil Crisis, September 1919 to March 1920

The ambitions of Royal Dutch–Shell and its allies had outstripped their means. With the Long-Bérenger accord signed, a law was to be submitted to the French parliament to continue the state import monopoly even after peace was signed with Germany. In the absence of renewed authorization, the monopoly expired with the peace treaty. France would use diplomatic authority to get oil for the combine abroad, while at home parliament would make a present of the domestic market.[150]

Standard Oil's war against the Shell combine jeopardized France's oil supplies far more than had the worst of wartime conditions. The company's decision to suspend oil deliveries germinated over five months. Concerned by inside information about the direction of French oil policy following the Long-Bérenger accord and the state monopoly renewal, Standard Oil's chairman A. C. Bedford met with Etienne Clémentel and Henry Bérenger at Clémentel's home for forty-five minutes on the afternoon of 10 May 1919. Bérenger informed

[150] The renewal of the monopoly was called the "projet Klotz," after the finance minister, who was not a major oil player but rather someone who did as he was told. Of his own finance minister, Clemenceau caustically remarked, "C'est le seul juif qui n'entende rien aux questions d'argent." The harsh prejudice characteristic of post-Dreyfus France is mitigated by Louis-Lucien Klotz's imprisonment for passing bad checks in 1929 (Jolly 1962, p. 2056).

Bedford of France's intention to have a "national policy" coordinated with that of Britain. Bedford asked for copies of the accords, but Bérenger referred him to the ministry of foreign affairs. Bérenger reproached Bedford for overpricing and complained about United States Shipping Board decisions during the war. He told Bedford that the oil monopoly was making money and the government wanted to keep it: "Mr. Bedford appeared surprised at this news, as if he had been induced into error, and remarked that Americans were for liberty of commerce and not for the state monopoly." Bérenger answered that the monopoly "was a matter of domestic policy of which each state was its own master." He told Bedford Royal Dutch–Shell had made "important contractual offers" that were important not just because of price but from "the point of view of oil organization in general." Bedford "was silent for a few moments" and asked questions about fuel oil. Clémentel intervened and suggested that Bedford meet with Bérenger in the senator's office at the Comité Général aux Essences et Combustibles.[151]

Bérenger and Bedford met again five days later. Bedford asked how the French monopoly would divide market shares and preserve brand names. Bérenger said only that a certain percentage of business could be reserved for Standard Oil. He reiterated that Royal Dutch–Shell had made "proposals" and asked why Standard Oil's principal client firms were not giving information about their business. The French refiner-distributors had said that Standard Oil forbade giving this information to the French government. Bedford agreed to send a company vice president to speak with Bérenger, but the interview was strained.[152] A month later Clémentel received a letter from Bernard Baruch, the American financier and statesman then with the American Commission to Negotiate Peace. He protested that "American oil producers have complained that they are shut out from free competition in French markets."[153] Writing to Clémentel, Bérenger denied any discriminatory intentions and complained that the promised Standard Oil vice president had yet to show up; he "was still waiting after thirty-two days."[154]

In October Bérenger was still waiting, but Bedford had concluded

[151] Quoted material from MAE Papiers d'Agents, André Tardieu, vol. 56, B: Comm. Gén. aux Essences, "Conférence entre Mm. Bedford et Henry Bérenger chez M. Clémentel," dated 10 May 1919.

[152] MAE Papiers d'Agents, André Tardieu, vol. 56, B: Comm. Gén. aux Essences, "Entrevue avec M. O. C. [*sic*] Bedford," dated 15 May 1919.

[153] MAE Papiers d'Agents, André Tardieu, vol. 56, B: Comm. Gén. aux Essences, letter dated 12 June 1919.

[154] MAE Papiers d'Agents, André Tardieu, vol. 56, B: Comm. Gén. aux Essences, letter from Bérenger to Clémentel dated 17 June 1919.

that further talks with Bérenger were pointless.[155] Standard Oil would take more forceful measures. On 26 September 1919 it told French representatives in New York that the company opposed the plan to keep the state-run monopoly after the war. The French replied by asking Standard Oil for $10 million in credit to buy oil for the same state monopoly.[156]

Standard Oil refused the credit. Its oil deliveries to France were cut off, even though American supplies accounted for over 80 percent of the country's total imports.[157] Bérenger did not worry: Royal Dutch–Shell was his ace in the hole. The Shell combine had opened a renewable six-month line of credit for France two weeks earlier; this would cover about one-half of imports from the Shell Corporation for six months.[158] The Shell combine had also contracted on 25 September 1919 to deliver, over the next eighteen months, 9,500 tons of oil per month.[159] Bérenger claimed that the Americans wanted too much money for their oil and that the poor exchange rate necessitated buying more cheaply elsewhere. Bérenger and the finance minister had agreed that "we should not buy any oil from the United States so long as they refuse to open up credits for us and maintain their exchange rates at catastrophic levels."[160]

Prices and the exchange rate were not the real issues, which included the state monopoly and Bérenger's favoritism for the Shell combine. Had he had credit, Bérenger would have bought Standard's oil, and in fact he ended up buying some anyhow, paying Standard Oil's price and then more to the Shell combine as middleman. Shell was compensating for part of the politically caused shortfall by buying from Standard Oil, cash up front. Standard Oil would sell to anyone with the cash to pay, which under penurious postwar conditions was a limited circle. France was like a prospective home buyer who has been denied a mortgage: the formal right to purchase the house remains, but for most people no mortgage means no house. France could buy oil from Standard Oil or anyone else, but without credit the right was

[155] MAE Etats-Unis, 1918–1929, vol. 250, "Ambassadeur Français Washington," dated 9 October 1919.

[156] MAE Etats-Unis, 1918–1929, vol. 250, "Obtention Crédits Amérique," dated 26 September 1919.

[157] MAE Papiers d'Agents, André Tardieu, vol. 56, B: Comm. Gén. aux Essences, letter dated 17 June 1919.

[158] MAE Etats-Unis, 1918–1929, vol. 250, "A Commissaire Française New York," dated 11 September 1919.

[159] Fin. B31828, "Note pour Monsieur Celier," 8 July 1920.

[160] MAE Europe-Russie, 1918–1929, vol. 526, letter dated 10 December 1919.

symbolic. Not all money had dried up, though, and France got some oil, though lacking the bulk of the American contribution.[161]

The French had needed and asked for the credits to buy oil. They had been refused. Bérenger's decision "not to buy" was a vainglorious attempt to make a virtue of necessity. The pro-Shell French government hoped other sources of oil would suffice. The task may not have been hopeless: although the Shell group's attempts to recover its assets in Romania were stalled, and would remain so for another year, there were old producers and new investors in the Caucasus who wanted to sell in France, if only conditions would stabilize. Bérenger contacted the diplomatic delegation from Azerbaijan, who said they could deliver one million poods of crude oil to the export reservoirs at Batum by the end of the year.[162] He also negotiated with several "independent" groups in the Caucasus: the Nobels, who had been the largest Russian producers before the war, as well as Armenian and Russian enterprises in the region—Mantacheff, Lianosoff, Adjenof, and Tchermoeff.[163] The French banking group Louis Dreyfus was trying to enter the oil business in the Caucasus; it controlled 12 percent of the production in the Baku region, against 40 percent for the Nobel group and 15 percent for the Shell group.[164]

Bérenger ran into unexpected opposition from the British government. The British in Batum systematically hindered oil exports from the Caucasus, pressuring the French during the prelude to the Treaty of San Remo. Bérenger wrote to the French finance minister: "This policy of stopping our purchases in the United States is possible only insofar as it can be immediately replaced with a policy of purchasing in Eastern Europe. The English have perfectly understood this, and it is for this reason that their agents want to get their hands on all the oil of the Caucasus, which is, for them, one of the principal keys to the oil of Persia and Mesopotamia."[165] In December 1919 Bérenger thought he had resolved this difficulty with the British; but export problems from the Caucasus persisted. Because of the revolutionary

[161] Fin. B31828, "Note pour M. Celier," dated 8 July 1920.

[162] MAE Europe-Russie, 1918–1929, vol. 526, "Télégramme Chiffre," dated 25 November 1919. A pood is a Russian unit of weight equal to 16.38 kilograms; the offer was about 16,000 tons, or half of one month's needs for France.

[163] MAE Europe-Russie, 1918–1929, vol. 526, "Note du Commissaire Général aux Essences et Pétroles," dated 26 October 1920.

[164] MAE Europe-Russie, 1918–1929, vol. 526, "Les Napthes de Bakou et de Grozny," dated 25 December 1919.

[165] MAE Europe-Russie, 1918–1929, vol. 526, letter dated 10 December 1919; MAE Papiers d'Agents, André Tardieu, vol. 56, B: Comm. Gén. aux Essences, telegram dated 20 December 1919.

turmoil, oil exports ceased altogether in a few months' time, making moot the question of British goodwill and Bérenger's success.

Bérenger saw clearly that the development of Mesopotamia and Persia was threatened by competition with oil from the Caucasus; the old question of Persian Gulf oil's competition with Russian exports from the Black Sea, which dated from before the war, had resurfaced in a new form. The Caucasus held only major rivals for Anglo-Persian, and the region's revival was unwelcome. As military occupiers of the Caucasus, the British used this oil source to pressure the French on other oil negotiations.

The troubled political conditions of the Caucasus made oil production difficult in any case, but the British military governors aggravated the problem. They imposed a five-ruble tax on each pood of exported oil; the French minister of foreign affairs complained to Ambassador Cambon in Britain of the "veritable British seizure" of the Caucasus, adding that even French naval cruisers in the area had trouble getting adequate oil.[166] The British requisitioned storage tanks at Batum and threatened to close the crucial pipeline from Baku if the local government did not comply; the threat was briefly carried out.[167] French firms paid high fees for the use of storage tanks in the area.[168]

Romania and the Caucasus remained beyond the reach of Bérenger's and Royal Dutch–Shell's ambitions. Yet even had oil been available, Bérenger had to find tankers to carry it. He counted on acquiring all or part of eight Standard Oil tankers that had been blockaded into German ports during the war. The tankers were classified "German" and turned over to the Reparations Commission, which was taking German ships to replace allied vessels lost in the war. The Allied Supreme Economic Council divided up the confiscated tonnage, liberally favoring France; Standard Oil and the United States State Department protested. During the appeal the tankers were assigned to British waters, and by December 1919 Bérenger had yet to get any, save one German tanker not claimed by Standard Oil.[169] With no oil from the Caucasus and Romania, the only way to bypass Standard Oil was to assign additional tankers to the long route to the Dutch East Indies. Failing to get Standard Oil's "German" tankers, Bérenger's favored

[166] MAE Europe-Russie, 1918–1929, vol. 526, letter dated 4 December 1919.

[167] MAE Europe-Russie, 1918–1929, vol. 526, "Rapport sur les pétroles du Caucase," dated 8 December 1919.

[168] MAE Europe-Russie, 1918–1929, vol. 526, "Les Napthes de Bakou et de Grozny," dated 25 December 1919; also "L'Industrie Pétrolifère et le Commandement Britannique à Batoum," dated 15 January 1920.

[169] Gibb and Knowlton 1956, pp. 269–274; MAE Europe-Russie, 1918–1929, vol. 526, letter from Bérenger to Berthélot, dated 10 December 1919.

Royal Dutch–Shell remained in the same position that had bedeviled it during the war: long supply lines and not enough ships. The fight with Standard Oil had come too soon: the Shell combine could not lease more ships at a reasonable price.[170]

By December 1919 shortages loomed in fuel oils and gasoline.[171] The Shell group's market share in France climbed from 15 percent to 45 percent, but this was a proportional shift owing to the market's diminishing size.[172] Standard Oil meanwhile aroused opposition to Bérenger in the United States and France. The American Petroleum Institute protested to the French ambassador that Americans abroad should enjoy the same privileges accorded foreign nationals in the United States, a thinly veiled reference to the Shell group.[173] The French ambassador commented that the Americans "think the monopoly will in practice be administered to their detriment and to the advantage of the British and Dutch oil companies."[174] He publicly maintained the neutrality of his country's legislation but admitted more delicately to his colleagues in Paris that the American representative "alleges, it is true, certain facts which seem to him to justify the fears of certain American exporters concerning the disposition of the French government regarding American oil."[175]

In January 1920 Bérenger learned that Standard Oil was carrying out "hidden negotiations" with French refiners (probably Fenaille et Despeaux) for long-term contracts. This was clearly illegal, since contracts had to be negotiated with the state monopoly.[176] But Standard Oil and the Banque de Paris et des Pays Bas were forging a new alliance, and the American company acted with impunity in France. A ministerial report observed: "The French monopoly project would have effectively limited the action of the respective [oil] groups and perhaps consolidated the policy of alliance with the English, and thus met with the opposition of Standard Oil. From this came the ardent press campaign in favor of a free market, a campaign obviously supported in France by the French banking group linked to Standard Oil, that is to say, the Banque de Paris et des Pays Bas."[177]

[170] Fin. B32314, "Le Monopole d'importation en France pendant la guerre et sa liquidation," n.d., prob. 1931. Cf. parliamentary accusations of intentional price hikes in world freight market in Nowell 1983, pp. 240–241.

[171] MAE Europe-Russie, 1918–1929, vol. 526, letter from Bérenger to Berthélot, dated 10 December 1919.

[172] Fin. B32314, "Le Monopole d'importation en France pendant la guerre et sa liquidation," n.d., prob. 1931.

[173] MAE Europe-Russie, 1918–1929, vol. 526, letter from American Petroleum Institute, dated 1 October 1919.

[174] MAE, Etats-Unis, 1918–1929, vol. 250, letter dated 7 October 1919.

[175] MAE Etats-Unis, 1918–1929, vol. 250, letter dated 9 October 1919.

[176] Fin. B31828, Bérenger to finance minister, dated 3 January 1920.

[177] MAE Etats-Unis, 1918–1929, vol. 250, "Note d'information au sujet de la politique française en matière de pétroles," dated 1 December 1920.

The Royal Dutch–Shell alliance lacked oil and lacked tankers. Before parliament in March 1920, the government representative, Antoine Borrel, responded to hard questioning by pro-monopoly deputies. If the government kept the state monopoly, there would be a "serious supply crisis."[178] The Shell combine itself was losing interest: French intervention had proved ineffective in helping Shell reclaim properties in Romania, and in the Caucasus the situation was worse. "At the beginning of 1920," a finance ministry document observed in July, Royal Dutch–Shell "was practically our only supplier of oil," but "these last months its deliveries have dropped off and the company provides only a part of our increased needs." France asked the combine for an additional three million pounds sterling of credit, but Gulbenkian offered only one million, conditional on helping the Shell combine gets its confiscated Astra Romana back from the Romanian government.[179]

Doing without Standard Oil had turned out to be more difficult than foreseen in the fall of 1919. The effort to give the French oil market to Royal Dutch–Shell ended quietly but ignominiously in April 1920, when an import license was issued to the Bedford Petroleum Company, a Standard Oil subsidiary. The import monopoly, barely two years old, remained on the books till 1921 but was effectively finished.

In the fall of 1919 a new parliament was elected, to assume power on 20 January 1920. Clemenceau was replaced by Alexandre Millerand as premier; Clémentel, failing to win reelection, lost his post in the ministry of commerce; and Bérenger lost his job in February 1920. His Commissariat Général aux Essences at Combustibles became a political football; it was transferred to the ministry of public works, where it became the Commissariat Général aux Essences et Pétroles.[180] It was a clean sweep of the Royal Dutch–Shell faction from office, and in March 1920 Standard Oil resumed normal supply procedures with France.

Other Factors Contributing to the Disintegration of the Anglo-French Oil Alliance

The Millerand government signed the San Remo treaty, but the conditions for its success had already been undermined in France and were to be attacked elsewhere as well. The Bolsheviks conquered Azerbaijan in May 1920, and oil exports dried up.[181] "In all their statements

[178] Déb.Ch., 26 April 1920, p. 1385; Déb.Ch., 10 July 1923, pp. 3321–3322, furnishes interesting details, including Standard Oil's defeat of a similar monopoly proposal in Italy. See also Nowell 1983, p. 243.

[179] Fin. B31828, "Note pour M. Celier," 8 July 1920.

[180] Nayberg 1983, p. 395.

[181] MAE Europe-Russie, 1918–1929, vol. 526, "Tiflis par Constantinople," telegram dated 28 May 1920.

[the British] made clear that Baku Oil was a secondary consideration, as compared with that of Southern Persia," Ullman writes.[182] In Romania, national domestic oil companies struggled against the international oil giants. The San Remo treaty established a firm French claim to the Deutsche Bank's former share of Mesopotamia; its other provisions were opposed by Standard Oil and the Bank de Paris et des Pays Bas. They launched a worldwide campaign against France and Royal Dutch–Shell, against Britain and Anglo-Persian.[183]

In the United States, the navy adopted measures restricting the ability of foreign firms to produce American oil from its reserves. Measures were introduced in Congress to deny oil exports to countries that discriminated against Americans.[184] Standard Oil not only broke the bond between the French government and Royal Dutch–Shell, it carried the struggle into the combine's backyard, forcing the Dutch government to let it into a productive concession in the combine's stronghold, the Dutch East Indies.[185]

To retaliate against Britain, Standard Oil aimed at Anglo-Persian's heart. Persian production, reaching nearly a million tons by the end of the war, had become commercially viable.[186] In attacking the British side of the Anglo-French alliance, Standard Oil based its strategy on three factors. Britain held each area militarily; Anglo-Persian was vulnerable to pressure in Persia in the struggle over Mesopotamia; and finally, the Standard Oil Company's president, Walter Teagle, was convinced that Persia's oil and Mesopotamia's oil were geologically the same field: "The future of the present Persian fields is particularly promising. There is every reason to believe that these Persian fields are large in extent and that they extend over into Mesopotamia. . . . In the settlement of the division of Turkey, consideration should be given to the oil possibilities. . . . undoubtedly the Euphrates Valley would pro-

[182] Ullman 1968, p. 322. With 67 percent of Baku oil in the hands of non-British oil producers and the Shell group and its holdings under the control of neither the British nor the French, the supporters of Anglo-Persian had little reason to commit troops to the defense of this area. Statistics on ownership in the Baku region are given above. The British had many other considerations in 1920 that argued against staying in the Caucasus: e.g., a populace tired of war, Irish independence, and an uprising in Iraq.

[183] Fin. B32313, "Rapport au sujet de la politique du pétrole en France," 24 May 1923, p. 21.

[184] Gibb and Knowlton 1956, pp. 274–275.

[185] MAE Etats-Unis, 1918–1929, vol. 250, "Bulletin périodique de la presse Hollandaise," with note from Ambassador Philips, dated 19 April 1921; also French analysis that ties the oil land leasing bill of 25 February 1920 to Dutch policy ("Extrait de l'analyse . . . du 20 mai 1921"); see also Gibb and Knowlton 1956, pp. 391–394.

[186] Nouschi 1970, pp. 46–47.

duce a large quantity of oil. I am wondering if there is any way we can get into the oil producing end of the game in Mesopotamia."[187]

Northern Persia also promised oil. Anglo-Persian formed North Persian Oils in 1920 in a bid to control the area. By May 1920 Royal Dutch–Shell counterattacked with its own request for a concession.[188] Standard Oil, not to be left out, approached the Persian government, which in September 1921 began negotiations with Standard Oil in return for a $5 million loan.[189] In March 1922 the Standard Oil Company advanced $1 million "without guarantee and without any conditions at 7 percent, payable next December, regardless of whether any accord is given concerning the Northern oil concession."[190]

In Persian politics at the time, $1 million was a lot of money to throw around without any guarantee. Before World I, loans of this kind were typically made against some form of security such as the revenue of a port or control over the public treasury. Loans were used to gain power over economic resources; they became a means to extend de facto imperial control, as in the competition between the British and the Russians. The debtor country wanted the loan money, and it balanced the loss of sovereignty by parceling out different parts of the country to opposed imperial interests. By dealing out $1 million as its first hand in Persia without asking for guarantees, Standard Oil was saying that there was more where that came from; a loan on paper, the $1 million was a bribe. The French ambassador in Tehran analyzed it thus: "It is truly a diplomatic success for the Persian government, because it is the first time that Persia has succeeded in getting money from anywhere other than England or Russia, without being [forced] to give up, as a price for this financial assistance, part of its political or commercial independence to the profit of its fearsome neighbors."[191]

The ambassador thought the loan a portentous sign of American ambition to extend its influence in the area, and the Persians saw it as a sign that the American government "would be disposed, in another way, to give its support to Persia to help safeguard its independence."[192] The American Sinclair Company soon thereafter also attempted to buy its way into the Majlis and hence into the Northern Persian conces-

[187] Gibb and Knowlton 1956, p. 285. The first ellipsis is the Standard Oil historians'; the second is my own. Geologically the oil fields are distinct, but Teagle was right—they both had oil.

[188] Fin. B32866, Deutsch de la Meurthe to minister of foreign affairs, letter dated 17 May 1920.

[189] Gibb and Knowlton 1956, p. 309.

[190] Fin. B32866, "Télégramme à l'arrivée," dated 26 March 1922.

[191] Fin. B32866, "Télégramme à l'arrivée," dated 26 March 1922. The bracketed word is missing in the original; "forced" or "required" is clearly intended.

[192] Fin. B32866, "Télégramme à l'arrivée," dated 26 March 1922.

sions.[193] American penetration appeared to be a genuine threat, and the British backed down. To cool the heated Persian oil politics, they came to terms with Standard Oil on major aspects of Middle East oil. They would renegotiate the Mesopotamian provisions of the Treaty of San Remo, on which France's future as an oil-producing power depended. All that was left of the world-encompassing Anglo-French oil alliance was Britain's promise to give France a share of Mesopotamia. The promise preserved the March 1914 Turkish Petroleum Company and kept Royal Dutch–Shell in the game.

TRANSNATIONAL STRUCTURING, 1914–1921

The view that German submarines caused the crisis that launched the French state's policy of foreign procurement in Mesopotamia and domestic regulation is bad history. The world struggle over oil resources ensnared states—states did not lead the struggle. The French finance ministry, having watched Bérenger and the Royal Dutch–Shell faction of the government get unceremoniously swept from power, was acutely aware that its etatist tradition had hit a limit in tackling Standard Oil:

> The national interest is quite clear and easily ascertained. France is a buyer. Unfortunately it will stay that way for a long time. The matter at hand therefore is to ensure the liberty of our supplies and not to be beholden to any foreign group, no matter which. . . . [A] state organization, with the power to buy and then divide these purchases among all groups, favoring as much as possible French producers, . . . is in theory the most attractive system, perhaps even ideal, . . . [but] in fact it cannot stand against the most serious objections. First, whatever may be the qualities and noble traditions of French administration, it is a dangerous principle to let these functionaries have the responsibility for making large contracts that have of necessity an arbitrary element. . . . by dividing up state purchases in the market, we hinder the play of competition, and prices are fixed perhaps at too high a level. Finally, the selling possibilities of the large producers are limited. Thus we arouse the hostility of extremely powerful groups. This hostility takes the form of sudden limitations on imports for which, in spite of everything, we have a need. And looked at from another viewpoint, this hostility is translated into press campaigns with a financial backing that is increased by the unfavorable exchange rate, becoming by that fact singularly dangerous and efficacious.[194]

[193] Gibb and Knowlton 1956, pp. 312–313.

[194] MAE Etats-Unis, 1918–1929, vol. 250, "Note d'information au sujet de la politique française en matière de pétroles," dated 1 December 1920.

The document shows extraordinary self-awareness: the "qualities and noble traditions of French administration," now called the "statist" or "etatist" tradition, had been clobbered, and the statists themselves knew it and wrote it. The French market was deregulated, the anti–Standard Oil alliance sundered. Two sovereign states, each ruling world empires and at the peak of their military power, were humbled by an oil company. States and corporations had fought over oil, ships, and markets. The goals of the states were not independently articulated objectives, but themselves the result of the international struggle among corporations. France's sudden interest in a national oil policy in 1918 was not the result of supply shortages but a by-product of Royal Dutch–Shell's problems: with Britain and Anglo-Persian, with oil supply shortages and a lack of tankers.

Transnational structuring is evident here, as in 1914. Plans to integrate Mesopotamian oil and national markets were forming in 1914, but the conflict made them obsolete. In 1918 and 1919 the Royal Dutch–Shell group took the mantle of power from the Deutsche Bank and tried to put together a similar arrangement, minus the Germans. The French monopoly plan and the Treaty of San Remo were parts of the new architecture for a world oil system.

Standard Oil had other ideas. By destroying the oil alliance between Britain and France, by threatening Anglo-Persian in Persia, by destroying French pretensions to regulate their own market, Standard Oil showed that it had learned how to play transnational politics.

The process of transnational structuring differs from "rent seeking." Standard Oil did not need a surplus profit or "rent"; it could make more money at a lower retail price if it could bypass French national interests with its efficient production and marketing. Its campaign against a regulated market was every bit as political, however, as Royal Dutch–Shell's campaign to put protective measures in place. French domestic firms were caught up in the struggle. Nonetheless, the "unregulated" market in which Standard Oil and the Banque de Paris et des Pays Bas' Omnium International would play increasingly important parts was a profound change from the unregulated market that prevailed earlier. Before 1914 the French domestic refiners had been the arbitrators between Standard Oil and the market. Now Standard Oil had a close alliance with a major French bank and was entering directly into large-scale retail operations. The French government had been beaten back: it could not envision future regulation of the market without Standard Oil's backing. As a result, French governmental power in oil matters declined: the Commissariat Général aux Essences et Combustibles, which under Bérenger had reached for greatness, for a time was to disappear as an effective political authority, and with it went French regulatory authority. But owing to the conditions just mentioned, we cannot say that this "free market" was the same as the cartelized, legally "free" market of pre-1914. There was, in short, a cumulative change owing to these activities. Transnational structuring, as described in the opening chapter, was proceeding apace.

I have made strong claims about the lack of a genuine strategic motive for French intervention in the oil market in the closing months of the war. That does not mean, however, that the "realist world" of power struggles among nations did not have a powerful impact on the world oil market. Britain's naval power forced Standard Oil's ships out of the German trade and unleashed on the world not a shortage of tankers, but a surplus. Turkey's and Germany's blockade of the Dardanelles forced Royal Dutch–Shell to dire extremities. Royal Dutch–Shell bided its time before launching its massive political campaign against Standard Oil, hoping that the Gallipoli campaign would be successful and that both oil and short supply routes would be restored. Royal Dutch–Shell's launch into "desperate strategies" of aggressive regulation corresponded with great events: revolution in Russia and an attempt by Anglo-Persian to profit from Britain's belated military success in Iraq to force the combine out of Mesopotamia.

However impressive the prowess of Germany's U-boats may have been, the factors I have described were much more important to the structure of the world oil industry and the major oil companies. I have not refuted realist notions of state conflict in the international system. I have improved on the "realist" appreciation of the war: Britain's blockade of Germany, Turkey's blockade of the Dardanelles, the fight over Iraq, were more important than France's losses to submarines. But I *have* refuted simplistic attempts to convert realist theory into an "explanation" for economic regulation in states, even states at war.

Transnational structuring affects the world of states as much as states affect the world of corporations. Standard Oil had to pull its ships from Germany; Socony surrendered ships to Great Britain's sovereign request; Royal Dutch–Shell could not get ships through the Dardanelles; whatever Standard Oil's impatience, the French market *was* regulated in 1918 and remained that way for over a year. But France had to deregulate in 1920, and Great Britain and France had to drop their worldwide oil alliance. In spite of the "systemic constraints" of the international system and the "anarchic tendencies of states in competition," power within the international system went as much from corporations to states as vice versa.

In the 1920s, the international market grew more complex. The French oil market fractured into even more factions: the oil majors among themselves; the oil majors against the French domestic refining interests; the slowly wakening coal interests; the financial interests that struggled to link Russian oil and Russian debt, alike thrown into turmoil by Bolshevism. These factors determined corporate goals and the evolution of transnational structuring in the world oil market over the next two decades.

Chapter Four

France and the Development of the World Oil Cartel, 1922–1939

In the late 1920s French oil policy took on the etatist form that it would keep through most of the century. The battle among oil interests was extraordinarily complex, and this chapter is only a partial treatment: it focuses on conventional oil extracted from the ground. Synthetic oils, hydrogenation, and their effects on the coal and chemical industries are discussed in the following chapter. France created a state-sponsored oil company, a law setting quotas, and an agency to enforce them. Though justified with a mercantilist rhetoric, these measures showed the powerful effects of transnational structuring, in which France was not unique. National oil firms did ally with banks to link oil to the settlement of Russia's debt; if French policy had a "local flavor," it was because of this.

The Compagnie Française des Pétroles was established in 1924; the state's equity participation was proposed in 1928. Standard Oil, which had been the target of earlier monopoly plans, would brook no more state-monopolistic insolence along the lines of 1914 and 1919–1920. The Compagnie Française des Pétroles would bring in oil from Mesopotamia, and quotas would ease the competition among international and French national companies. The Office National des Combustibles Liquides, which enforced the quota system after 1928, was originally created as a result of worries about how to integrate both Soviet and Mesopotamian oil into the French market. The French national companies leaned toward Soviet oil, while the international firms leaned toward Mesopotamian; their dispute brought into the fray French banks and their concerns about the outstanding debt of the Russian ancien régime.

Typically, Premier Clemenceau's appeal to President Wilson during the war crisis of late 1917, which supposedly convinced French elites to regulate the French oil market, is used to justify French interventionism. Premier Poincaré authorized the creation of a state oil company in 1923, and his national interest justification for this action makes it look as though French policy grew out of the far-sighted wisdom of these two statesmen. But the crisis of 1917 was exaggerated. The worldwide struggle for oil during World War I and the Standard Oil embargo of 1919–1920 have been lost from view because the "1917 crisis" version, publicly broadcast, covered up Bérenger's malfeasance in office.[1]

The realist-mercantilist perspective on French oil sweeps many facts under the rug. It cannot explain how France went from the alleged strategic crisis of 1917 to a deregulated market in 1920, only to pull a regulatory rabbit de novo out of the hat again in 1928. If strategic need was the motivation, why the vacillating domestic policies? The government's erratic behavior is explained by the rivalry of oil groups.

The letter Poincaré wrote to Ernest Mercier in 1923 looks quite "realist-mercantilist" as a public policy document.[2] The letter makes it easy to celebrate the state company as an instrument of the national will. But the stock membership lists, the real effective rates of taxation, and the difference between the state company as first elaborated before 1923 and as revised in 1928—all less "obvious" than Poincaré's calculatedly dramatic letter—show an entirely different version of the facts.[3] Poincaré's words about the Compagnie Française des Pétroles are more readily found than the addresses, banks, and subsidiaries of listed stockholders that show who the stockholders in the "state" company really were. Transnational structuring predicts that much of interest is concealed, because that is where private interests lurk. Poincaré's letter is suspicious because it has been so well publicized.

Transnational structuring warns us to look for other interested economic actors beyond the oil sector. In the 1920s banking factions viewed oil as an instrument to pay off Russian debt. One faction, expropriated oil landowners from prerevolutionary Russia, wanted the oil trade with the Bolsheviks to pay for their expropriated lands. Banks wanted the money to pay off bonds renounced by the Bolsheviks. Other oil firms simply wanted to buy or sell Russian oil because it was available.

[1] One broadcaster was Bérenger himself. See Bérenger 1920, a collection of parliamentary documents that includes the 1917 telegram from Clemenceau to Wilson.

[2] See p. 171–172.

[3] Realism also offers its examples of subterfuge. Morgenthau 1973, p. 537, provides an amusing example of realist diplomacy's potential for complexity: "When Metternich was informed of the death of the Russian ambassador at the Congress of Vienna, he is reported to have exclaimed: 'Ah, is that true? What may have been his motive?'"

The Soviet oil-debt-indemnity issue was a major part of European oil politics in this period.

The International Struggle for the Control of Oil

For the multinationals, failure to master every potential source of supply meant that a new oil discovery might drive prices down and threaten the stable income needed to finance worldwide operations. With no international cartel to replace the pre–World War I Epu and Asian agreements, the competition for control of foreign resources grew ferocious because of the need to protect established markets.

During the Great War production expanded in the United States and other parts of the world. Russian oil was taken off the market by the war and by revolution. If Russian oil reentered world markets, it would disrupt the oil giants' new supplies, beyond their indignation at the Bolshevik expropriations of "their" oil territory. Russia by itself threatened overproduction and falling world prices. Mesopotamia promised new oil fields as plentiful as those confiscated by the Bolsheviks, and still more falling prices. Neither region's oil was needed to supply world markets at existing price levels. Postwar oil politics revolved around an attempt to chart a course between the Scylla and Charybdis of the Transcaucasian and Mesopotamian oil-producing regions.

Russia was the immediate concern, and Mesopotamia the potential concern. Mesopotamia was undeveloped, Russia had already played a leading role. So long as no one knew what to "do" about Russia, Mesopotamia was best left locked up in long diplomatic negotiations. Russian oil might be rapidly put back into production by a revolutionary government whose politics, technical abilities, and marketing practices were unknowns. Russian oil first became an integral part of French national firms' attempts to defend their national market shares from the multinationals; Mesopotamia was the multinationals' response to their success.

Competition for Russian Oil

In 1911 the Royal Dutch–Shell group had purchased the Rothschilds' holdings in Russia. After the revolution, the company speculatively purchased additional oil titles from émigrés strapped for cash. Standard Oil purchased the Nobel properties, once the largest single producing and marketing group in Russia. Today these purchases appear wildly imprudent. But each company saw its rival as committed to Russian properties: if there were a small chance that capitalist property might be restored, neither group could afford to allow the enormous production of the region to slip from its grasp. Failing the near-term restoration of capitalism, the Soviet government, for want of capital, know-how, and access to downstream markets, might appeal to Western companies. Speculatively acquired property titles could be used to negotiate: the multinationals might offer to relinquish their legal claims in exchange for an exclusive relationship with Soviet suppliers. Buying ancien régime oil deeds was therefore justifiable and perhaps even necessary. The large firms could afford to lose money better than they could afford to lose a chance at Soviet oil.[4]

The Soviet government had repudiated czarist debt in 1918, but it retreated from this position in the early 1920s or, more accurately, tried to appear to do so with regard to oil.[5] There were good reasons: oil was the most readily available source of desperately needed revenue for the new regime, but it could not be sold worldwide without a proper distribution network. The greatest world oil distributors held titles to expropriated oil lands in the Caucasus and opposed Soviet exports. Soviet oil also had to pass through the Dardanelles. Turkey lay in the hands of the capitalist Allies, and the Soviets had to consider that the commercial antagonism of Standard Oil and Royal Dutch–Shell could translate into an oil blockade at this geographic chokepoint. This remained a major consideration even after the British withdrawal from the Caucasus, so the Soviets had to look "ready to deal" with the oil multinationals.

The Russian oil question remained moot as long as Turkey was disputed. No progress on Russian oil was made in international diplomatic conferences before and during most of 1922. Protracted negotiations over oil rights occurred at Cannes (January 1922), Genoa (April 1922), The Hague (June 1922), and Lausanne (November 1922 through February 1923, and again in April 1923). The last meeting, at Lausanne, came after the defeat of the Greeks and the reemergence of Turkey as a power with sovereignty. Britain accepted neutralization of the Dar-

[4] Gibb and Knowlton 1957, pp. 327–336, argue that the Bolshevik regime's collapse was considered imminent when Standard Oil bought the Nobel properties. But speculation continued after the British withdrawal. There were French purchases in the Caucasus as well, and also German, Turkish, and Italian "interest in opportunities created by the fluid political situation." French interests are surveyed in MAE Europe-Russie, 1918–1929, vol. 526, "Les Naphtes de Bakou et de Grozny," 25 December 1919.

[5] Fin. B32023, "Note sur le Règlement des Dettes de la Russie envers la France," by Léon Wenger, 16 March 1927. The Soviets wanted to trade partial recognition of old debts for diplomatic recognition by the *puissances bourgeoises*, but this offer was deemed insufficient by the latter.

danelles and lost control of the geographic oil spigot to Western markets. The pressure on the Bolsheviks to accept Western capitalists in the Transcaucasian oil fields was reduced.[6]

Banks in Great Britain and France were the principal creditors of the defunct czarist regime and therefore had an interest in the earnings and marketable assets of the new Soviet government. In 1927 Pétrofina calculated total Russian debt to France alone, exclusive of interest, at 27.5 billion gold francs, a staggering sum.[7] The problem for the creditors was how to foreclose on a national debt in default. French financial concerns wanted earnings from Soviet oil to go into a fund for debt repayment. The multinational oil firms were indifferent to the international bankers' problems and worried about the threat to their markets from Russian oil, and about who might regain control of Soviet nationalized oil lands. Turkey's victory and the neutralization of the Dardanelles were therefore major events in international oil markets, which weakened the bargaining power of the colossal financial and industrial concerns interested in Russian oil.

For the first time since the Germans closed the straits in 1914, the controllers of the Transcaucasus also had the geopolitical wherewithal to get oil out to world markets. The Soviet government might earn the money it needed to stabilize internally. Therefore not until 1922 did the major oil companies look for a serious "containment" policy that would use their oligopolistic control of downstream marketing activities to force favorable concessions from the Soviet regime as a price for market access.

One proposal would have had the major oil groups refuse delivery of Russian oil until compensation for, or restitution of, oil-bearing properties had been achieved. Exiled Russian property holders and oil companies ought to form a united front. In December 1921 Royal Dutch–Shell's Henry Deterding met to discuss the idea with Louis Pineau, then the second-ranking official in the reorganized Direction des Essences et Pétroles that had succeeded Bérenger's Commissariat.[8]

[6] Control of the Dardanelles was one reason Basil Zaharoff allied with the Royal Dutch–Shell group and supported the Greeks in the conflict with Turkey after World War I (Neumann 1935, pp. 168–181). Neumann contends that in the Greek-Turkish conflict Standard Oil and France backed Turkey, whereas Royal Dutch–Shell and Great Britain backed Greece. But he overreaches his evidence.

[7] Fin. B32023, "Note sur le Règlement des Dettes de la Russie envers la France," by Léon Wenger, 16 March 1927. Wenger headed Pétrofina's operations in France; the figure is his calculation. In 1926 paper francs had between one-seventh and one-eighth the value of inflation-resistant gold francs.

[8] Fin. B32308, "Note pour le Ministre," 6 December 1921. The reference to the "Syndicat International" appears in a section on Romania and is a precursor to the "Groupement des Sociétés Naphtifères Russes et Etrangères," which held its first meetings in the next nine months.

The December 1921 meeting was a bare beginning, and Royal Dutch–Shell, no longer at the center of French oil politics, may have been testing the new French government. Serious intercompany negotiations about the Russian oil problem began the following year.

The Franco-Belgian Coalition and Russian Oil

After World War I the British and French seized German oil holdings in Romania; Belgian capital, which had also been invested in Romanian oil before the war, expanded too. Pétrofina was founded in Belgium on 25 February 1920 and today remains one of Europe's leading oil firms. Backed by the French Crédit Mobilier, the principal French stockholder of the formerly German Spies Petroleum in Russia, Pétrofina was also owned by the powerful Société Générale de Belgique. In July 1923 the French Banque de l'Union Parisienne, which had backed Royal Dutch–Shell's 1919 attempt to monopolize the French market, also took a large holding in Pétrofina.[9]

Romanian oil, like Russian oil, passed through the Dardanelles. Almost equally close to Europe, the two regions were natural competitors: any major oil firm in Romania would fear Soviet oil competition. A good defense was to become the distributor of the Soviet production, which Pétrofina could use to expand in European markets. In addition, the French financial backers of Pétrofina were concerned with settlement of the Russian debt. The potent oil-finance alliance rapidly became a third force in French oil politics, taking its place alongside the long-standing rivalry between Standard Oil and Royal Dutch–Shell. The financial priorities of Pétrofina were writ large in the full legal name that is the origin of the contraction: Compagnie Financière Belge des Pétroles (Pétrofina = *Pétro*les + *Fina*ncière).

The Banque de l'Union Parisienne's addition as a shareholder redoubled the already potent backing. Disillusioned with Royal Dutch–Shell's failed monopoly attempt in the 1919–1920 collision with Standard Oil, the Union Parisienne found in Pétrofina a means to enter the oil business. As an ally of the Shell combine, the Union Parisienne had stood to become active in oil only if the combine's ambitious schemes for Mesopotamia and the French market succeeded. Since the schemes failed, the Banque de l'Union Parisienne looked elsewhere for

[9] MAE Europe-Russie, 1918–1929, vol. 528, "Le Président du Conseil . . . à Monsieur le Ministre des Finançes," 28 February 1924; Lepoutre 1923, pp. 82–83; see also Nouschi 1975, who stresses the importance of French and Belgian capital in the emergence of a Franco-Belgian alliance in Romania.

a direct oil participation. The move into Pétrofina complemented a separate move into the Compagnie Française des Pétroles (more about this below).

Even before the Pétrofina coalition, Russian oil imports as a means to settle the Soviet debt to France were considered so important that even when the Soviet Union was punished for nonpayment of bonds by suspension of its "most favored nation" status, the Direction des Essences moved quickly in 1921 to issue a decree continuing special tariff privileges for Russian oil, even though Russian exports at the time were almost nonexistent.[10]

In the spring of 1922 Louis Pineau, at the Direction des Essences et Pétroles, submitted to the finance ministry a plan to form a state-backed Belgian-French company to acquire oil land titles in the Caucasus. Pineau did not specify the kind of state backing, but the finance ministry evaluated the plan: "The company would have a relatively small capital, but would form subsidiaries which would be in charge of production. These subsidiaries would have the support of the state: political support at first, but financial support as well. The last would not be limited in extent; and the plan supposes, imprudently perhaps, that such support would be needed only 'in the event that political risks should increase above the level of normal commercial risks.'" The commercial part of the proposed company had in fact already existed for two years: Pétrofina. This firm sought government backing for its Russian ventures and attempted to link oil and Russian debt service: "The plan . . . is presented with great clarity as if for immediate application. It concerns putting together a financial company, Franco-Belgian . . . but majority French, to buy up the 'rights' of Russian proprietors on oil-bearing lands in the Caucasus and currently occupied by the Soviets."[11] Government backing would reduce risk and leverage the diplomatic and economic muscle of the group, which wanted "political support at first" but "financial support as well." The plan was pure Colbertian mercantilism, when government support was enlisted for risky trading in hostile waters. The finance ministry was skeptical about bypassing France's largest oil interests: the plan to favor Royal Dutch–Shell had already failed, and this proposal would have left out not only Standard Oil but also Royal Dutch–Shell, the French

[10] MAE Europe-Russie, 1918–1929, vol. 528, "Note sur l'importation du pétrole russe," 20 December 1924.

[11] Fin. B32308, "Note sommaire pour le ministre aux sujet d'une Note de la Direction des Essences," 20 June 1922. An outline of the original plan may be found in Fin. B32023, "Note complémentaire sur un plan de reconstruction de la Russie," 14 March 1922.

domestic firms linked with both, and their associated banks. The plan's objectives were therefore "uncertain and expensive."[12]

It was difficult for the French government to articulate a coherent policy on the Soviet oil question. The covert struggle for advantage was so intense and so fluid that the record of it is obscure; the overall picture is turmoil, high stakes, and inability of any one group, government or private, to prevail. Alfred Margaine, in a thorough if little known parliamentary history of 1920s oil politics, wrote that in the April 1922 conferences at Genoa "the Russian oil question was, even before the meeting, the pivotal point of the conference. Yet at no time do the minutes, or any of the diplomatic notes, mention oil. The United States was there as an observer, because of oil." Margaine adds, "The negotiations behind the scenes became so active that the representatives of the oil interests were infinitely more important than the official negotiators."[13] Some records confirm his impressions of "behind the scenes" action. Finance ministry documents refer to

> important scheming behind the scenes in which negotiators are trying, behind each other's backs, to put together combinations at the expense of states whose interests will be subordinated to the interests of the great powers. Turkey and Russia are the states in question, and the oil of Baku and Mosul has particularly drawn the interest of the great power organizers of the conference. Even the United States [an observer at the conference] showed exceptional agitation when these two magic words, Mosul and Baku, were pronounced. It would be useless to speak of the English passion, for England has conducted its policy of the last three years to cater to these fetishes, and has lavished millions of pounds on Denikin and Wrangel, and then on the Greeks.[14]

French oil representatives at Lausanne included Laurent Eynac, head of the Direction des Essences, his second-in-command Louis Pineau, and General Gassouin, who was on the board of directors of Standard Franco-Americaine. Also there for Standard Oil was H. E. Bedford, Jr., a relative of the American company's chairman and head of its operations in France. Margaine, writing five years later, concluded that "Standard Oil had acquired supporters in France more powerful

[12] Fin. B32308, "Note sommaire pour le ministre au sujet d'une Note de la Direction des Essences," 20 June 1922.

[13] Doc.Ch., Rapport 5170, 2ème séance, 6 December 1927, p. 365. Margaine's report, in spite of some errors, is the most comprehensive contemporary evaluation of oil politics of the 1920s.

[14] Fin. B32023, "Correspondance de signification spéciale," 6 December 1922. Anton Denikin and Piotr Wrangel were Russian generals who headed the "volunteer army" the allies sponsored against the Bolsheviks.

than those of Royal Dutch; the French and the Belgians followed Standard Oil."[15]

Margaine was right about the connection between France's Russian policy and Standard Oil, at least in late 1922 and early 1923. Louis Pineau's efforts to form a united front of exiled Russian oil property holders and oil interests began at Genoa in the spring of 1922 and continued through the Lausanne conferences into the following year. In March 1922 he hoped the French and Belgian banks would use "Russian oil to free us from the domination of the Anglo-Saxon trusts. But for that, France must, together with the Belgians . . . unite its interests to those of Russia. France must therefore wait for normal relations to be established with Russia so as to bring about direct negotiations with an eye toward obtaining special advantages in the Caucasus, advantages necessary for our supplies, which the national policies of Russia seem to favor."[16] Pineau dallied with the idea of an oil alliance with Italy but later broadened his approach to include the major international oil firms.[17] He had brave words about French independence, but he also urged an alliance with Standard Oil as the means to keep a foot in the door of Russian supplies—a step that could hardly be called independence from "Anglo-Saxon trusts," which meant Anglo-Persian, Royal Dutch–Shell, and Standard Oil.

The Standard Oil Company flirted briefly with Royal Dutch–Shell. The two companies met 29 September 1922, before the Lausanne conference, but the political situation was too fluid, and commercial rivalry too intense, to permit an alliance.[18] Pineau, pursuing his goal of a Franco-Belgian alliance on Russian oil, went to London in May 1923 and negotiated with Anglo-Persian's John Cadman and officials of Standard Oil for joint action on Russia. Their plan was vetoed by Royal Dutch–Shell.[19] Independence from the trusts was lower on Pineau's list of

[15] Doc.Ch., Rapport 5170, 2ème séance, 6 December 1927, p. 365. The French and Belgians wanted the Bolsheviks to restore full private property rights in the Soviet Union's oil fields; the British would have settled for a ninety-nine-year concession. Standard Oil worked closely with the Banque de Paris et des Pays Bas; the latter, through the Romanian Omnium International des Pétroles, was also involved with Belgian capital, through the group Jossé Allard and the Société des Pétroles Roumanie (Lepoutre 1923, p. 83).

[16] Fin. B32023, "Note complémentaire sur un plan de Reconstruction de la Russie," 14 March 1922.

[17] See MAE Europe-Russie, 1918–1929, vol. 527, "La Russie et le Pétrole Russe," by Louis Pineau, 24 February 1922.

[18] Fin. B32023, "Groupement des Sociétés Naphtifères Russes et Etrangères," 29 September 1922.

[19] Fin. B32023, "Memorandum," by Louis Pineau, 13 March 1926. Text of agreement in English is in Fin. B32023, "annexe 1—Texte redigé le 14 mai 1923 par Sir John Cadman," 14 May 1923.

priorities than a place in the sun for the "Franco-Belgian interests," Pétrofina.

The oil multinationals' attempts to forge a containment policy on Soviet oil were decisively set back in March 1923 when the Shell combine's Henry Deterding bought some, flying in the face of his own virulent anti-Bolshevism and his own earlier efforts to contain Soviet oil.[20] Perhaps the Soviets tricked Deterding into believing he would get permanent, preferential treatment. They did try to play on the rivalries of the international oil companies. Deterding's turnabout was a classic unannounced defection from a cooperation scheme under laborious negotiation. He feared that if he did not strike a bargain first, someone else would. Three years later he admitted to Pineau that his action had been a mistake; but his confession came too late to stop a pell-mell stampede of oil companies for Soviet contracts.[21] Royal Dutch–Shell's imprudent maneuver sundered the collective effort to wall off Russian oil exports. The Groupement des Sociétés Napthifères Russes et Etrangères, the organization in which the international oil interests had tried to come to agreement, split into three factions. The Shell combine dominated the Groupement, while the Belgian-French group formed the Syndicat Franco-Belge pour la Défense des Intérêts Français et Belges dans l'Industrie de Naphte en Russie. Standard Oil steered its own course.[22]

The split between Royal Dutch–Shell and Pétrofina's backers did not just reflect disagreement over importing "stolen" Soviet oil. Royal Dutch–Shell sought a special reimbursement fee on all exports of nationalized Soviet oil to compensate those who, like itself, held titles to expropriated land. The Franco-Belgian group had none and wanted to link any tax on Soviet oil exports to a general settlement of the czarist debt. Oil-land holders were not to be singled out as a special class for favored treatment.[23] It was a banker's view of how to handle Soviet debt, and Pétrofina was a bankers' oil company. Pétrofina uninhibitedly imported Russian oil that had formerly belonged to Royal Dutch–Shell or that Standard Oil had bought from the Nobels. As an oil company, Pétrofina saw Bolshevism's confiscation of its competitors' lands as an opportunity. As a financial group, it retained an oil-linked creditor's agenda with regard to Russia.

[20] Fin. B32023, Syndicat Franco-Belge to Président du Conseil, 16 June 1923.

[21] Fin. B32023, "Memorandum," by Louis Pineau, 13 March 1926.

[22] The Banque de Paris et des Pays Bas and its Omnium subsidiary discussed sending an oil mission to Russia in 1923, but it is unclear if this was coordinated with Standard Oil. See MAE Europe-Russie, 1918–1929, vol. 527, "Note pour le Président du Conseil," 6 February 1923.

[23] Fin. B32023, "Memorandum," by Louis Pineau, 13 March 1926.

The growth of Pétrofina, the international competition among the majors, and the intractable Russian debt problem directly influenced French policy. French domestic regulatory initiatives, and the French government's policies toward the majors, revolved around the conflicting demands on the Franco-Belgian financial interests, the majors, and French domestic firms.

Soviet Oil and French Domestic Regulatory Policy

In the 1920s the survival of French domestic oil interests depended on imports of Soviet oil. The international majors were trying to expand in the French market at the expense of the "ring of French refiners," Gulbenkian's felicitous phrase for the French national firms that for decades had dominated French wholesale distribution. In the prewar period the "ring" had defended the national firms against Standard Oil. The companies enjoyed tariff advantages and had a limited potential to switch to Russian and Romanian oil. Standard Oil chose not to disturb the prevailing political and economic arrangements, and the French national firms preserved their lucrative status as distributors of refined products. The war changed supply patterns and brought the ill-fated attempt to hand the French market over to Royal Dutch–Shell. Traditional sources of oil supplies, especially Soviet Russia, became less reliable.

Lamp oil was giving way to transportation fuel, which meant heavy new capital costs for gasoline pumps. Automobiles in France had increased spectacularly after the war, from 200,000 in 1920 to 800,000 in 1926.[24] The French domestic concerns found themselves beset with increased competition, decreased supplies, and insufficient capital to meet the requirements of a changing market. Under pressure, some firms sold majority shareholdings to the multinationals. By buying French firms the multinationals established wholesale-retail networks and access to the political connections of their owners. Royal Dutch–Shell bought its former collaborator, the firm Deutsch de la Meurthe, to found the Jupiter Company in 1921. This consolidated the collaboration between the Rothschild-Shell combine and Deutsch de la Meurthe that dated from before the war; Royal Dutch–Shell's initial equity holding of 49 percent in Deutsch de la Meurthe would expand to 75 percent by 1927. In 1920 Standard Oil bought the equally vener-

[24] See testimony of former Secrétaire Général Pierrot of the Chambre Syndicale de l'Industrie de Pétrole, in "L'Enquête parlementaire," *Revue pétrolifère,* no. 266 (21 April 1928): 545.

able Fenaille et Despeaux, with which it had collaborated closely during the war. The new firm was christened La Pétroléenne; Standard Oil also bought a 48 percent share of the Compagnie Générale des Pétroles in 1921. Anglo-Persian bought Paix et Lesieur in 1923 to form the Société Générale des Huiles de Pétrole.

Other prominent French firms, such as Desmarais Fères, Lille Bonnières Colombe, and the Raffinerie du Nord, received offers from the multinationals but opted for a more combative strategy.[25] In theory all sellers in the French market, even those recently bought by foreign firms, remained committed to long-standing cartel practices and the principle of negotiated market shares. Since the multinationals controlled the vast majority of oil arriving in France, including oil going to their nominally "independent" competitors, they wanted larger market shares that recognized their dominance. Cartel negotiations became acerbic. The independent French firms were severely handicapped by their lack of vertical integration.

The intense competition continued throughout 1923 and into 1924. Prices fell, and rebates to retail outlets became common. In June 1924 French marketers started negotiations for a *contingentement,* a cartel agreement. The offers of the multinationals were so unfavorable to French interests that these last threatened to withdraw from the Chambre Syndicale, the industry association that from the 1920s to the present has been the center of market division among major French oil firms. The majors backed off, and for 1925 an agreement was signed giving the French independent firms 44.5 percent of the French market up to the first million tons, with an entitlement to 23 percent of future market increases over that amount (see fig. 6). The international firms had captured most of the French distribution since the end of the war and had secured an even stronger grip on future market growth. Yet the French independent firms were not all small: the largest, Desmarais Frères, had 21 percent of the French market under the cartel agreement, more than Royal Dutch–Shell. In the mid-1920s Desmarais Frères spent FF120 million to modernize by installing gasoline pumps. The firm was as aggressive as its multinational rivals in meeting the challenges of the new postwar market.[26]

The French firms' inadequate access to "independent" sources of oil was their Achilles' heel. Romanian production was useful but not by itself sufficient to meet their needs. Romanian independents sold to

[25] See deposition of Robert Cayrol, director of Desmarais Frères, in "L'Enquête parlementaire," *Revue pétrolifère,* no. 267 (28 April 1928): 582; also Gibb and Knowlton 1957, p. 508.

[26] Figures from testimony of Robert Cayrol of Desmarais Frères, "L'Enquête parlementaire," *Revue pétrolifère* no. 267 (28 April 1928): 582–583.

LA REVUE PETROLIFERE

Diminution du pourcentage réservé aux Sociétés Françaises en fonction du tonnage par application de l'accord de 1925

(avant 1919, la totalité des sociétés était française).

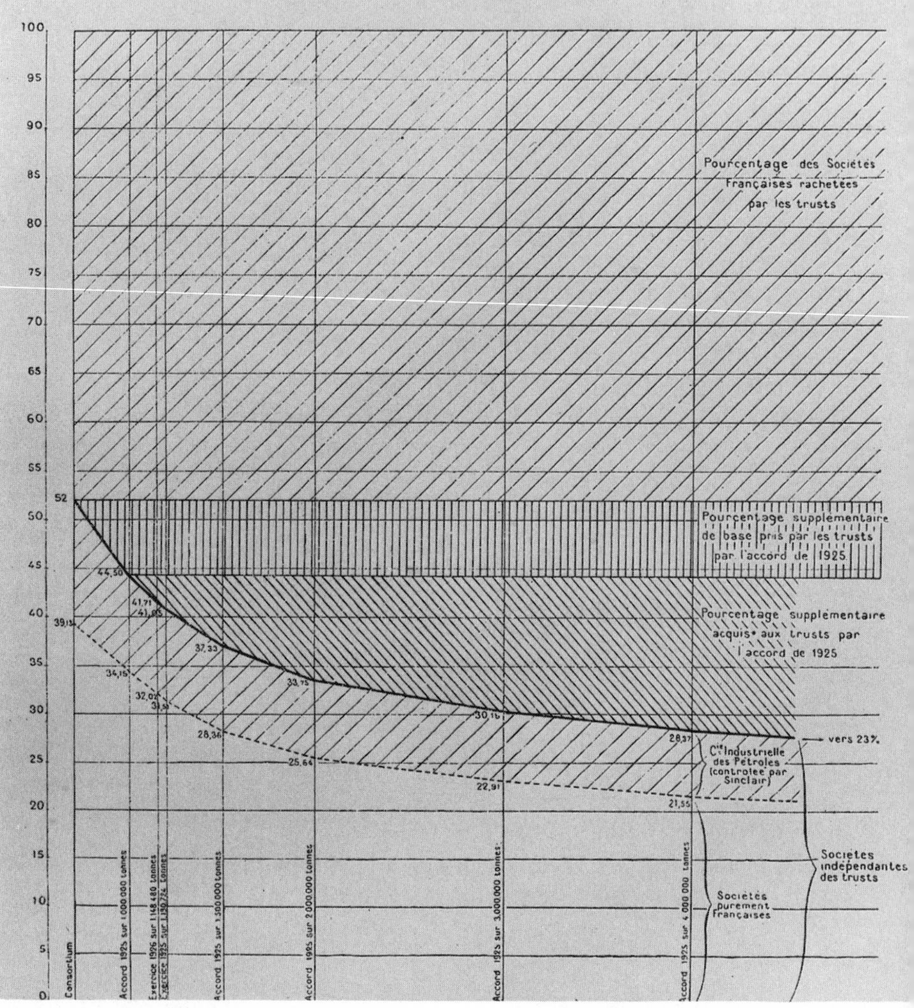

Figure 6. French 1925 cartel agreement. *Source:* Testimony of Léon Wenger (Pétrofina) before parliamentary committee of inquiry, shown as reproduced in *Revue pétrolifère,* no 268 (5 May 1928): 628. Phot. Bibl. Nat. Paris.

French independents, and some French firms had their own production in Romania; but the total still was not enough. True independence from the trusts required more sources of oil. The national firms embraced the Pétrofina group, whose Russian oil offered an alternative to the international majors. Led by Léon Wenger in France, Pétrofina worked with the Raffinerie de Pétrole du Nord and also became the designated representative *(mandataire)* for the leading French independent firms' dealings with the Soviet Union. This group included Desmarais Frères, Lille Bonnière Colombes, Pechelbronn, the Compagnie Industrielle des Pétroles, and the Société Française des Carburants.[27]

Creation of the Office National des Combustibles Liquides

The alliance between Pétrofina and French national producers created, from 1921 to 1925, a solid constituency for Bolshevik oil. Their lobbying created the Office National des Combustibles Liquides, the major pre–World War II oil regulatory agency. The French Left wanted to help the Soviet economy, and it found allies in the financial and oil interests that needed Soviet oil. As competitive pressure on the French national companies rose, they sought an energy agency with the regulatory power to make Russian oil imports easier. Their first efforts were directed at the Direction des Essences, a vestigial agency left over from Bérenger's days at the Commissariat Général aux Essences during and after the Great War. The first parliamentary proposal to increase its power was offered in April 1921 by the socialist Edouard Herriot.[28] In 1922 a similar measure to create an "Office National du Pétrole" was introduced: finance ministry officials knew it was linked to the Genoa conference and the "Franco-Belgian" oil group's desire to import Soviet Oil.[29]

Though beneficial to many French capitalists, the pro–Soviet oil posture of the proposed Office National was too pink for Poincaré, who became premier in early 1923 and whose oil policies were energetic in every way but this. The victory of the Cartel des Gauches in the spring of 1924 swept away Poincaré's hesitancy; the Office National des Combustibles Liquides was approved by parliament on 10 January 1925.[30]

[27] Testimony of Léon Wenger of Pétrofina, "L'Enquête parlementaire," *Revue pétrolifère,* no. 267 (5 May 1928): 623.

[28] Thomas 1934, p. 60.

[29] Fin. B32308, "Note sommaire pour le ministre au sujet d'une Note de la Direction des Essences," 20 June 1922.

[30] Labarrière 1932, pp. 122–127; Mertens 1926, pp. 245–260. Debate over the legislation began earlier; see Déb.Ch., 11 April 1924, pp. 2116–2118.

Louis Pineau, formerly head of the Direction des Essences, became the new agency's director.

Yet the Office National des Combustibles Liquides remained toothless. It enforced a three-month oil reserve requirement on oil importers, which had been required since 1921 by the Direction des Essences. It had licensing power for oil importers, but there were no provisions for controlling the overall quantity of imports. Licensing remained a strictly pro forma affair after the great crack-up of the state oil import monopoly in 1920. Official policy was described as *liberté contrôlée,* but this was a juridical nicety. The agency could not participate in commercial operations and at its inception was simply unable to be the arm of the interventionist French state. It had a mandate to help develop synthetic oil from coal (see next chapter), and the coal sector gave powerful support to the agency's creation. It reflected the aspirations of the corporate and financial supporters of Russian oil, who wanted to expand its authority. Six months after establishing the Office National des Combustibles Liquides, the French parliament sent a commission to the Soviet Union to study the possibilities for working oil there.[31]

The French national companies' alliance with Pétrofina caused tensions with the international majors to become even more acute; cheating on the cartel agreement became common. Russian imports grew from a combined figure of 56,600 tons for 1923 and 1924 to 90,000 tons in 1925, and ultimately to 220,000 tons in 1928, over 10 percent of the French market. In 1928 Pétrofina and its allies contracted with the Soviets for 2.2 million tons, threatening serious competition not only in France but all over Europe.[32]

For a few years Russian oil was a linchpin of the state's strategy to protect French national firms. Even with limited powers, Louis Pineau gained higher visibility at the Office National and persuaded the French navy to help the domestic French industry by buying Soviet oil. The details reveal how the government helped the French firms. Soviet refining produced a fixed ratio of heavy fuel and lubricating oils to lighter kerosenes and gasoline. The Soviets needed to sell their full range of products, but the French commercial market could absorb only the lighter ones; French purchases would have left the heavy Russian products unsold. Standard Oil had worldwide outlets and could place the Soviet's heavy oil products, and it tried to buy the full spectrum of production to deny this source to Pétrofina and its French allies. It negotiated with the Soviets in London and Paris, and in 1927

[31] Gargas 1927, p. 96.

[32] Doc.Ch., annexe 4327, 7 April 1927, p. 605; and Doc.Ch., annexe 5170, Sess.Ex., 6 December 1927, p. 368.

it nearly landed monopoly rights to sell Soviet oil in France, Portugal, Spain, North Africa, and Belgium.[33] Louis Pineau leaped into action on behalf of the constituents who had made his agency possible. He visited the Soviet embassy's oil representatives, reminding them that the special tariff exemption favoring Russian oil, which dated from 1921, "was simple toleration" and "could be ended at any time." This news, in a finance ministry summary, "seemed to cause [Soviet oil representative] M. Lomoff and his colleagues to reflect a bit."[34]

The Soviets complained of 50,000 tons of fuel oil stocks at Batum and Novorossisk, with more piling up each month. Standard Oil would buy them as part of the contract for light oil products: What about the Pétrofina group? Heavy fuel oils were the province of Royal Dutch–Shell, and the Pétrofina group did not want to challenge the combine. Who would take the heavy fuel oil to help the French group land exclusive rights on Russian light oils?

The French navy was planning to buy 100,000 tons of fuel oil in 1925. Pineau called the minister of the navy, who agreed to buy the Soviet oil.[35] The Russian Syndicat du Naphte and the French navy signed contracts in late February 1925.[36] Pétrofina eventually landed the Soviet oil monopoly for the French market, and the allied French firms were bolstered. The four-year campaign to secure government support for Soviet oil imports had paid off, and the French firms and Pétrofina were closer to their goal of a supervised, allocated market. The question of Russian debt remained unsettled, but however nettlesome that might have been to Pétrofina's banker owners, the oil company at least profited from Russian oil. The monopoly sales privilege was certainly preferable to no oil, no debt settlement, and only the thin hope that the Bolshevik government might collapse.

Pineau's intervention had shown that, unlike Bérenger's clumsier efforts, government action could shape the oil market without blowing up in its intended beneficiaries' faces. But survival was not prosperity: the multinational firms stepped up pressure, and even with Soviet oil

[33] MAE Europe-Russie, 1918–1929, vol. 528, "Note sur l'importation du pétrole russe," 20 December 1924. Many firms tried to negotiate regional, product-specific monopolies with the Soviet Union. Deutsche Erdoel sought a monopoly of Russian products for the German market; Pétrofina, in alliance with Desmarais Frères and Pechelbronn, sought a similar arrangement for France, for which it competed against Standard Oil and the independent known as PEN (Pétroles, Essences et Naphtes), an offshoot of the Bauer-Marchal Bank.

[34] MAE Europe-Russie, 1918–1929, vol. 528, "Note sur l'importation du pétrole russe," 20 December 1924.

[35] MAE Europe-Russie, 1918–1929, vol. 528, "Note sur l'importation du pétrole russe," 20 December 1924.

[36] MAE Europe-Russie, 1918–1929, vol. 528, "Note pour la sous-direction d'Europe," 20 February 1925.

the market share of Desmarais Frères, the leading French national firm, declined from 21 percent in 1924 to 17.95 percent in 1926. In June 1926 French cartel negotiations broke down, and in November price cutting and rebates abounded as the companies jockeyed for larger shares in the next round of cartel talks. Lacking the multinationals' deep pockets, the French national companies looked to the government for help. Without such help Soviet oil could only, in the words of the director of Desmarais Frères, "loosen the screw."[37]

Sympathetic to the French companies, in April 1926 parliament passed an amendment to a fiscal bill, accepting in principle the idea of more stringent market regulation. Importing "oil and its derivatives" was to be effected as of April 1927 by "the state or persons accredited by it."[38] The amendment needed clarification: if the state did the importing, the system would in reality be a state monopoly. If importing was done by "persons accredited by it," virtually any alternative might be envisioned, from a continuation of the existing license system to more intricate forms of state participation such as the *régie intéressée* or *société mixte.*[39] A complete reorganization of the oil industry could hardly be sandwiched into an amendment of a revenue bill: parliament's April 1926 vote was a warning to the oil groups that they should settle their differences or else parliament would. In today's parlance, it was a call for "voluntary" quotas.

The 1926 proposal had other objectives. One was to raise state capital

[37] Robert Cayrol, Desmarais Frères, before "L'Enquête parlementaire," *Revue pétrolifère,* no. 267 (28 April 1928): 581–583.

[38] Labarrière 1932, p. 112.

[39] In the *régie intéressée,* business is conducted on behalf of the state by a private person (or persons) who participates in the profit but is protected from losses and from capital risk. Profits are shared with the state; monopoly privileges are optional. In a *régie directe,* state administrative officers retain control of the enterprise, and the state receives the profits. In a concession, one has the right to participate in part or all of a market (monopoly concession) but must venture one's own capital and is exposed to potential loss. *Société mixte* may be applied to many different arrangements between the state and private capital. The *société mixte* functions like a private corporation, but the state owns shares and receives dividends; the degree of control will depend on the number of shares held by the state and on other charter provisions. In theory, private capital risks more in the *société mixte* than in the *régie intéressée,* but in practice the state would rarely allow a company in which it was a shareholder to collapse. The *régie intéressée* and the *société mixte* therefore converge in some characteristics. The *régie directe* connotes non-profit or limited profit status, as with municipally owned utilities in the United States, and usually is an enterprise organized by the state in the absence of a private service, as with water. Nationalized companies, *entreprises nationalisées,* as the *Grand Robert* notes, are incorrectly but commonly referred to as *régies,* this "misuse" extending even to legal entities such as the *Régie Renault,* a profit-seeking company created by private capital and subsequently taken over by the state. The many permutations of law, common usage, and case-by-case modifications make it unwise to expect linguistic precision where political practice blurs distinctions. See articles s.v. *régie,* and *exploitation* in the *Grand Robert* and *Petit Larousse.*

to modernize the French refining industry. Refining on French soil would let the domestic industry capture an additional part of the profits and would reduce the cost of imports, improving the balance of payments. Like the 1918 version, the state-run import monopoly would distribute through the French firms; oil in theory would be obtained from Texas Gulf independents, Romania, and the Soviet Union.

The project was linked to France's war debt to the United States. Talks on the allied debt to the United States were a feature of 1920s international finance, but linking a French oil monopoly with debt repayment was real enough for Standard Oil's A. C. Bedford to ask that the Bérenger-Mellon debt talks be used to pressure France to keep a "free market" for oil.[40] The debt talks brought together Standard Oil's commercial rivals. Bérenger's former links to the Shell combine and Rothschild interests were well known, and Andrew Mellon, United States treasury secretary, was from the family that headed Gulf Oil and substantial banking interests. But no substantive link of French oil refining and American debt emerged from the negotiations, even though after that meeting the movement to regulate the French oil market increased.[41] The April 1926 amendment to regulate the French market coincided with Bérenger's debt negotiations in the United States; the author of a major related proposal, Alfred Margaine, was a member of Bérenger's Radical-Socialist party. Deputy Margaine's figures indicate that refining oil in France would have saved FF500 million a year, about one-fifth of France's total 1925 trade deficit with the United States.[42] Mellon's Gulf Oil could have helped France supplement its Soviet imports. But Bérenger had left oil politics after the debacle of 1919–1920; whatever his historical connections with the Shell combine, he probably was closest to the Rothschild bank, and he wore a "banker's hat" rather than an "oil hat" during the negotiations. In his 1926 book *France and Her Capacity to Pay,* Bérenger argued that in corporate taxation banks "are burdened the heaviest."[43] His worry was finding sources of revenue and, perhaps wiser from his 1920 experiences, he never mentioned oil.

Labarrière calls the 1926 state monopoly proposal a device for the French national companies to sell out at a higher price than the multinationals would offer; they preferred to "get expropriated by the state

[40] Melby 1981, p. 98.

[41] The two men signed a debt agreement for nearly $7 billion to be paid over sixty-two years. Had the agreement been honored, France's World War I debt would have been retired the year this study was filed in dissertation form (1988). See front page of *New York Times,* 30 April 1926.

[42] Debt size estimated by Bérenger 1926, p. 69; Nowell 1983, pp. 248–249.

[43] Bérenger 1926, p. 58.

with an indemnity rather than be driven out of business after a struggle that would have drained their reserves."[44] But the state monopoly proposals were more important to Margaine's May 1926 proposal than the fiscal amendment of a month earlier.[45] Margaine talked tough to bargain tough, but he had no complete nationalization in mind. In 1927 he explained that he favored a *régie intéressée*: the state would monopolize the import business but would share refining and distribution with private interests.[46] He explicitly stated his desire to protect private interests: "I have simply tried to create a right for the state. The road to a concessionary system leads through a state monopoly. The state, once it has this monopoly right, can give it back [to the private sector] soon after."[47] Margaine would have used state resources to build up a domestic refining system and then reprivatize. The Soviets thought the plan would help them sell oil and tried to influence the outcome. Standard Oil chairman Walter Teagle noted that the French national companies and Soviet oil exporters had become a significant faction: "Mr. Teagle reviewed the situation in France as a whole, pointing out that different groups in France were interested along different lines *and that the Government had no definite policy regarding petroleum.* He mentioned the present difficult situation due to many opposition factions; that some people were opposed to Mercier & Cie. Nationale;[48] that the Russians had spent much money in France and, as a result, a number of people were trying to create an oil monopoly which they could turn over to Russia as an outlet for Russian oil."[49]

Margaine's proposal was useful as a threat against the international majors. When the April 1927 deadline set by the April 1926 amendment came, parliament had not prepared any policy, but the national companies' need for one had grown more acute in the absence of a satisfactory cartel agreement. Parliament upped the ante by authorizing a parliamentary committee of inquiry into the oil industry, which began its first deliberations in May 1927 and continued through the year and into the spring of 1928.[50] Many major figures in the French

[44] Labarrière 1932, p. 112.

[45] Doc.Ch., annexe 2903, 27 May 1926, pp. 372–375.

[46] Doc.Ch., annexe 5170, 6 December 1927, p. 355.

[47] "L'Enquête parlementaire," *Revue pétrolifère,* no. 267 (28 April 1928): 587.

[48] The Compagnie Française des Pétroles.

[49] Pat.Hearings, p. 3430; reprint of memorandum dated 31 August 1928. Italics added.

[50] The Commission de Pétrole's incomplete records in the Assemblée Nationale make reference to published government transcripts that Faure 1939, p. 87, already characterized as "difficult to consult, because they are out of print at the office of the Journal Officiel and not in published collections, and because they are not a part of the Annexes themselves." Melby cites a copy (1981, p. 98) but does not say where he found it. The *Revue pétrolifère* published some of the major testimony between 3 March and 19 April 1928 (see bibliography), for which information I am indebted to André Nouschi. Incom-

oil industry testified before the commission, and many witnesses stated that the inquiry itself had precipitated a truce among the oil rivals, who feared reprisals.[51] As Cayrol of Desmarais Frères testified in November 1927: "Right now the situation has modified to the point that there is a status quo, but don't think it's because we have succeeded in making the trusts understand the situation in France. If we are in the current situation, in which there are fixed percentages and no preferential increases for the trusts, let me tell you it's because they are afraid of the *gendarme.* And I dare say, M. Margaine, you probably count for something in that."[52]

Pétrofina and Desmarais Frères, along with other pro-Soviet oil firms, disliked Margaine's state monopoly proposal but favored assigned quotas. Quotas backed by state authority could not be cheated on as the cartel quotas had been.[53] Léon Wenger professed to like free markets, but he thought market controls necessary against companies that were "stronger than nations."[54] The French national companies had asked the Office National for quotas, but the government agency "was not equipped" to intervene effectively. Deputy Charles Baron asked for clarification:

> *Baron:* So, in a word you can say that you are asking for *l'état-gendarme* to set the trusts straight.
> *Cayrol:* Yes, for this reason the system of concessions would satisfy me in this regard.

Cayrol warned that Margaine's state monopoly would face "immense

plete handwritten and typed transcripts made by the committee secretary are in the archives of the Assemblée Nationale at Versailles, identified simply as "Procès Verbaux de la Commission des Pétroles, 1927–1928." This source includes committee deliberations from May 1927 and testimony of key witnesses such as Raymond Poincaré, Louis Pineau, and Louis Loucheur, none of which was printed in the *Revue pétrolifère.*

[51] Testimony of Léon Wenger, "L'Enquête parlementaire," *Revue pétrolifère,* no. 268 (5 May 1928): 629.

[52] Testimony of Cayrol, "L'Enquête parlementaire," *Revue pétrolifère,* no. 267 (28 April 1928): 586–587.

[53] Standard Oil's Bedford, whose reserved testimony perhaps reflected the need for restraint given his company's role as the bête noire of world oil politics, observed simply that "if you have a monopoly, you won't be thanked by consumers, because without free competition prices will rise. That is why I am opposed to a monopoly, whether it be public or private" (testimony of Bedford, "L'Enquête parlementaire," *Revue pétrolifère,* no. 269 [12 May 1928]: 666). Yet Bedford's company was routinely involved in cartel negotiations designed to raise prices, in France and elsewhere. Given the pitch of antitrust sentiment the Standard Oil Company aroused, the company would have been unwise to admit publicly that it favored a monopoly or quota system, whether in France or anywhere else.

[54] Testimony of Léon Wenger, "L 'Enquête parlementaire," *Revue pétrolifère,* no. 268 (5 May 1928): 629.

difficulties" in getting adequate oil supplies because of the opposition of the trusts. He thought the concession system would generate less antagonism and argued that with Margaine's state monopoly, France would have to work so hard to find alternative sources of oil that "your foreign policy will have to be aligned on your oil procurement possibilities."[55] Royal Dutch–Shell's representative resisted, arguing that private cartelization was preferable to government-enforced regulation: "[Government regulation] can have unexpected repercussions and bring an arbitrary element into the attribution of market quotas. Can you imagine how delicate, what a fearsome task it is, for someone who is not a part of the oil business, no matter how intelligent he might be, to attribute quotas? If the quotas are not fixed according to what each company can sell, the attribution of just 1 percent less to one party can take away six million francs from him and put them in the pocket of another."[56]

Royal Dutch–Shell was gaining a greater market share in France and preferred contentious private cartel negotiations to any state project. The combine's growth was threatened by the proposed oil legislation. The rival Pétrofina had in 1927 raised capital to build a refinery in France, and its chief officer had filed a detailed plan showing how Soviet oil could be used to finance the prerevolutionary debt and provide the basis for new investment to increase Soviet exports.[57] To cap off the suspicions of Pétrofina's rivals, the chairman of the parliamentary committee of inquiry, Charles Baron, publicly revealed that he was the son of a founding member of the Compagnie Industrielle des Pétroles, a French national company that had come under the control of Sinclair Oil and that collaborated with Pétrofina in importing from the Soviet Union. Baron's mix of left-wing politics and profitable business opportunities was particularly unpalatable to the international oil trusts.[58]

Royal Dutch–Shell grew progressively combative as imports from the Soviet Union to France and elsewhere increased. The Soviets made a contract with Standard Oil Company of New York (Socony) to supply markets in Royal Dutch–Shell's *chasse gardée* in the Far East; they had twice raised prices on deliveries to Royal Dutch–Shell, in 1925 and

[55] Block quotation and following citation from testimony of Cayrol, "L'Enquête parlementaire," *Revue pétrolifère,* no. 267 (28 April 1928): 586–587.

[56] Testimony of Brylinski, "L'Enquête parlementaire," *Revue pétrolifère,* no. 270 (19 May 1928): 696.

[57] Fin. B32023, "Note sur le Règlement des Dettes de la Russie envers la France," by Léon Wenger, 16 March 1927.

[58] Stated by Baron during testimony of Barbaudy (Cie. Industrielle des Pétroles), "L'Enquête parlementaire," *Revue pétrolifère,* no. 262 (24 March 1928): 393.

1926, and then cut the combine out of contracts altogether in 1927.[59] Henry Deterding protested the French navy's Soviet oil purchases and cried crocodile tears to Louis Pineau about how shortsighted his 1923 policy of importing Soviet Oil had been.[60] The following year Deterding embarked on a crusade against Pétrofina, contending that the Soviet oil was "stolen" from its rightful, prerevolution owners—that the firm was receiving stolen property. The Shell-controlled association of former Russian oil proprietors sent a warning letter to Pétrofina's Léon Wenger, who presented it in testimony. The letter threatened legal action to repossess imported Soviet oil or the money made from it, and in fact the combine did win a related suit to confiscate the Russian Syndicat du Naphte's deposits in French bank accounts, in a dispute over the Spanish market.[61] Even if unsuccessful, such suits promised expensive legal proceedings on a politically delicate topic: "The international group of oil companies of Russia has unanimously decided in its session of 15 March 1927 to advise companies and governments and third parties who have bought or who are going to buy oil products from the Russians that the said products are liable to be claimed by the members of the international group who are the legitimate proprietors of those products and who claim the right to prosecute, by all legal means, for compensation for damages caused by any dealings having as their purpose the purchase of petroleum products coming from the lands under exploitation in Russia that belong to them and have been confiscated by the Soviet government."[62]

Soviet oil sales were increasing in Asian and European markets; it menaced the combined markets of Royal Dutch–Shell, Anglo-Persian, and Standard Oil. The Soviets usually favored market "outsiders" wherever they did business. A Soviet official expressed sympathy for state-controlled oil initiatives in France, Spain, Italy, Greece, Yugoslavia, and Turkey. This exacerbated the oil industry's paranoia that the Soviet Union was exploiting its relationship with left-wing political parties to market its oil. For the oil man, the international red menace was a Bolshevik standing next to a gasoline pump.[63]

[59] Testimony of Branet (Jupiter, Royal Dutch–Shell's subsidiary in France), "L'Enquête parlementaire," *Revue pétrolifère,* no. 260 (10 March 1928): 334.

[60] Fin. B32023, "Memorandum," by Louis Pineau, 13 March 1926.

[61] MAE Europe-Russie, 1918–1929, vol. 529, "L'Attaché Commercial à Berlin," 10 January 1928.

[62] Testimony of Léon Wenger, "L'Enquête parlementaire," *Revue pétrolifère,* no. 268 (5 May 1928): 627. The letter was signed by Deterding and five others.

[63] On Spain, see Nowell 1985, pp. 114–117; also Fin. B32023, "Interview de M. Lomoff, Président du Syndicat du Naphte, au Journal Torgovo-Promislehayaya Gaseta," n.d., prob. 1925. Three years after its founding, the Spanish oil import monopoly, CAMPSA, suspended purchases of Soviet oil, citing "recent declarations of the Congress of Soviets

The Soviet Union was an outsider, a cartel spoiler, in the late 1920s. It sold oil to some at more favorable conditions than to others; it was conspicuous by the quantity of its reserves, but it was not alone. The parliamentary committee of inquiry pointedly invited the Phillips Petroleum Company president to testify, emphasizing the availability of other oil. Frank Phillips stated his desire to sell oil to France and broached the possibility of building a refinery there. An aggressive oil policy might yet lessen French dependence on the multinationals.[64]

From the tenor of committee hearings and from the coalition backing a reinforced Office National des Combustibles Liquides, one might have concluded in late 1927 or early 1928 that France was headed toward a fiercely independent oil policy. One might have anticipated sharp restrictions on the international majors in France and perhaps a *régie intéressée* or quota system that would have diversified suppliers, reduced the market shares of the majors, increased French imports of Russian oil, and contracted with Texas Independents. Pineau's Office National would have ushered in the new era.

But the international majors resisted. The quotas and enhanced powers of the Office National des Combustibles Liquides enacted in March 1928 did not bring about the independent policy describe above. On the contrary, imports from Russia and the Texas independents were reduced, while French national policy was locked to the interests of the majors. To understand why, we have to return to the early 1920s and see how the complex struggle over Russian oil became enmeshed in the struggle over Mesopotamia.

THE COMPAGNIE FRANÇAISE DES PÉTROLES

The pride of French etatism in the interwar period was the Compagnie Française des Pétroles (CFP). Its history is intimately linked to the struggle for Mesopotamian oil and is a major motif of the struggle for control of the French market. Rondot, once an employee of the company, opens his history with a caustic observation: "There is a rather widespread opinion that the creation of the CFP was decided a priori by French statesmen. This beginning, drenched in etatism, is a perma-

on the use of profits from commercial operations with foreign states for purposes of revolutionary propaganda" (Fin. B32023, "Ambassadeur de la Republique Française à Madrid à Monsieur le Ministre des Affaires Etrangères," 2 July 1930). More discussion is in Fin. B32023, "Note," 16 August 1928, and a translation of an *Izvestiya* article dated 1 December 1928.

[64] Testimony of Phillips, "L'Enquête parlementaire," *Revue pétrolifère,* no. 259 (3 March 1928): 281–286.

nent source of annoyance to the company's directors, an annoyance that one observer has rightly called an 'immaculate conception complex.'"[65] Unfortunately, Rondot's summary exposition of the company's origins does not vindicate his own sarcasm; but he aims true. The company's history was even less "drenched in etatism" than that of Anglo-Persian, whose origins were examined in chapter 2.

The "widespread opinion" that the company was created a priori by French statesmen originates in Premier Raymond Poincaré's 1923 directive to the company to assume control over France's shares in the Turkish Petroleum Company; troops had won the shares from Germany, and the Shell combine had won them for France from England. Poincaré's letter, dated 20 September 1923, was addressed to Ernest Mercier, who organized the company. This letter is one of two key documents in the "etatist" interpretation of French policy:[66]

> Following the interviews which you have had with the ministers of finance and commerce, I would like to confirm the general principles according to which the government would like a national oil group constituted. You will assume the presidency of this oil group, which you accept on condition that you adhere to government principles and on condition that your designation be accepted by the ensemble of the principal French companies in the oil industry. The group will be formed as soon as possible, in the form of a private company [Société Anonyme Française] whose statutes, in conformity with the principles mentioned above, will be approved by the government.
>
> *Government Objectives*
>
> The government desires a policy instrument capable of carrying out a national oil policy. The company must be essentially French and remain completely independent. It will try to develop oil production under French control in different productive regions. It will create any companies that it deems necessary for its prosperity. The company will, in addition, receive from the government the task of undertaking any works the government will deem opportune. If the company should undertake unprofitable operations, the state will accord it equitable compensation.
>
> *Company Objectives*
>
> The company's goals will be:
>
> 1) To organize and develop the resources and other advantages that the state will procure from diplomatic or other accords, now existing or to come, concerning oil.

[65] Rondot 1977, p. 5.

[66] The other key document is Clemenceau's December 1917 telegram to Standard Oil, cited in chapter 3.

2) To take participations or other interests in all other enterprises working in different oil regions, especially in Central America or South America, so as to diversify our source of supplies.

3) To take up, in due time and by the proper means, the question of Russian concessions, so as to avoid the dispersion of French efforts in the face of foreign competition.

4) To develop, with government support, whatever oil riches might be discovered in France, in the colonies, and in the countries under protectorate.

For each of these enterprises, the company will be able to work either directly or by means of a subsidiary that will have been created to this end, and over which it will retain control. The company will receive the support of the government and its administrators in its efforts and the works it will undertake, in conformity with programs established in accord with the government.

Organization of the Company

Any industrial or financial company whose purpose is the production, transportation, trade, distillation, refining, or distribution of oil will be able to participate in the company, either at the time of its founding or subsequently. In principle, the size of the participation will be determined by the importance of the national capital invested by each participating company. The case of certain companies that have significant foreign participation will be the object of special scrutiny. All necessary measures will be spelled out in the statutes to ensure that permanent control of the company will be effected by French capital. The board of directors will be French and approved by the government. The chief executive officer and other directors will also be French, unless exceptions are granted by the Government.

Relations with the State

As compensation for the advantages the state will give the company, the state will have the right to exercise control over its affairs. This control will be exercised by two government commissioners, to be designated respectively by the minister of commerce and the minister of finance. The role of the commissioners will be elucidated in the statutes.[67]

This is a determined executive giving a clear mandate to a subordinate. Mercier himself testified that the Compagnie Française des Pétroles was Poincaré's idea. The testimony and letter have propagated the idea of the Compagnie Française des Pétroles as a state-sponsored entity chosen to further the strategic and diplomatic interests of

[67] Text of letter is fully reproduced in another letter signed by Ernest Mercier, itself undated but probably from spring 1924, Fin. B32309, "Projet de Lettre à Monsieur le Président du Conseil."

France.[68] Mercier's prestige as company founder, and the later fact of state equity participation, make the *volonté d'état* the ready and easy way to understand the company's development.[69]

The true origin of the company lies in the struggle among the international oil majors, Russian oil interests, French banks, and the attempt to stabilize world oil markets in the 1920s.

Bérenger and Clemenceau's attempt to create a monopoly to favor Royal Dutch–Shell was defeated in the spring of 1920 by Standard Oil's worldwide counteroffensive, and afterward France became a cartelized but unregulated market. The British were also targets of Standard Oil's counteroffensive, and their agreement to admit American firms into Mesopotamia sealed the collapse of the Anglo-French oil alliance. The French signed a memorandum with the British in January 1921 dissolving the global oil alliance that had been envisioned in the San Remo agreement. The French retained one important right: participation in Mesopotamia.[70] The San Remo agreement's architects had thought that Royal Dutch–Shell would become France's principal supplier. The abrogation of the San Remo oil alliance put the last nail in the coffin of Royal Dutch–Shell's pretensions; when Standard Oil allied with the Banque de Paris et des Pays Bas, it was clear that the war booty of Mesopotamia would have to be divided in a new way.

When it put together the Société pour l'Exploitation des Pétroles, the Shell combine had enlisted the support of major French industrial groups. Their hopes for a share of Mesopotamian oil were dashed when the project collapsed. Even before the Shell combine's grandiose scheme, these groups had made independent overtures to the French government for concessionary rights in Mesopotamia. Separate requests were filed by the Banque Mirabaud, the Banque de Paris et des Pays Bas, the Banque Louis Dreyfus, the Banque de l'Union Parisienne, the "group of coal and steel interests" headed by Peyerhimoff, the Schneider industrial interests, and the Groupe Lilleois des Pétroles du Nord.[71] These groups wanted to get into oil: many of them also petitioned to participate in oil drilling in Palestine.[72]

The most powerful of these French firms was Omnium International des Pétroles, backed by the Banque de Paris et des Pays Bas, and after

[68] Testimony of Ernest Mercier, "L'Enquête parlementaire," *Revue pétrolifère*, no. 263 (31 March 1928): 435.

[69] Faure's account (1939, pp. 120–121) is representative of the étatist approach.

[70] CFP 82.7/1, "Note de M. Gulbenkian sur la constitution du capital . . . de la Turkish Petroleum Company," etc., n.d. but marked "note from before 1928."

[71] MAE Papiers d'Agents, André Tardieu, vol. 56 A, "Rapport sur les Pétroles," 6 May 1919, p. 9.

[72] MAE Levant, Turquie, 1918–1940, vol. 429, Pétroles, letter signed by Démachy to Ministre des Affaires Etrangères, dated 10 July 1919.

late 1919, allied in most international oil matters with Standard Oil. Omnium produced in Romania and in the early 1920s sought to expand internationally. In March 1920 the company commissioned an exploratory party in Turkey.[73] Significantly, this venture followed on the heels of Royal Dutch–Shell's failed monopoly attempt. A special exploratory dispatch to Syria was made in March 1921, and a five-month survey of Syria, Armenia, Mesopotamia, and Turkey was undertaken from February to August 1922.[74]

By September 1921 the idea of a state company was already circulating in government circles, explicitly stating that a state-authorized company for the French share of Mesopotamia would have to include Standard Oil. This was exactly what Poincaré authorized, and it most certainly was not his idea. The idea of a French company with Standard Oil was circulating well before Poincaré became premier and was a consequence not of etatism, but of the fact that French policy had been reversed by Standard Oil. After 1921 the whole thrust of French etatism was not to work against the American company, but to conciliate it. An official of the foreign affairs ministry said as much in dismissing a last effort by the Shell combine to hang on to France's Mesopotamian shares:

> If M. Bérenger formerly made promises, we don't know anything about them and have no documents on this subject in our files.[75] Most likely M. Bérenger had his intentions, but the agreement on oil was in its day just one project on which no commitment can be based. *As for the San Remo accord, it is impossible to concede advantages to one French company. The group to put together will have to include all the French companies, including even Standard Franco-Americaine.* One can easily imagine that the French state, as a supporter of the affair, could be one of the major participants in the enterprise, as shareholder or in some other manner, as expressed by M. Loucheur.[76]

Royal Dutch–Shell's Société pour l'Exploitation des Pétroles was to

[73] MAE Levant, Turquie, 1918–1940, vol. 429, Pétroles, "Président du Conseil . . . à Messieurs les Agents diplomatiques et consulaires française en Turquie et dans le Levant," dated 22 March, no year, prob. 1920.

[74] The 1921 Syrian expedition is in MAE Levant, Turquie, 1918–1940, vol. 429, Pétroles, two letters dated 26 February 1921 and 22 March 1921. The five-month survey is in MAE Levant, Syrie-Liban, 1918–1929, no. 344, A: Syrie-Pétroles, letter from Emile Aublé to Saint-Quentin, dated 14 February 1928.

[75] This is not true: material on Royal Dutch–Shell's plans is in the foreign affairs ministry. But it is true that Royal Dutch–Shell never had a signed contract to work the Mesopotamian concession for France.

[76] MAE Levant, Turquie, 1918–1940, vol. 429, Pétroles, "Note au sujet de la lettre et des documents remis par M. Deutsch de la Meurthe," 28 September 1921. Italics added.

have been a subsidiary with omnibus participation by many major French financial and oil groups, except for Standard Oil. The object was to land the concessionary rights without alienating the whole of the French international business sector. The Shell combine consortium, which might have included state equity, was the "state company" envisioned by Bérenger in 1919. The 1923 company envisioned by Poincaré in the letter quoted above was really not different in its general outline, except that Standard Oil was excluded from the Royal Dutch–Shell-sponsored project, and included in the second project.

Many firms floated the idea of an exploration or oil development syndicate whose eventual commerce would be supported by the state. I have described two in detail: Royal Dutch–Shell's attempt in 1919 and the 1920–1921 efforts by Pétrofina to get state sponsorship for control of oil lands in the Transcaucasus. The third group proposal came from Ernest Mercier and Omnium International des Pétroles. The Banque de Paris et des Pays Bas and its oil company formed, in early 1921, the Syndicat d'Etudes et de Recherches, whose objectives were nearly identical to Royal Dutch–Shell's. In the spring of 1921 this new syndicate sent its first exploratory mission to Syria, Armenia, and Azerbaijan.[77] Loucheur's September 1921 proposal (quoted above) for a group that would include Standard Oil directly follows the Banque de Paris et des Pays Bas' announcement of an oil syndicate. It was a mirror image of the Shell combine's efforts in 1918–1919. The difference was that this time Standard Oil and its French banking ally predominated. The Compagnie Française des Pétroles was not created by Poincaré, or by any particular brand of French "etatism" as such. Many competing initiatives emanating from the private sector had as a goal state help in acquiring foreign oil resources, especially but not exclusively in Mesopotamia. Standard Oil's aggressive policies and its astute banking alliance with the Banque de Paris et des Pays Bas was the winning combination out of many possibilities, and Poincaré's letter certified the victory.

Loucheur's desire for state equity in the Compagnie Française des Pétroles, discernible in 1921, was eliminated by Poincaré in the 1923 version. Poincaré's 1923 version of the Compagnie Française des Pétroles emerges as the least etatist proposal of the various options that had been proposed in the preceding five years. The mandate to form the Compagnie Française des Pétroles in 1923 did not purpose to create a strong state presence in the oil industry, which is the merely rhetorical flavor of the September 1923 letter, but to cut Standard Oil in on

[77] MAE Levant, Turquie, 1918–1940, vol. 430, letter from Omnium International des Pétroles to Ministre des Affaires Etrangères, 23 February 1922.

a piece of the state's action. Arguably, Poincaré perhaps wanted to foster an "oil peace" by accommodating the one giant company that had proved it could stop French plans. But Standard Oil and the Banque de Paris et des Pays Bas had been pushing Poincaré's supposedly etatist solution for at least two years.

Standard Oil got one of the largest slices of the new company, as shown by the original equity structure and the person Poincaré chose to run it. Ernest Mercier was president of Omnium International des Pétroles, the oil company owned by the Banque de Paris et des Pays Bas. He sat on the board of directors of other companies that were linked to the same bank and were, like Omnium, active in Romanian oil. These included Steaua Française and Steaua Romana. Mercier was not a disinterested person chosen to lead a policy of independence from the American oil giant; to the contrary, the banking group and oil industry members he represented could profit from a harmonious relationship with Standard Oil. Following Poincaré's instructions, Mercier furthered this harmony in the Compagnie Française des Pétroles, in which Standard Oil and the companies of the Banque de Paris et des Pays Bas held 27.6 percent of the founding shares, the largest single block (see table 2).

The dominance of Standard Oil and the Banque de Paris et des Pays Bas in the Compagnie Française des Pétroles did not perfectly mirror Bérenger's earlier monopoly ambitions. The Standard Oil faction had the lion's share, but Royal Dutch–Shell got a consolation share, unlike Standard Oil in 1919. Nor did the Compagnie Française des Pétroles envision, as had Bérenger for the Shell combine, getting the whole French market for Standard Oil and its allies. This was due, first, to the complicating presence of the Pétrofina group, which had not existed when Bérenger's plans were drawn up. Second, although Mercier's oil activities bore the stamp of the Banque de Paris et des Pays Bas and its alliance with Standard Oil, he had to take a broader view of oil policy because of the diversity of his own industrial enterprises. Mercier participated in a number of electrical ventures backed by major French banks, some of whom, especially the Rothschilds, were partial to Royal Dutch–Shell.[78] Mercier the oil man would have found it impolitic to collide with the major collaborators of Mercier the electric-

[78] Kuisel 1967, pp. 13–15. Kuisel justifiably defends Mercier against being the "man of the Rothschilds." But Mercier's "neutrality" about industrial factions was the neutrality of a man with his finger in many pies, not the neutrality of a man with his finger in none. Like Mazarin or Colbert, Mercier's services to the state brought him financial benefits, and this is in my opinion the continuing "French tradition" he best represents. But positioned near the center of French industry and finance in the first half of this century, he is a perfect subject to illustrate the development of French capitalism.

Table 2. Stockholders of the Compagnie Française des Pétroles, 1924 (Total initial capitalization: FF24,150,000)

Company shares[a]	Capital participation (FF)	A shares–B shares
Standard Oil–Banque de Paris et des Pays Bas[b]		
Banque de Paris et des Pays Bas	1,100,000	550A–1,650B
La Pétroléene	1,000,000	500A–1,300B
Standard Franco-Americaine	450,000	225A–675B
L'Economique	450,000	225A–675B
André et Fils	450,000	225A–675B
Cie Générale des Pétroles	350,000	175A–525B
Omnium International des Pétroles	475,000	475A–475B
Columbia	475,000	None–950B
Sté de Naphtes Limanowa	475,000	237A–713B
Pétroles de Dabrowa	475,000	237A–713B
Steaua Française	475,000	237A–713B
Sté Française des Carburants	450,000	225A–675B
Ernest Mercier	50,000	31A–69B
FF6,675,000, or 27.6% of total capitalization		
Pétrofina[c]		
Banque de l'Union Parisienne	1,100,000	550A–1,650B
Sté Générale	350,000	175A–525B
Pétrofina Française	475,000	475A–475B
Concordia	475,000	None–950B
Crédit Mobilier Français	900,000	450A–1,350B
Raffinerie de Pétrole du Nord	350,000	175A–525B
FF3,650,000, or 15.1% of total capitalization		
Desmarais Frères[d]		
Desmarais Frères	1,200,000	500A–1,800B
Aquila	350,000	None–700B
Sté Fse et Roumaine	175,000	287A–63B
Romano-Belge de Pétrole	50,000	None–100B
FF1,775,000, or 7.3% of total capitalization		
French National or "Independent" Companies[e]		
Cie Industrielle des Pétroles	875,000	462A–1,288B
Lille Bonnières Colombes	700,000	350A–1,050B
Consommateurs de Pétrole	450,000	225A–675B
Pechelbronn	450,000	225A–675B
Pechelbronn	350,000	175A–525B
FF2,800,000, or 11.6% of total capitalization		
Royal Dutch–Shell[f]		
S.A. Pétroles Jupiter	1,200,000	600A–1,800B
Raffinerie du Midi	700,000	350A–1,050B
FF1,900,000, or 7.9% of total capitalization		

Table 2. (continued)

Anglo-Persian Oil Company[g]		
Sté Générale des Huiles de Pétrole	875,000	437A–1,313B
FF875,000, or 3.6% of total capitalization		
Hausseman group[h]		
Omnium des Pétroles du Nord	50,000	75A–25B
Renastera Française	50,000	25A–75B
Sté Française de Silva Plana	475,000	237A–713B
Sté Fse des Pétroles Premier	475,000	237A–713B
Sté Financière des Pétroles	50,000	75A–25B
FF1,100,000, or 4.6% of total capitalization		
Miscellaneous Unidentified and Smaller Investors[i]		
Banque Nationale de Crédit	700,000	350A–1,050B
Crédit Commercial de France	500,000	250A–750B
Comptoir National d'Escompte de Paris	350,000	175A–525B
Banque de Mulhouse et Mulhouse	250,000	125A–375B
Sté Générale Alsacienne de Banque	250,000	125A–375B
Compagnie Algérienne	250,000	125A–375B
Sté Française de Gérance de la Banque d'Indochine	250,000	125A–375B
Sté Financière Française et Coloniale	100,000	50A–150B
Crédit Foncier Coloniale	100,000	50A–150B
Mallet Frères et Cie	50,000	25A–75B
Mirabaud et Cie	50,000	25A–75B
Huilerie Centrale de Saint-Ouen	25,000	12A–38B
Leprêtre	25,000	37A–13B
Cie Franco-Polonaise des Pétroles	475,000	237A–713B
Sté Pétrolea	50,000	75A–25B
Sté des Pétroles de Girabownica	50,000	75A–25B
Sté des Pétroles de Zagorz	50,000	75A–25B
Raffinerie Prédinger	50,000	None–100B
Victoria	50,000	25A–75B
Vulcanesti	50,000	25A–75B
Sté des Pétroles de Bustenari	50,000	25A–75B
Cie Commerciale des Pétroles	50,000	None–100B
Apostolane-Romana-Shela	25,000	NA–NA
Pétroles de Bellik à Grosni	150,000	75A–25B
Sté Civile de Recherches pour l'Afrique du Nord	400,000	200A–600B
Sté Alsacienne et Lorraine de Recherches Minières	200,000	100A–300B
Sté pour l'Approvisionnement des Consommateurs d'Huiles Combustibles	200,000	NA–NA
Sté Nouvelle de Sondages Bonne Espérance	100,000	50A–150B
Le Pétrole de France	100,000	NA–NA
Sté de Recherches d'Hydrocarbures	100,000	NA–NA
Sté de Recherches et d'Exploitations Pétrolifères	100,000	NA–NA
Cie Auxiliare de Navigation	450,000	225A–675B
FF5,600,000, or 23.2% of total capitalization		

Sources: Annuaire Desfossés (1925); Lender 1934, pp. 112–117; Fin. B32309, "Compagnie Française des Pétroles—Tableau de repartition de 25 millions," 5 March 1924, signed Mercier; MAE Papier d'Agents, André Tardieu, vol. 56 A, dossier Tardieu, "Rapport sur les pétroles présenté au conseil," 6 May 1919; Erard 1934.

[a]An A share, or preferred stock, carried more votes than a B share, or ordinary stock; the voting ratio was 20:1. Moreover, preferred A stock could be held only by French nationals, whereas ordinary B shares were freely negotiable. Both had a par value of FF500.

[b]Ernest Mercier, founder of the Compagnie Française des Pétroles, was president of Omnium and served on the board of directors of Steaua Française, Steaua Romana, and Columbia. Limanowa was founded as an independent venture between Fenaille and Despeaux and Desmarais Frères; after the war Fenaille and Despeaux were bought out by Standard Oil, bringing the company closer to the Standard Oil–Banque de Paris et des Pays Bas group. Mercier held shares in his own name.

[c]As a chief importer of Russian oil and an alternative supplier for the independent French firms, this group's political efforts were closely linked with those of French national companies, including Desmarais Frères.

[d]Largest French national or "independent" firm, entirely family owned. Its interests were closely connected with those of Pétrofina and the other French national companies.

[e]The Pechelbronn group is entered as stockholder in two separate places on Mercier's signed and hand-corrected document; I am not certain whether the double entry is an error or represents two separate stock purchases. Pétrofina's Léon Wenger served on the board of directors of Pechelbronn, the only firm producing oil in France (in territory recaptured from Germany in Alsace-Lorraine), as did Henri de Peyerhimoff, the prominent representative of French coal interests. Other French independent firms depended on Pétrofina as an alternative source of oil supplies.

[f]Royal Dutch–Shell's chief banking ally, if any, is not clear; it had worked with the Banque de l'Union Parisienne, but this relationship diminished after 1920. Internationally, Royal Dutch–Shell had an increasingly cooperative relationship with Anglo-Persian.

[g]Internationally, increasingly cooperative with Royal Dutch–Shell. Possibly may have had connections with Hausseman group.

[h]Here designated "Hausseman group" because the first four companies listed shared the same address at 69 boulevard Hausseman in Paris. The group played a role in oil investment in Poland, with the Société Financière and Société Française. These last two came under the control of Anglo-Persian after the collapse of speculation in Polish oil in 1928. Silva Planna later sold shares to Desmarais Frères.

[i]"Unidentified" in this category means that there are undoubtedly companies, especially banks, with ties to other oil groups named above, but that I have not established these ties. Mercier's notes indicate the Banque d'Indochine withdrew; it is listed here but not included in the capitalization figures. Companies listed in this group did not constitute an identifiable, coordinated set of interests; they were for the most part passive shareholders. Philippe Mullerfeuga in France is completing what promises to be a thorough doctoral study of ownership and concentration in the pre–World War II French oil industry, but his work is not yet available.

ity industrialist. He was thus a good choice to play a national role and to attempt to forge a unity of opposites by bringing rival factions into the Compagnie Française des Pétroles; but it is also clear, from the weighting of participation in the company at its founding, that the wind was behind the Banque de Paris et des Pays Bas and Standard Oil.[79] Royal Dutch–Shell's 7.9 percent equity was small compared with its former ambitions. In the first planning session for the Compagnie

[79]See table 2.

Française des Pétroles held by Mercier on 7 November 1923, Royal Dutch–Shell was not represented, nor did it find a place on the company's board of directors in March 1924.[80]

Mercier brought in the pro-Russian importers including Pétrofina, Desmarais Frères, and the other French national firms. Though not as close-knit as the Banque de Paris et des Pays Bas and the Standard Oil interests, this was a powerful group that agreed on the need for market quotas and for Russian oil imports. Together these interests bought 34 percent of the original capitalization of the Compagnie Française des Pétroles.

The Compagnie Française des Pétroles was formed in March 1924 after Poincare's government had been voted out. The interregnum of the Cartel des Gauches was about to begin. Poincaré and Mercier distrusted the Left and did everything possible to transfer the government's Turkish Petroleum Company shares to private ownership. The hasty transfer, at a purely nominal price, neglected to secure contractually the state's right to a share of the Compagnie Française des Pétrole's oil. The flaw in the company's charter would later be criticized by the Left and the Right, especially by Louis Loucheur.[81] Poincaré and Mercier were hardly securing their credentials as etatists when they signed away the government's rights to the oil.

The two feared the oil agenda of the left-wing government: the Office National des Combustibles Liquides could and did institute measures to help the Pétrofina group, catering to the constituency that had created it. Mercier had a different economic clientele and a different strategy: he wanted to promote Russian imports through the Compagnie Française des Pétroles, which would have forced Pétrofina to share its privileges with that company's other stockholders, especially Standard Oil. The Compagnie Française des Pétroles therefore acted quickly to send a commercial mission to the Soviet Union; negotiations for oil began shortly after the company was founded.[82] As Mercier and the Compagnie Française des Pétroles tried to bypass Pineau and get contracts for Russian oil, Mercier's Omnium International des Pétroles and the Banque de Paris et des Pays Bas, still working as a syndicate, directly competed with the Pétrofina group and the Banque de l'Union

[80] Fin. B32311, "Cie Française des Pétroles," 7 March 1924; Fin. B32309, Louis Tronchère to M. de Mouy, 22 March 1924; Fin. B32308, "Syndicat français d'Etudes Pétrolières," 7 November 1923.

[81] But Loucheur, also worried about the Left, signed the convention proposed by Mercier; the text was reprinted in the *Revue pétrolifère*, no. 262 (24 March 1928): 385–386.

[82] MAE Europe-Russie, 1918–1929, vol. 528, "Ambassadeur Français Moscou," 26 September 1925. This document tells of "eighteen months of negotiations," or just after the Compagnie Française des Pétroles began.

Parisienne for monopoly sales privileges in Turkey.[83] Thus, though both were members of the Compagnie Française des Pétroles, the two groups competed fiercely in important oil regions. When, in 1925, with Pineau's and the French navy's help, Pétrofina received exclusive marketing rights to Soviet oil in France, it was a blow to the interests Mercier represented.

Pineau struggled with Mercier for control of oil policy. The Office National was backed by the Pétrofina group and left-wing parties who favored strongly regulating the French market in order to favor Russian oil and protect domestic firms. By contrast, Mercier's freedom to act was restricted by the dominant banks in the Compagnie Française des Pétroles. The banks had stipulated that the company would never "have any monopoly or privilege, aside from the French part given to it in the business of Turkish oil" and that "participation in the Union National des Pétroles would leave intact the right of participants to pursue their activities in the other countries, especially Russia." They also stipulated that the "constitution of a national oil company will not restrict directly or indirectly freedom of oil commerce," although parliament's right to regulate the market was viewed as a separate matter.[84]

The champion of French oil independence was from its birth severely limited by the rivalries among its own stockholders, who sought to protect their existing prerogatives more than to create new ones for the state company. Mercier admitted that in the company the "trusts were very largely represented," but in spite of the Banque de Paris et des Pays Bas' large interests in Romania and their close ties with Standard Oil, he maintained that the banks were "independent" and that the Compagnie Française des Pétroles was "independent in the strict sense of the word."[85] This self-representation of independence has affected the historical image both of Mercier's company and of Mercier himself—what Rondot called the "immaculate conception" theory of the Compagnie Française des Pétroles.

Mercier's affiliation with the Banque de Paris et des Pays Bas swung the policy of the Compagnie Française des Pétroles and that of France away from Britain—where Bérenger had vainly tried to take it—and

[83] MAE Levant, Turquie, 1918–1940, vol. 430, "Omnium International des Pétroles," 10 January 1925; "Note sur la concession du monopole de vente des pétroles en Turquie," 14 December 1925; "Note pour Monsieur Seydoux," 27 February 1926.

[84] Fin. B32308, "Compagnie Nationale des Pétroles," minutes of meeting 26 October 1923. The Compagnie Française des Pétroles had not yet been named, and several alternative titles floated about until it had.

[85] Fin. B32311, "Rapport au Président du Conseil . . . sur la Compagnie Française des Pétroles," 18 December 1928, written and signed by E. Mercier.

back to "America," or Standard Oil. Privately, Mercier was clear about the dependency of France's "independent" initiatives on Standard Oil: "It has often been envisaged, in this business [the Turkish Petroleum Company], a Franco-American cooperation in which, without any illusion, America will have the predominant influence. If we bring about this cooperation in an effective manner, by an agreement on voting or an accord on commercial interests or some other manner, along with G[ulbenkian] . . . it is not to be doubted that such an association would be able to dictate its wishes in the T.P.C."[86] Poincaré had learned the lessons of the 1919–1920 embargo, and the selection of Mercier, as opposed to someone like Bérenger, was a clear recognition of the limits of France's power.

The unpublicized objective of allying with the Americans directly contradicted the etatist noise that interventionist policy would guarantee French independence from Standard Oil, "the trusts," or the "Anglo-Saxon trusts."[87] By decade's end the Compagnie Française des Pétroles became the means to stop the regulation of multinational oil firms in France. Mercier spoke plainly enough to French legislators on this point: "I do not believe," he said, "that the strong hand of the state on the importing and distribution of oil in France will help us in our task; I even think it would signal a veritable declaration of war against our coassociates in an enterprise that we have begun in common with them."[88]

In the 1920s the Compagnie Française des Pétroles could not protect the national market from the trusts that were anathema to the French national companies. The most dreaded of foreign companies, Standard Oil and its banking allies, held the largest block of the national champion's stock, and its president had been drafted out of the same milieu. The other large block of stock was controlled by the Pétrofina allies. Though mandated to explore other sources of oil, the Compagnie Française des Pétroles had to stick to Mesopotamia; the other likely oil sources, especially the Soviet Union, were fought over by powerful factions that could not reconcile their differences within the confines of a fledgling company, bereft of oil and market share. The state's national champion had no oil policy, and when it had tried to implement one for Soviet oil it had been whipped by Pineau and Pétrofina.

[86] CFP 81.1/62, "2ème Note," n.d., prob. early 1928. Gulbenkian is referred to only by the initial *G.*

[87] Going in one direction with vainglorious public fanfare while going the opposite direction in secret is an irritating but recurrent feature of French diplomacy. "Nuclear independence" from the United States was another such sham. See Ullman 1989.

[88] Testimony of E. Mercier, "L'Enquête parlementaire," *Revue pétrolifère,* no. 263 (31 March 1928): 438.

Mesopotamia was therefore the only other arena where Mercier might make a difference, and he would use it to enhance the power of his company and the groups he represented.

The Struggle for Mesopotamia

The Compagnie Française des Pétroles vigorously entered the diplomatic fray to turn its shares in the Turkish Petroleum Company into an active, oil-producing asset. Here company representatives coordinated their efforts with the foreign affairs ministry and were indeed *étatique.* French participation in Mesopotamia required the conjoined effort of private interests and the state, a necessary guarantor of security in the international arena.

If we go behind the diplomatic façade of unity, the conflict of private interests within the Compagnie Française des Pétroles tarnishes the luster of statism. From 1924 to 1928 the state internationally represented a private entity with no state equity; this may have benefited France, but parliament later asked why the state needed to turn its Mesopotamian shareholdings over to private interests. Why not, adversaries of the company would later ask, have the government form a state company and keep the profits for itself, producing oil for its own strategic needs and selling whatever surplus it had to the private sector? Poincaré and his followers replied that the state ought not to be involved in the expensive risks of oil drilling and exploration. The investment burden should be borne by the private sector. Yet some state support was necessary because French capital was too timid to embark without it on such a difficult commercial venture.[89]

The claim that French capital was too timid to venture into oil was nonsense. When the Compagnie Française des Pétroles was formed, French capital was simultaneously working up an enormous speculative bubble in the Polish oil fields. The bubble collapsed in 1928, at a cost to French investors of two billion francs, ten times the investment in the Compagnie Française des Pétroles even *after* subsequent increases in its capital, or about one hundred times its initial capitalization.[90] Moreover, the "risk" in Mesopotamia was never whether oil was there. By 1924 Mesopotamia was regarded as the most certain future oil supply in the world. That oil had already played a major role in the

[89] Melby 1981, pp. 67–69.

[90] On Poland, see Lender 1934, p. 116. The company's capitalization was raised to FF50 million in 1926, FF75 million in 1927, FF150 million in 1928, and FF205 million in 1930 (Labarrière 1932, p. 146). French capital was also heavily invested in Romania.

strategy of the First World War. That oil had preoccupied one of the world's most prestigious banks, the Deutsche Bank, which lost its claim only by force of arms. That oil had helped occasion even more Turkish brutality against the Armenians, whose claims to independence threatened control of oil-bearing regions. That oil had been coveted by Anglo-Persian and the British Admiralty since 1912. That oil had precipitated the world's second largest oil company, Royal Dutch–Shell, into an audacious postwar maneuver to corner the French market and gain control of the Mesopotamian shares. That oil had been the object of Turkish decrees and intense negotiations in the Ottoman Empire since the 1890s. That oil had precipitated Standard Oil into an unprecedented worldwide offensive against the exclusionary policies of its archrivals, leading to the embargo of 1919–1920. That oil had for decades preoccupied one of the world's most prominent oil investors, Calouste Gulbenkian, and in France had already been solicited by many private groups, including the Banque de Paris et des Pays Bas, which was not short of capital for the world's richest oil field.

The risk in Mesopotamian oil was not whether there was any to be found, but the terms under which it would be developed. The real risk was for the French government: if it kept the shares for itself, it did not have sufficient autonomy to face down the wrath of private investors who wanted to be in on the profits. Poincaré had as examples of corporate power the Shell project of 1918–1919, Pétrofina's solid backing of the Office National, and Standard Oil's heavy influence in the deregulation of the French market in 1920. Poincaré knew he could not keep the shares for the state and face the combined power of these companies. In dividing them up, he diluted potential opposition by giving everyone a slice of the pie, even if some received more than others. The only risk assumed by French shareholders was that the French state might fail to turn the paper claim to rights in Mesopotamia into a workable black gold bonanza.

The Compagnie Française des Pétroles insulated interested parties from their respective risks: Poincaré's government placated the oil business groups by amalgamating them into a common pool, never mind how "French" they might be. Shareholders bearing the important "A" stocks had to be French nationals, but it was a cosmetic requirement. A registered subsidiary of a foreign firm in France could hold shares in the Compagnie Française des Pétroles. The state exerted strenuous efforts to help the anxious stockholders, but their interests were far from united. The main arena of contention was between the pro-Soviet faction, whose bankers strove to link Russian oil to debt payments, and the Mesopotamian faction, whose success would depend on pushing Russian oil out of the European market.

The struggle among the Anglo-Persian, Royal Dutch–Shell, Compagnie Française des Pétroles, and Standard Oil companies (which dominated the American consortium, the Near East Development Company)[91] was a replay of the even longer struggle among the original participants in the Turkish Petroleum Company, Royal Dutch–Shell, Anglo-Persian, and the Deutsche Bank. Now the French substituted for the Germans, and the Americans wanted in: the pie had to be sliced four ways instead of three.[92] Haggling over how to slice the pie and other details accounts for the principal parties' failure to reach any agreement before 1928. The Federal Trade Commission in 1952 listed the six major areas of disagreement.[93] The resolution of these disputes enmeshed French domestic and international policy in the second world oil cartel, whose origins coincide with the resolution of the Mesopotamian oil dispute.[94] Since they are necessary to understanding the development of the world oil cartel, the six areas of dispute are summarized here.[95]

1. The open door plan. The open door plan was the American diplomatic thrust of Standard Oil's worldwide campaign against exclusion from foreign oil resources, especially the Anglo-French oil alliance of 1920. The United States government wanted equal legal treatment for its nationals in other countries; assurance that economic rights granted in mandate regions would not exclude other foreigners; and guarantees that no monopoly would be granted to any one group.[96] In theory the "open door" meant an equal chance for oil development in Iraq

[91] FTCI, pp. 53–54. In 1928 the Near East Development Company comprised Standard Oil of New Jersey, Standard Oil Company of New York, Gulf Oil and Refining, Atlantic Refining, and the Mexican Petroleum Company. By 1934 only Standard Oil of New York and Standard Oil of New Jersey remained. In 1931 Gulf Oil dropped out of Iraq in a move that must have been related to the functioning of the Red Line agreement and Gulf's participation in the Kuwait concession; but secondary and primary sources are mute on the subject. Two other participants in the NEDC, Sinclair and the Texas Company (Texaco), dropped out in 1921 and 1922.

[92] The San Remo agreement had also promised the Iraqis up to a 20 percent participation, which was not so much a charitable consideration on the part of the vanquishers of the Ottoman Empire as a practical means to finance the region's new government (Penrose and Penrose 1978, p. 43). This inconvenience to the great powers was deftly resolved by eliminating the Iraqi share altogether.

[93] FTC1, p. 55.

[94] The Federal Trade Commission report is often cited in most American works on the oil industry. It is widely available in American libraries in the censored version of 1952; additional pages of deleted material are available through the Truman archives. The first, published version of the report, widely available in government depository libraries, is here referred to as FTC1. The deleted material recently made available by the Truman archives is referred to as FTC2; a small fraction of the classified material was published in 1975 (see bibliography).

[95] Blair 1976 is also a standard reference.

[96] FTC1, p. 51.

by Italian, Japanese, Soviet or other nations' companies. In practice the open door applied only to the United States; once the Americans were admitted, the door swung shut. Gulbenkian sardonically comments in his memoirs,

> Immediately after the participation of the Americans had been decided, other difficulties cropped up. The American groups, presumably to carry their Government with them had evolved a formula known as the "Open Door" which was translated into obtaining for the Iraq Petroleum Company the confirmation from the New Iraq Government of a concession over what was formerly the vilayets of Baghdad and Mosul, but with the obligation to keep for their own use 24 plots of 8 square miles each with the further obligation to put similar plots up for auction to the whole world at regular intervals. This "Open Door" policy as I pointed out must have been put "en vedette" to convince the governmental authorities of the non-monopolistic and broad-minded policy of the oil groups. In itself, it was an eyewash that the whole world would participate in the exploitation of Mesopotamia.
>
> It is unnecessary to go into the details but soon after this redundant "Open Door" was hermetically closed and no one talked about it any more. There could not be a more closed "Open Door" policy than the one followed by all the oil groups to collar every possible concession from the Iraq Government in order to prevent other competitors to apply for same.[97]

The twenty-four plots were alloted to the Turkish Petroleum Company. Other likely oil plots would be put up for bid; but the agent of the Iraqi government was the Turkish Petroleum Company, which rigged the bidding rules against any outside competition.[98] Gulbenkian, whose vast profits from the scheme did not cloud his often caustic appraisal of his collaborators, correctly calls the concession plot scheme "eyewash." Eyewash or not, it was difficult to devise a concessionary agreement that made a bow to Iraq's symbolic sovereignty and that also kept the prerogatives of the oil companies. These negotiations helped delay ratification of the concession until March 1925.[99]

2. *The self-denying clause.* The self-denying clause kept the companies participating in the Turkish Petroleum Company from competing with it. The clause dated to the Company's founding 1911 and 1914 agreements; it prevented consortium partners from outbidding one another

[97] Gulbenkian 1945, p. 25.

[98] FTC1, p. 56. Money for the plots was to be paid to the Turkish Petroleum Company; Turkish Petroleum could outbid any competitor, and since it paid the money to itself, it got back the money and obtained the parcel for no real outlay.

[99] FTC1, p. 56.

for oil outside the immediate concession. Hence in 1914 it was confined to the Ottoman Empire: members of the Turkish Petroleum Company could not claim oil outside the concession but were limited to the confines of the Ottoman Empire, and then they competed against their concession partners. An acerbic dispute arose between the French interests and Royal Dutch–Shell, which had overstepped the 1914 protocol by developing oil found on the Farsan Islands in the Red Sea.

Gulbenkian and the French interests in the Compagnie Française des Pétroles insisted on the self-denying clause for the same reason that the Deutsche Bank had first written it. The clause protected undercapitalized firms from overcapitalized firms. If, for example, the Turkish Petroleum Company worked area A and oil was found in adjacent area B, a Turkish Petroleum Company member, X, might bid for area B's oil and develop it, getting the whole profit on B for itself. Then company X would use its shareholders' influence in the Turkish Petroleum Company to hold back production in the original area A. This was rational for company X as a minority participant in area A but a complete owner of area B. The highly capitalized members of the company could, without a self-denying clause, collude to expand their operations at the expense of the Turkish Petroleum Company, eventually gaining control of Iraqi production.

With the self-denying clause, oil in an adjacent parcel would have to be developed by the whole Turkish Petroleum Company, allowing the weaker, undercapitalized members some protection from the better-financed members. Gulbenkian and the French had to threaten legal action as minority shareholders to force Royal Dutch–Shell to their view. These maneuvers, intricate in their details but straightforward in their motivation, consumed a great deal of legal, business, and diplomatic energy in the mid-1920s. The Standard Oil group moved to the French-Gulbenkian position when it realized that the self-denying clause would help control world overproduction; the clause also complemented the bidding system already in place.[100]

Growing Soviet exports, as well as oil development in Venezuela and elsewhere, created a need for some kind of "faucet" on Mesopotamian

[100] FTC1, pp. 58–61. Under the bidding system of the Turkish Petroleum Company as assented to by the Iraqi government in 1925, if company X wanted to bid for area B independently of Turkish Petroleum, it would have to pay the Turkish Petroleum Company for the land. The money would be paid to Turkish Petroleum and therefore would be divided among the members: 23.75 percent each to Royal Dutch–Shell, Anglo-Persian, and the Standard Oil group, and 5 percent to Gulbenkian (who used these intricacies to preserve the legality of the 1914 concession, and with it his personal share). Setting aside the arithmetic, the result was this: three member firms would always be able to bid for a new plot at one-third the outlay of one firm working alone. Group cohesion was therefore in all the members' interests.

oil, to slow down production when prices weakened. Standard Oil's position on this changed: in June 1927 the company still opposed the restrictions of the self-denying clause, fearing that Anglo-Persian and Royal Dutch–Shell would veto oil development that would compete with their oil in the Far East.[101] Assenting to the principle that its share in the Far East would neither grow nor diminish made this objection vanish, and the self-denying clause in Iraq became both the result and a reinforcing element of the world cartel of 1928. France and Gulbenkian, worried about holding their own solely within Iraq, supported the self-denying clause for a different reason; nonetheless, they also were folded into the world cartel.

The self-denying clause had to apply to a defined area. The original 1914 agreement covered the Ottoman Empire. Following that lead, the self-denying clause was applied to what had formerly been the Ottoman Empire. The French presented a map in October 1927, the basis for the "Red Line agreement," which was their legal, surveyor's version of what the amorphous Ottoman Empire had been and thus their exact version of what territories the self-denying clause included.[102] The Red Line encompassed the Middle East, extending around what would today include Turkey, Iraq, Syria, Jordan, Israel, Saudi Arabia, Oman, and the Emirates, excluding Kuwait. Royal Dutch–Shell won an exemption for its oil on the Farsan Islands. The ghost of the dismembered Ottoman Empire lived on in the Turkish Petroleum Company, not for the glory of the sultan, but as a boundary for the ambitions of corporate shareholders. The spirit and form of the consortium, first negotiated in 1914, was reincarnated in 1928.

3. The working agreement. After the all-important Red Line agreement, other elements of corporate operations were written out in a document known as the "working agreement." These elements included how the company would be taxed, how it would set the price for its oil, whether the new company would directly market its oil at the retail level or would simply sell to members, and so on. Major issues therefore had to be resolved. These included the fourth area of dispute.

4. The American group's share of the operation. Who would have to give up how much in order for the Americans to participate in the Turkish Petroleum Company? Anglo-Persian gave up its 50 percent share of the Turkish Petroleum Company, derived from the 1914 agreement, to give the Americans a quarter of the company. The four major companies ended up with 23.75 percent each, the 1.25 percent difference

[101] CFP 81.1/62, letter from Piesse to Wellman, 14 June 1927.
[102] FTC1, pp. 58–59.

between that and 25 percent going to make up Gulbenkian's share of 5 percent. Anglo-Persian came out of 1928 with half of what it had in 1914. Unhappy, the company demanded "compensation" for what it had never really owned, which was the next dispute.

5. *Anglo-Persian's overriding royalty.* On the company's original twenty-four plots Anglo-Persian would receive an overriding percentage of 10 percent. Thus the other companies really got 23.75 percent of the 90 percent of production left after Anglo-Persian got its royalty. Anglo-Persian had an effective revenue stream of 29 percent on the first twenty-four plots, but its voting block was set at 23.75 percent. Limiting this compensating royalty to the original twenty-four plots was the real reason for the awkward plot formula, not the cosmetic appearance of bidding, which fooled no one in Iraq or abroad. Gulbenkian observed: "One cannot refrain to remark that this 10% royalty, although granted in compensation, is in reality a gift because none of the Companies had any rights or concessions in Mesopotamia."[103] As a coparticipant in world imperialism, Gulbenkian wrote with rare candor.

6. *The controversy with Gulbenkian.* The Armenian sat as a peer with the world's largest corporations and their diplomats. The corporations tried to buy him out, which he refused, and to force him out, which was difficult, since it could not be done and still respect the legality of the 1914 lease. The fiction of this lease, which covered a territory no longer under the same sovereignty, was useful to the major parties for many reasons. Anglo-Persian liked the document because the 1914 version gave it the lion's share, which allowed the company to pressure the others into giving it an overriding royalty. Royal Dutch–Shell, as a 1914 signatory, liked the document because it prevented Anglo-Persian from pushing it out of Mesopotamia. The French members of the Compagnie Française des Pétroles liked it because without it the shares they had won in the war were worthless. Standard Oil liked the agreement once it was included. All four groups and Gulbenkian liked the document because it enabled them to present the new Iraqi government with a fait accompli, avoiding negotiations for a new concession.

For the majors, almost everything was right with the 1914 agreement except that Gulbenkian was in it. To throw out Gulbenkian, they had to throw out the agreement, and in the end they liked the agreement more than they disliked Gulbenkian. But he still presented special problems. The companies had decided that the Turkish Petroleum Company would not become an oil company with refineries and corner gasoline stations: that would compete against the member firms and require unnecessary investments in service stations. Instead, the com-

[103] Gulbenkian 1945, p. 27.

pany would sell crude to its member firms. The member oil companies wanted their consortium to sell them crude at a nominal price to reduce their tax liabilities to the British Crown (under which the company was incorporated). Gulbenkian, as a passive investor, had no use for cheap oil; turning Turkish Petroleum into a nonprofit company meant no return on his invstment. He adamantly opposed the agreement until the French agreed to buy his share of oil.[104]

In addition to simply being in Iraq, France's major victory was the self-denying clause. The French oil interests in the Compagnie Française des Pétroles had succeeded in maintaining the protection won by the Deutsche Bank in 1914. It is not clear that they could have done so otherwise, and this is an example of how the politicking of one phase of transnational structuring can carry over into a later phase.

The structure of the Compagnie Française des Pétroles, though ostensibly a purely "internal" matter for France, also reflected the cumulative effects of clashing international interests. Standard Oil, Royal Dutch–Shell, and Anglo-Persian each had shares in France's national champion, in addition to their own direct holdings in the Turkish Petroleum Company. For the French national firms and the Pétrofina interests, shareholdings in the Compagnie Française des Pétroles were the only way into Mesopotamia. When the Mesopotamian concession was signed on 31 July 1928 the French state had no direct benefit in Mesopotamia at all and had been unable to clear the way for French capital without first making generous concessions to the trusts, from whom it ostensibly sought independence. But the state's desiderata were not the issue, for the whole idea for the Compagnie Française des Pétroles had its origins in the syndicate of interests sponsored by the Banque de Paris et des Pays Bas. These interests sought collaboration rather than competition with the multinationals, many of which were commercial allies in other endeavors.

The capacious Ottoman Empire was only a corner of a much wider world. All the members of the Turkish Petroleum Company except France had other oil fields. By boosting production in Romania, Iran, Venezuela, the United States, Mexico, or Indonesia, the other companies could make money without having to give a 23.75 percent share to French interests and an additional 5 percent to Gulbenkian, as they had to do in Iraq. The international oil majors sought to limit the role of Mesopotamian oil in their world markets. They had been trying for some time to form a group where they could meet to talk about global conditions. In 1925 they tried to form a syndicate in Romania to serve

[104] FTC1, pp. 64–65. Gulbenkian's passivity as an oil investor is illustrated in the name for his operations in Iraq and around the world: "Participations and Investments."

this function, called OPQ. According to deputy Alfred Margaine, who cited minutes of the Compagnie Française des Pétroles, "The principal interest of this organization was to bring together the chief oil personalities in the world, but it will lose some of its importance once the Iraq Petroleum Company achieves the same objective through the membership of its board of directors."[105] France's presence in the Turkish Petroleum Company, in 1928 rebaptized the Iraq Petroleum Company, guaranteed that Mesopotamia would not develop to its full productive potential. This was a simple outcome of France's being there. The Mesopotamian oil fields would become a swing producer whose output constantly came under pressure from partners who had alternative oil sources.

By July 1928 the Compagnie Française des Pétroles had been formed and the participation agreement for Mesopotamia was also signed and ready. The alliance of Pétrofina producers and French domestic companies, their pressure for oil import quotas, the growth of Soviet imports and how much of the French market they might claim were unresolved issues. The only certainty was that Mesopotamian oil could not find an outlet if Soviet oil exports kept increasing. Just as the original 1914 concession agreement in Mesopotamia precipitated state monopoly proposals to guarantee market access in France and Germany, the Iraq settlement of 1928 precipitated a new attempt to rationalize world markets. The attempt of 1914 never came to fruition; the far more comprehensive agreements of 1928 were successful.

The World Oil Cartel

Not coincidentally, the July 1928 Iraq settlement followed hard on the heels of major French oil legislation in March 1928. Two decades later, even the French oil industry officials acknowledged that the Iraq Petroleum Company and its accompanying agreements marked the "beginning of a long term plan for the world control and [world] distribution of oil [from] the Middle East."[106] The claim may seem hyperbolic; but it was too modest, reflecting the limited view of French officials whose role in the world cartelization effort was limited to their exiguous jurisdiction, their nation-state. For the multinational companies, conditions in any one nation were a "local" affair; France and the

[105] Déb.Ch., 12 February 1931, p. 632.

[106] FTC2, p. 306. FTC2 here cites a Compagnie Française des Pétroles memorandum prepared in 1947. Bracketed words are supplied to clarify meaning.

Middle East were important parts of the 1928 accords, but world oil leaders had in mind the systemic functioning of the whole.

In the late summer of 1928 the top officials of the world's three largest oil companies met at Achnacarry castle in Scotland and negotiated an end to the epic international struggle for markets and supplies that had pitted them against one another since World War I. The conference is sometimes called the Achnacarry agreements, but is also named from the chief provision of its settlement: Standard Oil, Anglo-Persian, and Royal Dutch–Shell would divide up world oil markets in the proportions they then had. The market was to be left "as is." Few familiar with these accords have failed to be impressed by their depth and comprehensiveness.[107] The oil market did not become engraved in stone after the 1928 accords. These accords bore the same relation to the world oil market that the American Constitution did to the early United States as a polity: it was a set of principles that left ample room for disagreement but retained a coherence that determined the shape of events to come, providing a focus even for those who opposed it.

The industry's consensus about an oil-laden Iraq was officially confirmed in June 1927 with a major oil strike. Probably because inside traders wanted to reduce their exposure in higher-cost oil areas such as Poland and Romania, the news was kept quiet till 15 October 1927.[108] The new well produced 12,000 tons of oil a day; a typical Romanian well produced about 250 tons a day. Increases in North American and Venezuelan production were already threatening world oil prices; rising Soviet oil exports and the veritable flood represented by Mesopotamia aroused painful visions of price wars for company leaders with a world vision. Royal Dutch–Shell's Henry Deterding wrote that "the new production will give rise to discussions involving big views and principles."[109]

"Big views and principles" meant a review of the world oil market from top to bottom. Russian oil remained the biggest bone of contention. By 1927 the Shell combine had purchased on speculation the greater part of the ancien régime oil-bearing properties in the Soviet Union. Of these titles, 71.5 percent were in "British" hands, that is, in the London office of Royal Dutch–Shell; another 20.2 percent were

[107] The best history remains FTC1. Blair 1976 was a participant in the committee that wrote FTC1. Popular works on the history of the oil industry rely heavily on FTC1, but Hexner 1946 wrote well before. His book is extremely valuable, and perhaps for that reason it is ignored today.

[108] Thomas 1934, p. 95; Labarrière 1932, p. 150; Melby 1981, p. 86; Kuisel 1967, p. 35; Yergin 1991, p. 204. These authors all date the oil discovery to 15 October 1927. But a 14 June 1927 letter in the Compagnie Française des Pétroles' archives says plainly, "Oil has been found in Iraq" (CFP 81.1/62, Piesse to Wellman, 14 June 1927).

[109] FTC1, p. 59n.

owned by French nationals, including the Pétrofina group's banks and the Bauer and Marchal group, which backed the French independent company PEN. Belgian and German speculators divided the rest.[110] The Shell combine continued to seek reimbursement from Pétrofina for imported Soviet oil.

The political position of Royal Dutch–Shell had dramatically improved in Great Britain. Long opposed to Anglo-Persian in Mesopotamia and Iran, the combine had repeatedly tried to swallow up Anglo-Persian and its Burmah Oil parent. By the late 1920s all this had changed: The new threat in Mesopotamia was Standard Oil, whose sister company Socony was, like Pétrofina, a major vendor of Russian products. Standard Oil and Socony were emerging as coparticipants in Mesopotamia and would be advantageously positioned to supply Far East markets. Socony was also an active marketer of Russian products in the Far East. Royal Dutch–Shell's dominance of Asian markets was under siege. Anglo-Persian, also a big marketer in Asia, was similarly beset. The two companies, which had competed bitterly for twenty years, now had converging interests.

Their concordant interests turned into collaboration and partnership in the development of Mesopotamia, underpinned by changes in equity holdings and marketing. In 1928 Anglo-Persian's parent company, Burmah Oil, bought a million shares of Shell Trading and Transport, which remained, however, under the control of its dominant partner, Royal Dutch. Anglo-Persian, in which Burmah Oil had large holdings, formed a joint subsidiary with Royal Dutch–Shell's Asiatic subsidiary. This new holding company, Consolidated Petroleum, assumed control of its two parents' marketing in the Sudan, Egypt, Palestine, Syria, the coast along the Red Sea, Ceylon, and other parts of eastern and southern Africa.[111] Consolidated Petroleum augmented cooperation already under way in the Asian market: a year earlier, Burmah Oil's operations in India had merged with the combine to form Burmah-Shell Ltd. These world-spanning consolidations were scrutinized in the French ministry of foreign affairs: "This change in relationship between Royal Dutch–Shell and Anglo–Persian can have important consequences . . . indeed, the new Anglo-Dutch and American combinations are of a nature that will help bring about new ententes among the various trusts, ententes that could bring about

[110] Goulévitch 1924, p. 320; 1927, p. 242.
[111] MAE E-Levant, Irak-Mésopotamie, 1918–1929, vol. 34, "Les trusts et le pétrole russe," 25 October 1928; Ferrier 1982, pp. 463, 512.

restrictions in production or, alternatively, lead to the adoption of a common policy with regard to Russian oil.[112]

Through the working agreement in Iraq, exchanges of equity, and the creation of joint marketing companies, Royal Dutch–Shell had been "anglicized." It cooperated with Anglo-Persian; yet control remained at The Hague, as it has since the beginning of this century. The era of wasteful competition was over, and the two firms now jointly rationalized their operations to save freight costs. Oil from Persia no longer threatened the Shell combine's Asian markets, and oil from Indonesia no longer jeopardized the Burmah markets.[113] Each market was supplied from the closest source.

In their new harmony, Royal Dutch–Shell, Anglo-Persian, and Burmah interests became equally peevish about Soviet oil exports to Asia via the Suez Canal.[114] In France, Soviet oil was sold by Pétrofina and several other national companies and would become hotly debated in the March 1928 petroleum law. In Italy, Soviet oil was also sold through national companies, but as elsewhere 1928 marked a turning point. Italy had earlier recognized the Albanian government of Ahmed Zaku, who received loans and direct financial support from Great Britain; he granted concessions in Albania to Anglo-Persian, which agreed to sell 40 to 50 percent of the oil pumped there to Italy. On 30 and 31 July 1928, as the working agreement for Iraq was signed, the Anglo-Persian and Italian companies, including the state-owned Agip and two other Italian independent companies, met in Paris and, the next day, in Lausanne to sign agreements that phased Soviet oil out of the Italian market. Anglo-Persian became the principal supplier.[115]

[112] MAE E-Levant, Irak-Mésopotamie, 1918–1929, vol. 34, "Rapports des grands producteurs de pétrole entre eux," 3 December 1928. The translation of *entente* in French business language is subtle, the English "understanding" being rather weaker than the alternative translation, "cartel." This passage shows the senses of the word, here left in the French: it means "understanding," in the sense of policy collaboration that goes beyond production control (on the matter of Russian oil) and also includes production control, the chief tool of cartels.

[113] MAE E-Levant, Irak-Mésopotamie, 1918–1929, vol. 34, "Les trusts et le pétrole russe," 25 October 1928.

[114] MAE E-Levant, Irak-Mésopotamie, 1918–1929, vol. 34, "Les trusts et le pétrole russe," 25 October 1928.

[115] MAE E-Levant, Irak-Mésopotamie, 1918–1929, vol. 34, "Accord entre l'Agip et l'Anglo-Persian," 10 November 1928. "Agip" is Azienda Generale Italiana Petrolii. See also the extremely detailed article by Baldacci 1926. The other Italian companies were the Società Nazionale Olii Minerali and Benzina Petroleum. Fear of turning Italy back to Soviet oil may have prevented League of Nations oil sanctions from being imposed on Italy when it invaded Ethiopia. The British probably also promised Agip an entrée into Iraq in return for the latter's offtake of British production in Albania and its cooperation in not taking Soviet oil. A mixed British-Italian consortium, the British Oil Development Company, was allowed to bid successfully for oil lands in Iraq in the 1930s. The British Oil Development Company sold out to the Iraq Petroleum Company in 1935 without

Worldwide, except where Pétrofina had monopoly import rights, Soviet oil was sold through the Standard Oil Company of New York. Socony dominated sales in Turkey and carried the oil as far east as the Pacific. There were two 1927 contracts between Socony and the Soviet Naphtsyndicat. In the first, Socony agreed to construct a refinery at Batum and would be paid back with lamp oil to market in Asia. In the second, Socony received exclusive marketing privileges for bunker oil in Constantinople, Port Said in Egypt, and the port of Colombo in Ceylon.[116]

Through the Near East Development Company, Socony had signed the working agreement in Iraq. Royal Dutch–Shell and Anglo-Persian hoped to curtail Socony's sales of Soviet oil through a combination of incentives and intimidation. In Great Britain an energetic "press campaign" against "stolen" Soviet oil was launched by the *Daily Mail,* which the French ministry of foreign affairs noted was "undoubtedly inspired by Anglo-Persian, Royal Dutch, and Burmah Oil, the last two being particularly hard hit on the Indian market by American competition."[117] Consumers in the British Empire were discouraged from buying Standard Oil products. Standard Oil of New Jersey found itself getting bad press along with Socony. The latter would have ignored the Shell combine's negative campaign, but the chairman of Standard Oil of New Jersey, Walter Teagle, wanted to placate Deterding.[118]

Teagle could secure Socony's cooperation with the Shell combine because by 1927 John D. Rockefeller, Jr., had enough of Socony's shares and outstanding debt to influence company policy.[119] This is a contro-

ever having produced oil. It began operations in the late 1920s under the direction of British admiral Lord Wester Wemyss and the Italian companies, Agip and the Società Nazionale Olii Minerali (MAE E-Levant, Irak-Mésopotamie, 1918–1929, vol. 34, "Consul de France en Irak à son excellence," etc., 17 October 1929). A French foreign affairs ministry memorandum from 1928 describes the British Oil Development Company as "probably patched together by Anglo-Persian" (Fin. B32310, "Ministère des Affaires Etrangères à Londres et Washington," 13 August 1928).

[116] Fin. B32023, "L'Ambassadeur . . . à Moscou à Monsieur le Ministre des Affaires Etrangères," 1 September 1927.

[117] MAE E-Levant, Irak-Mésopotamie, 1918–1929, vol. 34, "Note," 16 August 1928; also Fin. B32023.

[118] Fin. B32023, "Note," 13 August 1928.

[119] Fin. B32023, "Note," 13 August 1928; that the French were unaware of the shareholdings is shown in Rock.Arch. JDR Jr. Business Interests, box 126, folder SO Indiana, 1919–1928: a 9 February 1928 document lists shareholdings in Standard Oil of New Jersey and Standard Oil of New York among family holdings; same archive series, box 137, folder "SONY," letter dated 3 February 1926, tells of the purchase of 561,573 new shares; same file, letter of 8 February 1927, mentions the purchase of $2,375,000 of five-year debentures, and a letter of 10 February notes that of $450,000 more of twenty-five-year debentures; and same archive and series (87.1.S91) shows the purchase of $4,000,000 of total $20 million debt issue on 10 February 1928; same archive series, box 137, folder Vacuum Oil Company, letter of 18 December 1929 from G. B.

verted point because legally after 1911 there was no formal connection between the firms; but John D. Rockefeller, Jr., was the power behind the throne that forced Socony to toe the line on Russian oil, a key objective of the Achnacarry accords.

Deterding, virulently anti-Soviet, also needed his arm twisted.[120] Teagle insisted that Socony would market Soviet oil until contracts expired; but Royal Dutch–Shell and other former proprietors could get compensation. Socony arranged for Royal Dutch–Shell to have options on Soviet oil in a quantity equal to its own, and Deterding, as he had in 1923, dexterously subordinated his political inclinations to his desire for crude oil. For the next four years Socony paid a 5 percent fee to owners of nationalized Soviet oil properties, a fee collected almost entirely by the Shell combine. The Soviet Naphtsyndicat, aware that the trend in world markets was against it and that the anti-Soviet propaganda in Great Britain could hurt sales, directed its marketing company in Great Britain also to pay the 5 percent fee. To maintain good relations with Socony, whose world market distribution it needed, the Soviet company paid a 5 percent rebate, calling the payback preferential treatment to a favored customer. Socony then promptly turned the fee over to the expropriated Shell combine interests as "compensation."[121]

These side payments are as close as the Soviet Union ever got to paying anyone anything for properties seized during the revolution. Oil companies succeeded, however modestly, in getting compensatory payments that had eluded the diplomats of the capitalist nations. The Soviets did not win Socony's gratitude, however. As the 5 percent fee arrangements were being made, Socony agreed with the other oil companies not to renew its Soviet contracts when they expired. The "situation of the great world trusts has considerably changed since the time when an 'oil war' had Standard Oil of New York and Royal Dutch at each other's throats," observed the French finance ministry.[122]

The Soviet Union had to cooperate with the "as is" agreements. It was weak in distribution and vulnerable to aggressive pricing by the

Waley to Thomas Debevoise, shows that the Rockefeller family was intimately involved in the export and foreign market strategy of Standard Oil of New York.

[120] Deterding's anti-Sovietism earned him an interesting portrayal by an American Communist party member, Glyn Roberts (1938).

[121] Arch.Nat. Desmarais Frères 130AQ1, "Le Pétrole Russe, Sir Henry Deterding et les Anciens Propriétaires," newspaper clipping dated 4 May 1929; Fin. B32023, "Note," 13 August 1928; Fin. B32023, "Note," 16 August 1928. Corley 1988, 2:31–45, does not mention the details of the 5 percent rebate, but his discussion of Socony and the cartel agreement in India is related.

[122] MAE E-Levant, Irak-Mésopotamie, 1918–1929, vol. 34, "Rapports des grands producteurs de pétrole entre eux," 3 December 1928.

other multinationals; to maintain its profits, it had to go along to get along. As the Federal Trade Commission wrote, by "1929 . . . the Russians had entered into cooperative arrangements with the 'as is' group of companies in Great Britain, and by 1932, in Sweden. Having acquired small portions of many national markets by 1931, the Russians tended thereafter to be cooperative with the cartels."[123]

Opposition to the world oil cartel came from regional producing or retail companies. Political fights over production controls were featured in Romania and in Texas, while import quotas characterized consuming countries that had independent distributors, as was the case in France, Spain, and Japan.[124] In countries such as Iraq and Venezuela there were no national business interests large enough to dispute the quota allocations of the world cartel. Teagle of Standard Oil and Deterding of Royal Dutch–Shell agreed on methods to limit Venezuelan production in 1928. Gulf Oil joined them later as a junior partner.[125] The multinationals established their procedures for controlling production with very little reference to local concerns.[126]

"Local arrangements" were the world cartel's vocabulary for national markets.[127] Federal principles guided its actions: top management established market percentages and general guidance, but the day-to-day management of the cartel, the adjustments to local conditions, the exchange of information on market shares and prices, the compensatory payments or "fines" for selling too much, and the allocations of crude oil all were to be worked out as "local arrangements."[128]

"Local" interests, however, appealed to their national governments for mediation with the world cartel. These "local" regulatory fights were handled within the "local" structure of government. National concerns would square off against the "local" national managers of worldwide companies whose political goals reflected global market objectives. Repeated across many countries, these conflicts constituted a world regulatory fight in which the large corporations sought to fit local regulation into the overarching strategy of their international cartel.

Many national leaders, including some in France, thought their "local" laws "regulated" the multinationals. For the world corporations, local laws fit into a common mold; through those laws, they orches-

[123] FTC1, p. 278.

[124] On Japan, see Samuels 1987, pp. 173–179. The Japanese version of the French law of 1928 was floated in 1934. Spain is discussed in note 63 above.

[125] FTC1, pp. 44, 178.

[126] MAE E-Levant, Irak-Mésopotamie, 1918–1929, vol. 34, "Les trusts et le pétrole russe," 25 October 1928.

[127] FTC1, p. 234.

[128] FTC1, p. 234; on fines, see FTC1, p. 232.

trated and regulated state behavior to enforce their strategy. By its size, scale, and international regulatory ambitions, the world cartel clearly went beyond mere price fixing, and the Federal Trade Commission tried to express the size of the undertaking:

> The case studies . . . illustrate the widespread development of local cartel arrangements following the conclusion of the Achnacarry or "As Is" agreement of 1928. . . . Its principles, which were elaborated in subsequent agreements in 1930, 1932, and 1934, proved applicable in all kinds of markets, in large countries and small, in industrialized economies and in agricultural and even "undeveloped" economies.
>
> There can be no doubt that most of the local cartel arrangements were guided by the international agreements. In nearly all petroleum markets of the world, outside of the United States, subsidiaries and affiliates of the principal parties to the international agreements—Standard (New Jersey), Royal Dutch-Shell, and Anglo-Iranian[129] were predominant forces. Cartel agreements would have been unworkable without their leadership and cooperation.
>
> . . . .
>
> The assertion that local customs and usages determined the growth and form of the marketing cartels must . . . be questioned. In this regard the significance, effect, and form of the local marketing arrangements of the 1930s is much the same as that of the international "as is" agreements. The international agreements largely determined the character of local cartelization. Local cartel arrangements pursuant to the "as is" agreements were themselves largely determinative of the customs of the trade.[130]

A 1933 French finance ministry report made the same point: "We should remember that the solution adopted [in 1928] was in some ways imposed on the state by forces greater than its will . . . it had to recognize, juridically, the existing 'acquired rights' ['as is'] and make them the basis of all policy developed as a result of the law of 1928. No one can call those rights into question without bringing down the entire edifice."[131]

The world oil cartel fostered production controls and joint marketing contracts similar to the one between Anglo-Persian and Royal Dutch–Shell.[132] "As is" members swapped crude to avoid "cross haul-

[129] Anglo-Persian's name after 1932.

[130] FTC1, pp. 346–348.

[131] Fin. B32314, "Comment se pose la question du monopole d'importation des pétroles," October 1933.

[132] FTC1, pp. 136, 162. E.g., in Kuwait, oil jointly produced by Anglo-Persian and Gulf Oil was sold to Royal Dutch–Shell, which worked with Anglo-Persian in still other marketing deals. Royal Dutch–Shell, Gulf Oil, and Standard Oil coordinated the sale of Venezuelan crude and so forth.

ing," establishing by common interest what allied regulatory efforts had striven for in the face of corporate resistance during World War I.[133] The "as is" agreements guided the dominant producers, as members of the American Petroleum Institute, in their lobbying for production controls in the United States, enhancing the regulatory authority of the Texas Railroad Commission and similar state entities.[134] Under "as is" Anglo-Persian, Royal Dutch–Shell, and Standard Oil divided European markets and negotiated with "local" firms in the various countries. "As is" restricted the number of newspaper advertisements and other publicity firms might use to compete. "As is" secretariats assembled statistical information and coordinated overall policy on a world scale.[135] In France and elsewhere, "as is" led to compulsory auditing of balance sheets of small and large firms in the cartel.[136] "As is" members monitored the market share of small firms: "No firm was too small or insignificant to escape their attention, control over supplies being an especially potent weapon." Under "as is" the three major firms acted harmoniously, as a "joint venture," when dealing with European governments on regulation of the oil market.[137] "As is" set up a world basing-point system, so that prices paid for oil would be equivalent to the price of oil shipped from the Gulf of Mexico, even if its actual trajectory was much less, as with Mesopotamian oil shipped to France. Through World War II Allied governments dutifully paid for oil shipped from nearby Iran and Mesopotamia as if it had come from the Gulf of Mexico. Romania was exempted from the basing-point system, but to little effect. The Soviets knuckled under and charged Gulf rate freight prices on oil shipped from the Black Sea.[138]

[133] FTC1, pp. 204–205. E.g., Standard Oil might deliver oil to Royal Dutch–Shell in the United States in return for an equal quantity of Royal Dutch–Shell's oil from Indonesia, delivered to Standard Oil in the Asian market. Standard Oil would save freight costs from the United States to Asia, Royal Dutch–Shell would save freight costs from Indonesia to the United States. Extremely cost effective, this arrangement increased barriers to entry for newcomers.

[134] FTC1, p. 217, cites the Small Business Committee on the series of state-level initiatives enacted in the 1930s to circumvent national antitrust statutes: "No single item [state regulation] is in itself controlling; taken together they form a perfect pattern of monopolistic control over oil production and the distribution thereof among refiners and distributors, and ultimately the price paid by the public." Also cited is Frankel 1969, pp. 116–117: "There can be little doubt that the American counterparts of this international set-up ['as is'] were 'conservation' and 'proration' as we knew them in the thirties." Export associations formed under the authority of the Webb-Pomerene Act to coordinate American oil exports with "as is" principles were less successful (FTC1, pp. 227–228).

[135] FTC1, pp. 230–231, 241, 247, 264.

[136] Texaco refused this provision; the practice was standard in France (FTC1 p. 243n).

[137] FTC1, p. 279.

[138] FTC1, pp. 352–360, discusses the basing-point system.

Mercantile States and the World Oil Cartel

No major country's oil policies can be examined without reference to the global context of the "as is" agreements. Independents' struggles against the "as is" companies to enlarge their market were foreseen and were the reason for enforcement principles and regulation. We cannot attribute oil policies of this period to particularistic causes, such as cultural traditions of government in any one country, for the reason that there is no set of traditions or state institutions that applies to England, Texas, France, Italy, Sweden, Germany, Japan, Venezuela, Iraq, and all the other countries of the globe where market rationalization resulted from the "as is" agreements. The world oil cartel of the "as is" agreements grew and developed through the 1930s even as other forms of international cooperation, from the League of Nations to the world monetary system, came unraveled.

France's "independent" state policy reflected the "as is" strategy of the world cartel. France's domestic oil policy is one local instance of the global pattern. What is surprising about the world oil cartel is not that it faced resistance in various countries, but that it worked at all and that it worked for so long. France's domestic policy as a "piece of the world oil cartel" shows the attributes that gave the world cartel longevity, flexibility, and stability.

French Domestic Policy and the World Oil Cartel

The French state was one of many national arenas where firms fought for supremacy. The Shell combine had sought to take over the entire French market; then Standard Oil and the Banque de Paris et des Pays Bas, though never trying to win the entire French market, insinuated themselves into every aspect of French policy, as is evident in the creation of the Compagnie Française des Pétroles. Other competitors such as Pétrofina also tried to bend state policy to favor their interests.

When "as is" and agreement over Mesopotamia created a growing truce (cartel) in international markets, the French state, whose policies once generated conflict, became the arena for reconciliation. Cartels are the business equivalent of peace; but as with any "anarchic" system, maintaining such peace entails enforcement. The tendency toward disorder among nations has its commercial equivalent in the tendency toward competition among firms.

The world oil cartel solved the enforcement dilemma without a central authority by obtaining from many different governments regulations that tended toward the common goal. Anglo-Persian's entente with Royal Dutch–Shell was commercially beneficial but also was backed by at least the British government, as chief stockholder in the Anglo-Persian. Iraqi oil became a joint venture of the world's major corporations, their rights as stockholders contracted and enforced under British law. Local state law regulated overproduction in Texas, under the guise of "conservation." In France the government perfectly mirrored the "as is" agreements as it parceled out market shares.

The major oil companies caused the international convergence of regulatory policies. Where "local" interests objected to the policies of the multinational originators of "as is," the multinationals enlisted domestic allies such as banks, making compromise a frequent outcome. There was a world system for the production and distribution of oil, a system that had very few people at its head and that had an articulated set of goals.

French policy elites knew of the world market cartel; many world cartel meetings were held in Paris.[139] The stances taken by French elites reflected their financial and political alliances. Their ability to influence policy varied, creating the blend of objectives that became French oil regulations. The major French banks wanted domestic policies that harmonized with international oil policies. The international oil companies, heading toward general agreement on such key production areas as Iraq, Venezuela, and Russia, and on consuming areas such as the Asian market, wanted French policies in line with the trend. But there were obstacles: French managers of multinational subsidiaries were less willing than their parent firms to divide the domestic market; it was their fiefdom, and they would have preferred sovereignty to suzerainty. Independent French national companies had no multinational parent to hold them in check as they defended their market shares. Ernest Mercier observed in December 1928,

> The trusts now have as spokesman Henry Bedford, a distinguished, courteous, and understanding man; they defend, as they understand them, their very special interests and do not hide that fact. On the one hand, the great foreign oil leaders, of like mind with M. Bedford, have an evident desire to reach an understanding [*entente*]; on the other hand, they are not particularly excited about the repercussions of our projects on their French industrial enterprises, which represent only a few percentage points of their total activity. However, they attach a great deal of importance to establishing a foreign commercial agreement [i.e., a cartel] with us; and for these reasons I have always thought that, even today, if we could get the majority of independent firms to agree with us, one face-to-face conversation in London and Cadman [Anglo-Persian], Deterding

[139] FTC1, pp. 238–239.

[Royal Dutch–Shell] and Bedford [Standard Oil] would have brought the trusts around to the fait accompli.

The French representatives of the trusts are of an entirely different mind and are more passionate, even violent; they see only their immediate checkerboard and have staked everything on the policy debate.

The banks have taken a position in favor of international agreements.[140]

The Mesopotamian settlement and brewing world cartelization in 1928 affected French policy. The international accords had two clear counterparts in French regulations. The first was in the oil import law of 1928: Article 5, called the Loucheur amendment, took the highest market share in the previous five years as the baseline for allocating quotas: it was the legislative expression of the "as is" goals of the multinational firms. Article 8 provided incentives for refining, and by increasing capital requirements, restrained the growth of independent importers who would have bought their oil from independents in Romania, the United States, and elsewhere.

The second counterpart was in the structure of the Compagnie Française des Pétroles. The first draft of the law of 1928 favored the interests of Pétrofina and Desmarais Frères and raised the possibility of increasing imports from Russia and Texas. The drafts violated the "as is" sentiments of the majors and were torpedoed by them. France was a partner in the Iraq Petroleum Company, and the world market division would not work if France collaborated with the majors on production but worked against them domestically. Keeping market shares "as is" for the multinationals while introducing Mesopotamian production meant eliminating some major current source of French oil, not adding more from Russia and Texas.

Louis Pineau used the early drafts of the 1928 oil law to bargain with the multinationals in the Iraq Petroleum Company negotiations. In a 1927 document Mercier recorded Pineau's "adroit use of the trusts' worries about the parliamentary discussions on the oil regime" in a meeting with Sir John Cadman.[141] Ironically, the same tool worked to constrain French independence. The world cartel, Mercier knew, would limit French policy to the needs dictated by the settlement in Iraq:

[140] "International agreements" is here used to translate "*combinaisons extérieures.*" Fin. B32311, "Rapport au Président du Conseil . . . sur la Compagnie Française des Pétroles," 18 December 1928. The Pétrofina group and its banks were not part of Mercier's "banks"; limiting Russian oil imports was not in its interest.

[141] CFP 81.1/63, "Note sur la situation actuelle de la participation française dans la T.P.C.," 8 September 1927.

> [Cadman] has already advanced (perhaps to keep us interested in him) certain ideas concerning accords for sales and refining.
>
> Well, it is certainly the case that such accords will be the logical and necessary consequence of current agreements, and especially of the working agreement. These accords will be of a rather long duration, at least for the first phase, in conformity with France's interests.
>
> . . . .
>
> In this respect a major revision of the current oil regime in France will certainly gravely and almost immediately compromise all our work and the results we have obtained with such difficulty. It would certainly make much more difficult an effective cooperation with the coassociates the French state has given to the Compagnie Française des Pétroles.[142]

This document is remarkable in explicitly linking the Mesopotamian settlement with broader cartel agreements. Mercier meant that the French import law could not be Margaine's 1925 monopoly project—could not bear the stamp of the French national interests or of Pétrofina, which had backed it. In late 1927 and early 1928, as the parliamentary hearings dragged on, the Pétrofina group and French national companies were poised to win with their "left-wing" proposal to import more Russian oil. The left-wing parties wanted to help the Soviet economy; their business backers wanted access to the oil, pure and simple. The right wing resisted. Mesopotamian oil meant less Soviet oil sold in France; less Soviet oil meant less revenue for the Communist regime, and financial pressure on the Soviet government: the Catholic church had, to escape taxation from anticlerical French parties, placed much of its prewar wealth in Russian bonds and was a major loser to Bolshevism.[143] Pétrofina's bankers were also interested in the viability of the bonds, but as the decade wore on and the oil company grew, this consideration took a backseat to oil, especially after Pétrofina became the exclusive vendor of Soviet products in France and much of Europe.

To head off the pro-Soviet proposal the government proposed a rival plan, which catered to the desire of the pro-Soviet oil importers for quotas, but in a way that discriminated against Russian oil. Poincaré proposed import quotas that would encourage crude oil to be brought into France and refined; oil importers would have to be licensed for specific quantities.[144] Regulatory power would be in the Office National

[142] CFP 81.1/63, "Note sur la situation actuelle de la participation française dans la T.P.C.," 8 September 1927. In this document Mercier also mentions Finaly, of the Banque de Paris et des Pays Bas, and his personal relationship with Walter Teagle of Standard Oil, as having laid the groundwork for the working agreement.

[143] See Zeldin 1973, p. 332.

[144] Déb.Ch., 7 March 1928, p. 1271.

des Combustibles Liquides, the favorite of the Left. But the agency's favoritism for Russian oil was erased by a provision that the amount assigned to companies for the first two million tons of annual sales had to equal their highest market share of the previous five years. Soviet oil imports had been limited in that period. When the law was passed, Pétrofina's imports from the Soviet Union were scheduled to increase dramatically, but the law choked off this threat to "as is" principles. Instead, French independents would have to get additional oil from the Compagnie Française des Pétroles, whose production would be controlled by the international majors.

Poincaré's project wrote into law the 1925 private cartel whose favoritism to the international majors had generated such bitter opposition. Soviet oil imports for a time increased with the rest of the French oil market but could not grow as a percentage of the total. This was a defeat for French national companies such as Desmarais Frères, and it limited Pétrofina's prospects in France. The market share restrictions were the same as the "as is" principles agreed to by the oil majors only a few months later. The oil import law's discrimination against Soviet oil earned it the enmity of the French Left; the communist deputy Alexandre Piquemal denounced the chamber of deputies as the "chamber of the cartel."[145] The government project passed 335 to 185.

Still, the defeat of Pétrofina and the domestic French firms was only partial. They had wanted quotas; the argument was over which faction to favor. They had, moreover, kept their options open by backing the Compagnie Française des Pétroles. Robert Cayrol, head of Desmarais Frères, was one of the principal French negotiators for the working agreement in Iraq. Mercier even praised Cayrol's "tenacity," which "greatly contributed to earning the respect of our partners and a certain esteem for their French colleagues."[146] Mesopotamian oil could compensate Pétrofina and the French firms for their loss of Soviet imports.

For the national firms and Pétrofina, much of what had been lost on the quota issue might be recouped if they could control the Compagnie Française des Pétroles. The conflict with the majors shifted into this new arena. The company had been on the margin of domestic oil politics for its first five years. Now that oil from Mesopotamia seemed likely to come to France, there was room to dispute shares of the market's growth, though the first two million tons annually were parceled out under the quota law. The French national companies saw a chance

[145] Déb.Ch., 6 March 1928, p. 1218.

[146] CFP 81.1/63, "Note sur la situation actuelle de la participation française dans la T.P.C.," 8 September 1927.

to salvage their independence in the details of the new firm's operations. Who would refine its oil? Who would sell it? Would it compete like any other company, or would it have special privileges? Different answers to these questions caused a struggle over clauses of the French national company's "convention," or charter.

Thus far major oil questions had been disputed outside the confines of the company. Now the battle erupted among its principal shareholders. Robert Cayrol of Desmarais Frères parted ways with Pétrofina and brought together a group of French independent companies with the idea of "combining, in the Compagnie Française des Pétroles, all the possibilities for independent distribution."[147] If the independents could use the refineries to be built by the Compagnie Française des Pétroles, they could get a larger share of the market than their individual quotas under the 1928 law; they would have sole right of access to France's Mesopotamian oil. Cayrol wanted to control the marketing end of the same oil that the majors sought to control at the production end: whatever the international majors decided to produce, the French domestic interests would have the right to sell. Cayrol also wanted a 2 percent rebate on oil bought from the Compagnie Française des Pétroles, guaranteeing crude to the French companies at a lower price. Since the firms he led were "giving" their distribution facilities to the new company, the rebate would compensate them for their trouble. The international majors[148] and other French firms[149] rejected Cayrol's proposal. Under the protection of the 1928 law, they wanted to build their own refineries, and they said Cayrol wanted "a commercial privilege based on oil from Mesopotamia."[150] Several other proposals for the domestic distribution of Mesopotamian oil were floated.[151]

The banks in the Compagnie Française des Pétroles wanted the costly refining company, which they would finance, to be profitable. By contrast, Desmarais Frères and other French firms wanted discounted oil. Cayrol would have turned the refining company into a nonprofit utility, giving the national firms the benefits of vertical integration—a refining operation and direct access to Mesopotamia—at very low cost. Knowing they would earn little from distribution, the banks wanted to

[147] Fin. B32311, "Note," 20 December 1928.

[148] Royal Dutch–Shell, Standard Oil, and Anglo-Persian.

[149] The Compagnie Industrielle des Pétroles, controlled by Sinclair Oil; Pétrofina's Raffinerie de Pétrole du Nord; and the Bauer and Marchal Bank's Société des Pétroles, Essences, et Naphtes.

[150] Fin. B32311, "Note," 20 December 1928.

[151] See Nowell 1988, p. 352, for details. Archival sources are Fin. B32311, "Note," 20 December 1928; Fin.B32311, "Note remise par la Banque de Paris et des Pays-Bas," 19 December 1928; Arch.Nat. 324AP77, André Tardieu, "La Compagnie Française des Pétroles et l'Etat Français," 20 March 1930, p. 33.

shift profits to production and refining. They also wanted to limit their day-to-day involvement in oil company management.

Solving these conflicts gave the Compagnie Française des Pétroles its particular character. In the parent company there figured all the major members of France's oil industry and finance, including the influential Banque de Paris et des Pays Bas, which was important by itself and also for its role as intermediary with Standard Oil. The international luster of the parent Compagnie Française des Pétroles suited the elite oil-company-only members of the Iraq Petroleum Company and the world oil cartel: in the Compagnie Française des Pétroles, the international majors of the world cartel sat in common council with the principal national oil interests of France. The foreign majors intended to build their own refineries in France and steered clear of infighting over the "French" refinery, whose output was governed by the "as is" agreements.

In the Compagnie Française de Raffinage, the principal shareholders were the French national companies that desperately needed modern refining technology to compete against the majors. The national firms used the Compagnie Française de Raffinage to organize against the foreign majors; it was also a way to raise capital for a joint venture. But their prominence in the Compagnie Française de Raffinage was not control; the Compagnie Française des Pétroles owned 55 percent; the national firms shared 35 percent of the remaining equity.[152] A 10 percent share invested by the Office National des Combustibles Liquides provided some "free" state cash for the capitalization and allowed the agency head, Louis Pineau, who set market shares, to have a seat in a refining company where domestic French interests were represented. The five participating French firms are listed in table 3.

The French firms in the Compagnie Française de Raffinage got the benefits of vertical integration without its full costs. They sat on the board of directors of the refining end of the business. The largest French national firms also sat on the board of directors of the parent Compagnie Française des Pétroles, where they dealt directly with the multinational companies about Iraqi production. The Compagnie Française des Pétroles became a Janus-like creature: one face included the multinational concerns interested in harmonizing and integrating French oil developments with their world strategies; the second face comprised the contentious independent national companies, who consistently favored pressuring the Compagnie Française des Pétroles to produce more oil, so that it had to push for more oil from the Iraq

[152] Fin. B32311, "Compagnie Française des Pétroles," 30 April 1929; Arch.Nat. Desmarais Frères 130AQ1, newspaper clipping dated 4 May 1929.

Table 3. Shareholdings in the Compagnie Française de Raffinage (Founded 30 April 1929, capitalized at FF100,000,000)[a]

Shareholder	Percentage of total
Compagnie Française des Pétroles	55
Office National des Combustibles Liquides	10
Desmarais Frères	15.8
Sté Lille Bonnières Colombes	6.5
Sté Française des Carburants[b]	3.4
Consommateurs de Pétrole	2.4
Pechelbronn	1
Sté pour l'Approvisionnement des Consommateurs d'Huiles Combustibles	0.5
Ernest Mercier[c,e]	0.00017
Jules Mény[c]	0.00017
Henri Desprez[c,g]	0.00017
Louis Tronchère[c,e]	0.00017
André Pellissier[c]	0.00017
Paul Desmarais[c,d]	0.00017
Robert Cayrol[c,d]	0.00017
Alexandre Palliez[c,f]	0.00017

Sources: Fin. B32311, "Compagnie Française des Pétroles," 30 April 1929; Arch.Nat. Desmarais Frères 130AQ1, newspaper article dated 4 May 1929.

[a]The cited sources for initial capitalization and shareholdings conflict in some cases. These figures are therefore approximate.

[b]Linked to the Banque de Paris et des Pays Bas.

[c]Individuals with minimum shareholding requirement legally necessary to sit on board of directors.

[d]Of Desmarais Frères.

[e]Of the Compagnie Française des Pétroles.

[f]Of Lille Bonnières Colombes.

[g]Of the Compagnie Auxiliaire de Navigation.

Petroleum Company and make the oil market more "French." Because of this dual nature, the Compagnie Française des Pétroles was disliked by the international majors, who saw in it domestic agitation for what they most wished to avoid: increased oil allotments from Mesopotamia. The firm was also disliked by domestic concerns, who saw international coordination of what they wanted to avoid: restraint of exports from Mesopotamia. Bringing Iraqi production into France for refining increased the tensions between multinational companies, domestic firms, and the banks in the national company. The banks and the national firms fought even as the Iraq Petroleum Company increased capitalization to develop its oil fields. The Compagnie Française des Pétroles had to match that capitalization, but the banking interests, uncertain about refining arrangements, went on a capital strike. They feared profits would evaporate in overly generous concessions to French distributors. The Compagnie Française des Pétroles was legally obligated to match the capital outlay of its partners in Iraq, and its position there was jeopardized by the capital strike. Tensions were heightened by suspicions that difficulties with choosing a pipeline route in Iraq were really a trick by the majors to slow down development.[153]

Through most of 1928, the Compagnie Française des Pétroles' chances for success, in Iraq and as a domestic company in France, looked tenuous. Mercier observed that "a sufficiently important group had not emerged" with the structure of the Compagnie Française des Pétroles to guarantee efficacious decision making.[154] The logjam between domestic firms, French banks, and the multinationals left him unable to raise the required capital. He asked for Poincaré's arbitration and offered to resign.

Poincaré refused the resignation and gave Mercier a new negotiating tool. He offered the company a new charter; this time the state would take an equity participation of 25 percent. This new participation, by expanding total shares, would reduce the return of shareholders who had already invested in the company. To mollify them, he reduced by half the schedule for taxation on profits that had been agreed to in 1924. The state's equity would boost investor confidence by signaling government commitment to the company's operations. The government would not let disagreements over refining, or attempts to slow the pace of Iraqi oil development, threaten the company's viability. The Banque de Paris et des Pays Bas, which had organized previous increases in capital and refused more because of the refining dispute, was now content. Its investment risk was reduced, its tax liability was halved, and the government contributed the extra money needed in the face of the bank's reservations about the profitability of the refining.

Poincaré's reduction of the taxation schedule scandalized some. A highly critical 1930 report sent to Premier André Tardieu observed: "The state, incomprehensibly generous, has abandoned exactly half of its share of the excess profits in the Compagnie Française des Pétroles. It may seem incredible—but there it is."[155] Another feature of the taxation schedule, which perhaps because of its "technical" nature attracted less attention than the schedule itself, was the clause on surplus profits. "Surplus profits" were defined as any profits earned after the first 10 percent "profit" returned on the total invested capital. For any year when profits fell below 10 percent, if in some later year profits exceeded 10 percent, then the excess of that year could be applied to

[153] FTC2, pp. 145, 147.

[154] Fin. B32311, "Rapport au Président du Conseil . . . sur la Compagnie Française des Pétroles," 18 December 1928.

[155] Arch.Nat. 324AP77, André Tardieu, "La Compagnie Française des Pétroles et l'Etat Français," 20 March 1930, p. 38.

the earlier year before constituting a taxable income. Thus if in one year the Compagnie Française des Pétroles earned an 8% profit, and in another year it earned 12 percent, the 2 percent above 10 percent, which was taxable as a "superprofit" under the charter, was not taxable if it was disbursed to shareholders to make up for the earlier deficit. Oil did not begin to flow until six years later, so there were six years of no profits that had to be compensated for out of "surplus profits" above 10 percent. The first 10 percent of declared *profit*, not gross earnings as a percentage of invested capital, was tax free. Some of the companies' critics felt that the tax talons of the French "strong state," when it came to collecting revenue from its oil interests, were made of papier-mâché. And in this vein it is not surprising that a parliamentary investigation nearly fifty years later discovered that in France the oil companies "pay practically no taxes.[156]

Fair taxation is of course a highly subjective concept. The revised, halved taxation schedule seemed excessively generous to critics, the more so since the first taxation schedule had been drawn up in 1924 by two businessmen, Ernest Mercier and Louis Loucheur, both members of the Right. From today's perspective the package of tax incentives given the Compagnie Française des Pétroles, given the size of the undertaking, is probably not excessive, but it is a matter of record that the tax cut became political ammunition against the company. The tax schedules appear in table 4 for the 1924 charter and table 5 for the 1928 revised charter. Both tables give effective tax rates assuming different levels of profits on an invested capital of FF100 million.

Poincaré's offer to purchase shares in the Compagnie Française des Pétroles constitutionally required parliamentary approval of the company's charter. For five years the charter had passed unchallenged by both left- and right-wing governments. The attention of the Left had been turned toward better trade relations between the French national companies and Soviet oil companies; French naval purchases of Soviet oil had been one consequence. After the 1928 oil import law and the parallel "as is" agreements, the long-term prospects both for Soviet oil imports and for the companies dependent on them were compromised. Desmarais Frères and other French national companies hoped that oil from the Compagnie Française des Pétroles and its subsidiary, the Compagnie Française de Raffinage, would make up for this loss.

But there was another group of firms that were not members of the fledgling national champion and did not have amicable relations with the international majors. For them parliamentary review of the national champion's charter was a chance to rewrite, indirectly, the law of 1928. If control could be taken away from the interests that dominated the

[156] Schvartz 1974, p. 17; see also pp. 33–35.

Table 4. Tax Schedule of the CFP, as established in 1924 (assumes FF100 million invested capital)

Amount of profit (millions)	Marginal tax rate (%)	Amount paid (FF) (median)	Real rate as percentage of profit
up to 10	0	nothing	0
12	10	200,000	1.7
14	15	500,000	3.6
16	20	900,000	5.6
20	25	1,900,000	9.5
30	30	4,900,000	16.3
40	35	8,400,000	21.0
50	40	12,400,000	24.8
60	45	16,900,000	28.2
70	50	21,900,000	31.3
80	55	27,400,000	34.3
90	60	33,400,000	37.1
100	65	39,900,000	39.9
110	70	46,900,000	42.6
111 and above	75	—	approaches 75

Source: Calculated from table in Arch.Nat. 324AP77, André Tardieu, "La Compagnie Française des Pétroles et l'Etat Français," 20 March 1930, p. 38.

Note: Calculations are made on the upper limit of each tax bracket. The tax rates are marginal tax rates and do not apply to the whole amount of profit in excess of the first 10% of invested capital, but only to the amount in that bracket. Real rate of taxation on total is shown in far right column. The marginal tax rate, however, determines the incentive for new investment, because it determines the profitability of the next franc invested.

Compagnie Française des Pétroles, the law of 1928 could yet be turned to work in favor of the pro-Russian oil groups. The company charter became the place for these die-hard excluded groups to stage a last stand.

Thus began the debate over the equity, and as a corollary the control, the state should have over the Compagnie Française des Pétroles. By pushing state ownership to 51 percent or more the Left hoped to secure state control over the principal tool of national oil policy. Since the company could force its competitors to buy a percentage of its oil, the firm could have been a step toward the nationalized industry with private distribution and private capital participation proposed in 1925 by Margaine. Mercier wrote, "It is quite clear that the CFP, under the complete control of the state, would be, in the hands of a socialist or socialist-leaning government, a marvelous instrument for the realization of an oil monopoly in France. . . . clearly the CFP could not, from one day to the next, replace the entire French industry, but it would undoubtedly be a serious start, as a provisional step in the establishment of a complete monopoly."[157]

[157] Arch.Nat. 324AP77, André Tardieu, "L'Attribution à l'Etat de la majorité dans la CFP serait l'acheminement au monopole," n.d., prob. 24 March 1930.

Table 5. Tax schedule of the CFP, as revised in 1928 (assumes FF100 million invested capital)

Amount of profit (millions)	Marginal tax rate (%)	Amount paid (FF) (median)	Real rate as percentage of profit
up to 10	0	nothing	0
12	5	100,000	0.8
14	7.5	250,000	1.8
16	10	450,000	2.8
20	12.5	950,000	4.8
30	15	2,450,000	8.2
40	17.5	4,200,000	10.5
50	20	6,200,000	12.4
60	22.5	8,450,000	14.1
70	25	10,950,000	15.6
80	27.5	13,700,000	17.1
90	30	16,700,000	18.6
100	32.5	19,950,000	20.0
110	35	23,450,000	21.3
111 and above	37.5	—	approaches 37.5

Source and notes: Same as for table 4. Final tax bracket approaches 37.5% as profits approach infinity.

Supporters of the charter protested that the Left, even when it had commanded a majority in parliament, had not raised the objections that came up in 1928. That meant tacit approval of the company, they argued, and the new objections to the company's structure in 1928 were thus a policy reversal.[158] This was true: but the Left was not following a Marxist principle of state ownership. It was responding to the commercial circumstances of its allies from the business sector, as in earlier years. Margaine, author of the 1925 proposal for a state monopoly, showed a purely commercial concern when he objected that Mesopotamian oil would flood French and world markets and reduce prices, exclaiming to his colleagues, "You will reduce the value of oil in Romania to nothing!"[159] That was the worry of independent importing firms faced with competition from the companies in the Compagnie Française des Pétroles.

These business lobbies gave the resistance to the Compagnie Française des Pétroles surprising parliamentary strength, unsuspected even by the Left. In March 1930 a preliminary committee vote on ratification of the company's bylaws (the "convention") went against the Com-

[158] For example, Déb. Ch., 12 February 1931, p. 635; also Arch.Nat. 324AP77, André Tardieu, letter signed by Mercier, n.d., prob. March 1930.

[159] Déb.Ch., 12 February 1931, p. 631.

pagnie Française des Pétroles and resurrected instead the defunct 1925 monopoly project, by a vote of eight to seven. This was reported to Premier Tardieu:

> Yesterday afternoon the committee on mines . . . took into consideration Charles Baron's opposition project for an oil monopoly, by a vote of eight to seven. Following this surprising vote the committee reporter, Monsieur Charlot, resigned; the majority was composed of four socialists and four moderates who backed the Saint Gobain position. Monsieur Charles Baron and all the other members of the commission of mines, rather astonished by the results achieved, asked [Charlot] to reconsider his resignation, declaring that their vote was only a vote on principle and that they would be content, in place of a monopoly, with the construction of a state-owned refinery capable of supplying the needs of the army and the navy . . . in reality, this surprise vote caused the greatest amount of confusion, and it will clearly be necessary for the government to intervene soon in this discussion.[160]

This note states the alliance of the Left with the Saint Gobain chemical company. Baron, a socialist who was preeminent in oil matters and whose family wealth was tied to independent oil importing, backed away from his own public goal of a state monopoly when confronted with his group's unexpected success. Even the prospect of a refinery for the army and navy, which might be interpreted as the French Left's solicitude for the national defense, was tied to the Saint Gobain interests. This surprising development resulted from Saint Gobain's position in the chemical industry and its desire to expand into oil.

Saint Gobain, a premier member of the French industrial establishment, was a newcomer to the oil business. Historically the chemical industry had made its products from coal derivatives—tar and coke—produced in the reduction of coal to gas for the gaslight industry or in the manufacture of steel. Today the primary feedstock for the chemical industry is oil, a more convenient form of the hydrocarbon stock also available from coal. Saint Gobain's desire to enter the petroleum field reflected the desire to adapt to a new generation of technology.

The 1928 law gave Saint Gobain a chance to get into the oil business. The market quotas and tariff advantage for refining incited the international majors, Pétrofina, and a few other large French national oil firms to build their own refineries. Construction began in great haste, because whoever finished first would have a greater claim to import

[160] Arch.Nat. 324AP77, André Tardieu, "Note pour le Président," 21 March 1930. The Procès verbal of the meeting in question is in Ass.Nat. Procès verbaux de la Commission des Mines et de la Force Motrice, 20 March 1930.

allocations made beyond the two million ton level for the French market. Up to the two million ton level, the market had already been apportioned out to the highest shares of existing firms during the previous five years.[161] Saint Gobain targeted a medium-sized refinery for the market growth beyond two million tons, hoping to sell fuel to independent distributors while using secondary products for its chemical operations.[162]

Saint Gobain's refinery was underway when the worldwide "as is" agreements, and the charter of the Compagnie Française des Pétroles, threatened to turn it into a white elephant. The multinationals were planning their refineries' capacities at about 75 percent of their expected needs; they would make up the difference by importing refined products. The calculation was as follows: a firm with refinery that made 75,000 tons of oil, plus 25,000 more in refined imports, could run its refinery at 100 percent (75,000 tons) and could buy an additional 25 percent of its total sales from the Compagnie Française. But a firm that had only a refinery that ran 75,000 tons annually and did not import and sell refined products would still have to buy 25 percent of its total sales, or 18,750 tons annually, from the Compagnie Française. The refinery would never run at more than 75 percent efficiency, and possibly even less—a disastrous limitation. Saint Gobain fought the 1928 Compagnie Française charter because it correctly ascertained that, if passed without modification, the provisions would effectively modify the import quota law and discriminate against French firms that wanted to refine without being retailers.

Saint Gobain allied with the Left against the Compagnie Française des Pétroles. The company rounded up allies, forming an association of like-minded but smaller firms that Mercier characterized as "newcomers who want to build refineries in France." The Saint Gobain group threatened to interfere with the Iraq Petroleum Company working agreement by making an alliance with Calouste Gulbenkian to take his oil and refine it for sale in France. Louis Pineau, Mercier's longtime adversary in the Office National des Combustibles Liquides, consid-

[161] Arch.Nat. Desmarais Frères 130AQ1, newspaper clipping dated 4 May 1929.

[162] Ass.Nat. Procès verbaux de la Commission des Mines et de la Force Motrice, 20 March 1930. Saint Gobain and other groups assumed that awards of market share over two million tons would be proportional to investment. This is not unusual. Krueger 1974, p. 292, offers this general rule: "When licenses are allocated in proportion to firms' capacities, investment in additional physical plant confers upon the investor a higher expected receipt of import licenses. Even with initial excess capacity (due to quantitative restrictions upon imports of intermediate goods), a rational entrepreneur may still expand his plant if the expected gains from the additional import licenses he will receive, divided by the cost of the investment, equal the returns on investment in other activities." Saint Gobain did not get additional import licenses, however.

ered limiting, to favor Saint Gobain, the annual import allotment to the Compagnie Française des Pétroles. Mercier complained to the French premier Tardieu that Pineau's machinations were "known to everyone in the oil business" and that "various parties are afraid that their established rights will not be respected," which meant the oil companies and banks who were Mercier's collaborators.[163] Mercier and others of the Right were convinced that cash payments by Saint Gobain to the socialists explained the stiff opposition the Compagnie Française des Pétroles encountered. Léon Blum was implicated by name: but hard evidence is not to be found, unless the alliance between Saint Gobain and socialists in the Chambre is itself sufficient. Mercier declared to a stockholders' meeting of the Compagnie Française: "Certain private interests that are out of government favor have taken the expedient of attacking the work we have striven for six years to accomplish and to present this work in a manner I will not bother to characterize. These private interests have procured for themselves an expected degree of support by means that it would be, someday, very interesting to bring to the light of day." Léon Blum publicly denied charges of having been sold to Saint Gobain.[164]

To save Saint Gobain, the socialists demanded a refinery for the army and navy whose specifications matched the Saint Gobain project. Charles Baron envisioned "of course, a monopoly, mixed-enterprise system," but when pushed to clarify, he wanted one refinery to be owned and operated by the state. He cited cost figures for a refinery of 400,000 tons annual capacity, then revised the estimates to match a 300,000 ton refinery. Saint Gobain had initially demanded a 400,000 ton import license for its refinery and had gotten assurance it would have a permit for 300,000 tons.[165] Saint Gobain wanted an adequate import license or else wanted to cut its losses and sell off its refinery, which the socialists hoped to pass on to the army and navy. When the Compagnie Française des Pétroles' convention was ratified over the opposition's objections, negotiations began for the Saint Gobain refinery to be sold not to the army and navy, but to the Compagnie Française des Pétroles.

Saint Gobain dropped its opposition to the Compagnie Française des

[163] Arch.Nat. 324AP77, André Tardieu, "Gulbenkian," 24 March 1930. Mercier lists the refining "newcomers" as "Berre [St. Gobain], Brest Pétrolier, Bordeaux, etc."

[164] Mercier's accusation is in Déb.Ch., 12 February 1931, p. 631. Blum's denial is in "Deux Mots à Coty," *Le Populaire,* 8 July 1930. Clipping is in Fin. B32309, folder 7. Others had also accused Blum; hence the "denial" is dated before Mercier's remarks.

[165] Ass.Nat. Procès verbaux de la Commission des Mines et de la Force Motrice, 20 March 1930; Saint Gobain figures are from Robert de Vogue, president of Saint Gobain's Compagnie des Produits Chimiques et Raffinerie de Berre, same dossier, meeting dated 26 March 1930.

Pétroles in exchange for being able to sell off its refinery. But the negotiations begun in 1931 collapsed, and in 1933, in the wake of that collapse, the socialists initiated a last effort to bring state control to the oil industry and talked again of an independent refinery for the army and navy. By then these issues, their accompanying ideological arguments, and their underlying causes were tiresomely familiar to everyone. The Compagnie Française des Pétroles had been ratified, oil from Mesopotamia was about to flow, and Saint Gobain was saddled with an unprofitable refinery, a victim of its own bold attempt to outwit the major players of the world oil cartel in regulatory politics. As a last insult to the Saint Gobain cause, in 1934 the Compagnie Française des Pétroles began to build its own refinery almost next door to Saint Gobain's.[166] The Saint Gobain refinery remained a millstone around the company's neck until Royal Dutch–Shell reluctantly bought it in 1951.[167] A Desmarais Frères' memorandum comments, "Shell seems less interested than Berre [Saint Gobain] . . . in the projected merger." It was the import licenses that went with the refinery, not the refinery itself, that interested Shell.

Saint Gobain and its allies had hoped to beat the Compagnie Française des Pétroles by gaining control of it. They wanted 51 percent state ownership of the stock, but they only managed to raise the state share from the 25 percent proposed by Poincaré in 1928 to 35 percent; with preferred stock, the state's voting rights were 40 percent. Mercier fought continuously to keep state participation and state control to the bare minimum. The increase in state share to 35 percent was of no consequence in the overall control of the state over the company and did not achieve the objectives of the business interests who wanted the company reoriented to other sources of oil. The equity participation did not reflect the mercantile power of the state.

So why the great struggle? There were three reasons. First, even for the losers, a show of strength increased the compensating favors, especially import licenses, that would be thrown their way. When the Left's campaign against the Compagnie Française des Pétroles began, the government froze long-term refining licenses, holding them hostage to the ratification process.[168] Under the 1928 oil import law this was illegal, but it was done anyway. The tactic put the opposition under enormous pressure, but it also meant that nothing had been decided: the squeaky wheel got some grease. Saint Gobain's 300,000 ton annual

[166] Fin. B32312, "Rapport au Ministre," 3 February 1934.

[167] Arch.Nat. Desmarais Frères 130AQ3, "Note à l'attention de Monsieur Cayrol," 30 August 1951.

[168] Ass.Nat. Procès verbaux de la Commission des Affaires Etrangères, 9 February 1931.

import license was almost certainly a bone thrown its way that it might well not have had.

Second, although the Compagnie Française des Pétroles maintained the theoretical right to make its competitors buy 25 percent of their sales from it, the company never did so. It marketed through the member companies of the Compagnie Française de Raffinage. This was an important boost to Saint Gobain and a practical payoff for the Left's intense opposition.

Third, in a political fight as complex as this one, outcomes are uncertain. The opponents of the Compagnie Française des Pétroles were permitted to hope that one day they would win. A simple increase in the state's minority shareholding in the Compagnie Française des Pétroles placed them that much closer to getting majority control for the state if, for example, the Left won a sweeping victory in an election. And socialists of this time liked state equity: as Edouard Herriot explained, "So long as we live in a world of corporations, the influence of the state will be measured by the size of its participation."[169] His rhetorical support for state influence handily fit the needs of business interests like Saint Gobain.

Interest group pressures created both the Compagnie Française des Pétroles and the Office National des Combustibles Liquides, which gained considerable power from the oil import quota legislation of 1928. The Compagnie Française des Pétroles, founded in 1924 with no state equity, resorted to state backing only in 1928 when interest group conflict threatened its ability to raise capital. And then, because some private interests were unhappy with the effect of state participation on profits, additional tax concessions were made. The French Left allied itself with businesses that opposed charter provisions in the Compagnie Française des Pétroles, which they furiously attacked in a desperate effort to gain control of national oil policy. They wanted to increase the quota of Russian oil and give independent interests a larger share of the French market.

French oil legislation domestically reinforced the cartelized world market embodied in the "as is" agreements. Mesopotamian production would be tightly coordinated within the limits of those agreements, whose three principals not only held the controlling interest in the Iraq Petroleum Company but also were significantly represented, along with their banking allies, in the Compagnie Française des Pétroles. Russian oil's share in the French market would be reduced proportionately to increases from Mesopotamia, declining as this oil started to

[169] Déb.Ch., 25 March 1931, p. 2191.

Table 6. French imports from the Soviet Union, 1925–1939

Year	Tons	Value (FF)
1925	176,800	109,500,500
1926	168,300	203,808,000
1927	373,100	221,777,500
1928	425,900	261,620,000
1929	384,100	212,823,000
1930	565,190	289,394,000
1931	884,640	250,079,000
1932	1,102,900	258,366,000
1933	860,300	155,659,600
1934	671,090	103,560,000
1935	258,936	43,132,000
1936	260,418	40,032,488
1937	83,338	27,793,972
1938	214,620	70,681,655
1939	150,000	52,850,000
TOTALS	6,579,632	2,301,078,215

Source: Fin. B32313, Léon Wenger to Robert Coulondre, 15 May 1939.

reach the French market in 1934 and after (see table 6).[170] The legislation worked to prevent the market entry of powerful firms, as evinced by the unhappy fate of the Saint Gobain refinery.

Conclusion

In 1914 Mesopotamian oil nearly led Germany and France to pass oil laws to make room for the new production and to enforce locally the broader provisions of the cartel known as the European Petroleum Union. In 1928 the principle of 1914 was tried again on a larger scale; and this time it succeeded. To use a cosmological metaphor, the world oil cartel of 1928 was a "big bang" that created the basic structure of the oil universe. Long after the cartel itself ceased to police markets on a world scale, many institutional remnants, or pieces of the "big bang," continued to have an impact on the world oil market. The policies and resources of states were also affected.

France's Compagnie Française des Pétroles owed its birth, its equity, its Mesopotamian oil, and its rights in the domestic market to the outcome of political strife among the principal actors in banking, the domestic French industry, and the international industry. The origins

[170] Russian import figures are in Fin. B32313, Léon Wenger to Robert Coulondre, 15 May 1939, and are reproduced in table 6.

of the company considerably antedate Poincaré's official letter of commission in 1923; in fact the whole company was the joint revenge of Standard Oil and the Banque de Paris et des Pays Bas for Royal Dutch–Shell's ambition to corner the French market. The company's structure showed French weakness in international oil. The desire was to conciliate the Standard Oil Company and form a pro-American oil alliance, in direct contrast to the anti-American oil alliance that had collapsed in 1920. In ordering the company's creation, Poincaré was neither prescient nor aggressive: the policy had been rammed down France's throat by the Standard Oil embargo of 1919–1920.

The company's founder, Ernest Mercier, represented the banking milieu most closely aligned with Standard Oil. As Walter Teagle, the president of Standard Oil, said in 1928: "What we want and hope for in France is a unification of the various French interests behind a policy satisfactory to the majority"—with an explicit guarantee that Standard Oil's market share be assured.[171] That was what Mercier delivered, and Standard Oil memorandums show that the relationship between Mercier and Standard Oil was close. Mercier himself wrote that the structure of the Compagnie Française required France to participate in international cartel agreements, which governed the regulatory structure of the domestic market. State participation was not at first envisioned by Poincaré, and when it was finally adopted, it had nothing to do with the "need" for the state to supervise or exercise control over its "national champion." The Banque de Paris et des Pays Bas and other banks were worried about the French refiners' designs on the company and sought a reduction of risk in the form of state participation. As if apologizing for taking a share of the company, the state offered a 50 percent tax break to private shareholders whose profits would be diluted. When the state's shareholding was increased beyond the original 25 percent, it was done over Poincaré's and Mercier's objections. The French Left, the Saint Gobain group, and would-be importers of Russian oil raised the bitter parliamentary fight to increase the state's share to 35 percent, and higher if possible. Their first purpose was to get de jure and de facto control of the company, to favor their particular oil interests. Their second, more successful purpose was to blindside the government with a show of force that would win them concessions on the quotas in the law of 1928, regardless of what was done with the Compagnie Française des Pétroles.

After seven decades, this company still exists and still sells oil; in 1975 the Schvartz report found that it was still engaged in actively colluding or—if one prefers a weaker term—cooperating with the

[171] Pat.Hearings, p. 3430, Memorandum of meeting, 31 August 1928.

goals of the international oil companies. Today the company is undergoing privatization, a development that is scarcely surprising—the original state participation was to support private capital. But the greater point is this: regardless of what the company does today, its very existence after nearly seven decades is a continued expression of the process of transnational structuring and demonstrates the durable effect this process has on the international political economy.

The import quota law of 1928 was also the product of transnational structuring. It perfectly reproduced within France the broader outlines of the world agreements between Anglo-Persian, Royal Dutch–Shell, and Standard Oil. The oil import quota law of 1928, though now modified to accommodate the European Community, likewise became a long-standing feature of the French market and is another durable fragment of the "big bang" in the international oil industry. The Office National des Combustibles Liquides was another lasting creation: unlike the oil import quota law of 1928 and the Compagnie Française des Pétroles, it did not originate in the political desiderata of the international majors. The French domestic firms and the international finance and oil interests of Pétrofina sponsored its growth. The Office National's interests worked in opposition to the intentions of the Compagnie Française des Pétroles until the world cartel agreements of 1928 and the convention for the national company made it clear that Mesopotamian oil would prevail over Russian oil. The desire to support the Pétrofina group aligned French naval oil purchases with Soviet oil needs, which lasted until World War II. Rival interest groups "split the difference," and as French national policy swayed toward Standard Oil and pro-Americanism, limited side agreements were made in a contrary direction. The Office National was renamed the Direction des Carburants ("La Dica") after World War II and is another fragment of the "French state" that owes its existence to transnational structuring.

Pétrofina is not strictly speaking an "institutional expression of the state," but it too is a durable product of the international oil politics of the period. Though often advertised in Europe today simply as "Fina," the full name of the company, from *pétroles* and *financières,* still proclaims to the world the bankers' preoccupation with Russian oil and Russian debt that played a major part in the politics of the 1920s. The czarist debt has long since vanished as an issue, but the company has become a long-term member of the international political economy.

The "big bang" of 1928 had durable effects on Iraq as well as France, though as a "peripheral" country literally created by the world's greatest empire this is less surprising. Nonetheless, the principal agreements on its oil production quotas and shares among the four major companies (now called Royal Dutch–Shell, British Petroleum, Exxon, and the

Compagnie Française des Pétroles) lasted until 1973, affecting Iraq's revenue, its development, and its people's experience of foreign control, weakening the central government,[172] and creating a diplomatic unity between the French and the Iraqis, whose shared desire for more oil production was continually frustrated by the international majors. The unity of purpose between France and Iraq lasted through the Iran-Iraq war of the 1980s until Iraq's catastrophic war with the United States and allied powers in 1991.

Transnational structuring had other effects. The British, unwilling to help the non-British firms that dominated the Russian oil fields before World War I, may have exited the Transcaucasus a little more eagerly than they would have otherwise. Russia's withdrawal from world markets in the 1930s was due less to rising internal demand than to the cartel agreements that kept out of its hands the refining technology needed to boost production. The 1928 cartel agreements eased Russia out of France, out of Italy, and out of Japan; in other areas, such as Sweden and Great Britain, the Bolsheviks toed the line and obeyed the "as is" agreements. Anxious to defend itself from the exclusive tendencies of the international oil companies, the Soviet Union supported oil monopoly laws that it hoped would work to its benefit in France, Italy, Spain, Greece, Yugoslavia, and Turkey. Its efforts to create its own "transnational structure" redoubled the unity of purpose that guided its adversaries in the world oil cartel, who saw Mesopotamian development and Russian oil as incompatible. To placate these redoubtable adversaries, the Soviet regime was even willing to pay 5 percent of its gross to Socony, which paid the money to the expropriated landowners of Russia. Since oil has been Russia's principal means of earning foreign exchange in this century, the limitations of "as is" no doubt hindered the country's development in the 1930s.

Some think that Russian oil exports fell worldwide in the 1930s because production could not keep up with growing internal demand—that there was not enough to export, which caused the Soviets to "withdraw from the markets."[173] But this "withdrawal" correlated with rising Mesopotamian production. One may cautiously predict, in the hope that in the post-Soviet era the right records will be opened, that evidence will be found showing that the decision to sell oil to the Nazis in the late 1930s was in part motivated by the Soviet Union's inability to place oil in world markets controlled by "as is."[174] Current explanations in any case are contradictory: one cannot affirm that Russia "with-

[172] See Nowell 1991.

[173] FTC1, p. 278.

[174] Polanyi 1957, p. 248, offers a similar explanation about Soviet agricultural exports.

drew" from world markets because of internal demand, then suddenly had enough to export to the Nazis in the late 1930s. Stalin probably calculated that the need for foreign exchange overrode all other considerations, and if he could not sell in world markets, then he would sell to the Nazis.

Oil import legislation was introduced in Japan and Spain as well as in France. In fact the Spanish monopoly was much more *étatique* than France's, reflecting the fact that it was backed not by a domestic oil industry but by a group of Spanish banks eager to claim the oil industry as their own exclusive province. In Japan the process was similar to France, though domestic oil companies counted for more.

Oil production agreements were reached in Mesopotamia and, for a brief time, the entire Saudi Arabian peninsula under the Red Line agreements; Venezuela had similar production arrangements worked out by the majors; and the Texas Railroad Commission, which had been created earlier, received its greatly enhanced regulatory powers of oil production as a result of the "big bang" of 1928. The Red Line agreement is gone; the Texas Railroad Commission still exercises its regulatory authority.

The transnational structuring of the 1920s and 1930s created durable corporate and state institutional entities. In countries such as the Soviet Union and Iraq, transnational structuring also affected developmental capabilities by, in effect, creating a spigot on the potential to earn revenue and then shutting it off. Seemingly "realist" or military decisions, such as to fuel the French navy with Soviet oil, were consistent with the desires of specific interest groups. Even the pro-Americanism of French oil policy is an artifact of interest group struggle that demonstrated not the strategic lessons of World War I, as France supported the Shell combine for the first two years after the Great War, but Standard Oil's ability to influence sovereign decisions. The Saint Gobain refinery's failure to achieve adequate import quota licenses to be profitable is not a mere footnote: as we shall see in the next chapter, the chemical companies were the only firms outside the world hydrocarbon cartel with the technological capability to forge ahead with such projects as synthetic oil and synthetic rubber. France, and other nations alike, lacked commitment to firms such as Saint Gobain. As a result, these countries were denied access to the chemical high technology of the era.

The international basing-point pricing system was not, perhaps, profoundly structuring; but it enhanced the financial earnings and hence the political prowess and weight of the great oil companies, which succeeded in getting the great powers of the world to respect this lucrative fiction even through World War II. The international oil majors showed no more concern over the financial drain they caused their home governments than they did for the Soviet Union. For those who believe that the strong French state "defended the national interest" in joining the international oil cartel, it is interesting that before World War II this "strong" government paid for its Mesopotamian oil as if it had been pumped and shipped from Texas, a costly fantasy to which all the great powers of the era, capitalist and noncapitalist alike, duly adhered.

The process of transnational structuring did not create a gigantic regulatory juggernaut that permanently held sway over the globe. It affected the distribution of revenues, of capabilities; it created institutions and practices around the world that proved remarkably durable even though price competition sometimes occurred. To judge the world oil cartel as a mere price-regulating device is beside the point; it was much bigger, much broader, a major international political economic event of the century, on a scale comparable to the monetary agreements among nations. Unlike "politics among nations," as Morgenthau would describe international politics,[175] this was politics *through* nations. State regulatory powers were drawn upon by international actors for market-related purposes that were orthogonal to the power-maximizing drives posited by realism-mercantilism. Transnational structuring in the oil industry consolidated even as other facets of international cooperation, such as the League of Nations or the gold standard, were losing their grip. Transnational structuring is a thing apart from these better-known kinds of international cooperation and agreement; to "see" it at all, one must concentrate on business interests on a world scale.

Even so, this chapter's description of the world oil cartel remains parochial, describing only the trifling collision of international firms with each other and their local adversaries, who all played with national power, national institutions, and regulatory structures the way children play with blocks. The truly cosmopolitan view can be had only by looking at the world *hydrocarbon* cartel, which was an expression not of oil's struggle with oil, but of all oil with coal.

[175] Morgenthau 1973.

Chapter Five

The World Hydrocarbon Cartel, 1922–1939

The agreements of 1928 did not just cover oil extracted from the ground; they durably affected the chemical and coal industries that had a serious potential for producing in the developing fuel market. Though seemingly arcane today, liquid fuels derived from coal had the potential to enhance, at a cost, national independence and self-sufficiency. The world *hydrocarbon* cartel extended the limited market-sharing objectives of the "as is" agreements to the evolution and control of technology. This chapter examines the world hydrocarbon cartel and its implications for the worldwide transition from coal to oil, concluding with an assessment of the principal achievements of French oil and hydrocarbon policy in the decade before World War II.

It is easy to believe, wrongly, that synthetic oil from coal never took off because of its expense relative to gasoline. That is not so. All the major European countries were willing to pay a price, a price higher than oil, to protect their coal economies. Coal was simply too politically and economically significant to do otherwise. The question was, How high a price? Oil from coal could have competed against oil at a higher price with the necessary protection, but not at an impossibly higher price. The efforts of the world hydrocarbon cartel were directed at making coal's difficult situation impossible. This could be done only by keeping chemical firms from using new hydrogenation technologies to make chemicals, synthetic rubbers, and fuels.

The point can be reduced to two equations:

1. Possible synthetic fuel program = [(sales of gasoline made from oil from coal) + (sales of all chemical by-products from the manufacturing process) + (sales from products made from related applications, especially synthetic rubber) + (sales of gasoline made from oil at same facility)] − (cost of coal feedstock over equivalent oil feedstock).

2. Impossible synthetic fuel program = (sales of gasoline made from oil from coal) − (cost of coal feedstock over equivalent oil feedstock).

Equation (1) was the strategy of any company or state authority eager, for whatever reason, to move the coal industry toward the production of synthetic oil and hence the new transportation era. This strategy would not have produced synthetic oil at a price equivalent to that of oil imports; but it would have been close enough for states to adopt the strategy on a wider basis than in fact happened. Equation (2) was the strategy of the world hydrocarbon cartel, and it was the one that won: outside Germany, those who made synthetic oil did not market the by-products, made no synthetic rubber, and did not blend conventional oil with oil from coal.

Energy autarchy offered to insulate national economies from such massive disruptions as the blockades of World War I (e.g., of the Dardanelles and of Germany); from the heavy-handed politics of the international companies; and from the social dislocations that would follow from a massive conversion from coal to oil. But the road to autarky was not taken. The worldwide failure to develop the alternative fuels offered by coal reflects the effects of the world hydrocarbon cartel. The world oil cartel of the previous chapter was but one component of the greater whole.

France's diminutive synthetic oil program was again typical of the broader world pattern. The outcome reflected the preeminence of the oil companies and banks that backed the world hydrocarbon cartel; an aggressive, mercantilist etatism disguised an insipid reality. An ancillary pattern of neglect took root across a broad range of energy policies in France, where the government might have improved the French strategic posture even if we set aside the question of synthetic oil from coal. I could describe this as typical of the Third Republic's many failures; but many countries of quite different character did the same, and the reasons are not to be found in particular qualities of individual states. The principal force at work was transnational structuring.

HISTORICAL REASONS FOR THE DECLINE OF COAL

There are two reasons why the coal industry failed to adapt to the evolution of the petroleum era. The first is historical and market based: in the nineteenth century, the oil industry developed in markets that were marginal to coal, and this explains why, when coal first confronted petroleum as an adversary in the fuel market, it faced large, fully developed corporations rather than weak new entrants.

The second reason is political: in spite of the failure to jump into the oil market in the nineteenth century, the coal industry had another opportunity in the 1920s and 1930s. The attempt was made, but it failed because of the 1928 patent accords. In fact, the "as is" agreements were but a part of the more significant patent accords, which is why I distinguish here between the world petroleum cartel and the world hydrocarbon cartel, which included oil, but went beyond it. In this second argument I explain why the strategy summarized in equation (1) failed to materialize.

The term "coal complex" used in this chapter means the mines, the chemical companies, the coal gasification companies, and the steel companies that consumed large quantities of coal and produced from it a host of salable products. This coal complex ought to have resisted the development of the oil market, but surprisingly many histories of international oil companies and international oil politics are mute on this subject.[1]

Retrospectively, the decline of coal seems "easy to explain." Oil is intrinsically a better fuel than coal for most purposes: it is cheaper (oil extraction uses less labor than coal mining), is easier to handle, packs more caloric potential per unit of volume, and burns more cleanly. Had free markets operated, European coal use would have declined much earlier than the 1950s and 1960s. Large industries, however, do not operate in free markets because they have the political power to change those markets. The long decline of coal indicates in fact that there was a long political struggle. Nowadays we usually think of coal's decline as a matter of labor unions and government subsidies. A major question has been left unasked, however: Given the coal complex's massive presence in nineteenth-century industrial economies, why did it fail to buy its way into the oil business, capturing the infinitely smaller oil industry before it had time to overwhelm coal?

The historical, market-based answer. In the 1860s and 1870s the oil industry catered to the illumination market. The coal complex's illumination product, coal gas, was used in large cities where the high-density market justified the expensive infrastructure needed for manufacture and distribution. Worldwide, hundreds of millions of people did not live in urban zones. Oil was the premier illumination fuel, distributed cheaply in individual containers far beyond the limits of coal gas systems. Oil lamps were superior to candles in illumination power, reliability, and price. Even in cities with coal gas, lamp oil was the poor man's illumination fuel. High installation costs for coal gas kept it at the upper end of the market.

Until the 1880s and 1890s coal was the unchallenged fuel of choice

[1] Gordon 1970 does not discuss synthetic oil from coal programs.

for the world's railroads and ships. The large market of coal-consuming ships and railroads would have guaranteed a lengthy decline for the nineteenth century's premier fuel in any circumstances. Coal people should have seen themselves as transportation fuel producers, and some at least should have sought entry into oil production as an adaptive strategy. Oil's potential as a shipping fuel became evident toward the end of the nineteenth century.

During oil and coal's "age of illumination" the two fuels not only did not compete, they were complementary, one lighting the cities, the other the countryside. The oil multinationals, servicing the worldwide rural illumination market, grew to unprecedented size compared with the nationally fragmented coal industries.

When at last oil challenged coal in the transportation market, it did not do so as a weak new entrant with a novel product. Sometime between 1878, when the first oil tankers were launched by the Nobels, and 1911, when the British Admiralty adopted fuel oil and anyone could see this was the wave of the future, coal company executives would have had to "wake up" to the competitive challenge of oil.[2] But their oil competitors were by then already powerful, worldwide corporations, with two or three decades' advantage in finding, developing, and exploiting oil resources. The established oil industry was tough to take on, and this is one reason we do not find coal companies that entered the oil business. In a later era a maker of mechanical cash registers, the National Cash Register Corporation, changed itself into a major computer producer to keep up with the new generation of technology; but none of the coal complex industries made the analogous transition to oil.

There were other elements to the seemingly insurmountable fait accompli that the coal complex faced. The automobile and truck industries are now "obvious" competitors to coal-fired trains. But they did not begin that way. The automobile started as a "luxury item" for Sunday driving; trucks began as distributors from train stations. When at last internal combustion vehicles threatened rail transportation, the Sunday drivers and local delivery trucks had built up a distribution network that was already a sizable entry barrier. And yet the advantage of this built-up distribution network was acquired over several decades in market niches that did not at first threaten coal.

Additionally, electrification would eradicate both industries from the lighting market. Oil executives saw one stark imperative in electrification: concentrate on the development of transportation fuel or face shrinking markets. The coal complex responded differently: electrifi-

[2] Henriques 1960, p. 271.

cation worried mostly the coal gas companies. The other members of the coal complex (the mines, the chemical companies, the steel companies) saw lighting as only part of their market; politically, the "front line of defense" was left to the coal gas makers. Such markets as transportation, home heating, and supplying the raw feedstock of coal tar to the chemical industry seemed "more secure." Electrification helped the oil industry to focus its objectives; the more dispersed members of the coal complex reacted in a desultory manner to a more amorphous threat.

Consequently, as a matter of timing, it makes sense that the members of the coal complex would not challenge oil until about the 1920s. But many new oil companies were founded in the decades after the 1920s, when oil made great and irreversible strides in coal's traditional transportation markets. Banks started oil companies: Why did coal concerns not do the same? The 1920s and 1930s were the last years when the major world coal industries were robust enough to challenge the rise of the oil market. The reason for their failure in this period was political.

The World Hydrocarbon Cartel, Synthetic Fuels, and the Political Reasons for the Decline of Coal

In the 1920s and 1930s there were two strategies that permitted members of the coal complex to eye with interest the development of a liquid fuel market. The benzol market offered limited opportunities to enhance the profitability of traditional coal-processing technologies. By contrast, the new technologies of the synthetic fuels market could have revolutionized the entire coal complex, gearing it up for massive liquid fuel production and oil refining, and raising possibilities for the production of other materials such as synthetic rubber and other synthetics.

Benzol was condensed from the gases released by burning large quantities of coal, especially by steel and coal gas companies. The rich, high-octane benzol fuel was suitable for airplane engines and often was mixed with either gasoline or alcohol to lessen its volatility and make it more suitable for automobiles. In France a benzol-alcohol or benzol-gasoline mix was known as a binary *(binaire)* fuel. Sometimes all three liquid fuels were mixed in a trinary *(ternaire)* fuel that had a higher octane rating than pure gasoline.

Benzol production techniques developed during the late nineteenth and early twentieth centuries. The extraction of by-products from coal burning was the heart of the nineteenth- and early twentieth-century chemical industry. Benzol production was routine for French firms that burned large quantities of coal, chiefly the steel industry and the gaslight industry.[3] Coal burning also produced coal tar, a residue that large industrial firms either processed themselves or resold to the chemical industry as a feedstock. The historical association of steel and chemicals is explained by the large quantity of coal tar produced by blast furnaces.

When the gasoline market developed for automobiles, coal-burning firms commonly either distributed benzol directly to customers, as retailers, or made wholesale deliveries of the product to other distributors, who made binary or trinary fuels. Benzol was a true synthetic fuel, though not what we usually understand today by the term "synfuel."

Synfuel dates from the early twentieth century, when industry's ability to manipulate hydrocarbons improved dramatically. These technological advances developed in part because of the oil industry's need to produce the maximum of marketable products from crude petroleum. Increasing the gasoline yield relative to the "residual" fuel oil was a high priority as automobile engines became the dominant consumers of oil.[4]

The chemical industry, which made all manner of synthetic materials—chemicals, explosives, dyes, and fertilizers—also wanted to get the most from its coal tar feedstock. As a result, hydrocarbon processing technology advanced similarly in the oil-refining and chemical industries. The German chemist Friedrich Bergius took out the first pathbreaking patents for the liquefaction of coal—that is, turning coal into a usable petroleum product—in 1913.[5] He took out patents in a number of countries and established the International Bergin Company in Holland to handle licensing. To simplify, Bergius' high-pressure "hydrogenation" was designed to "refine" coal, as if coal were an extremely thick crude oil. The oil produced could then be refined in a manner similar to refining ordinary petroleum. Hydrogenation was so named because high pressure forced hydrogen to mix with coal derivatives without the need for expensive, slow-acting catalysts.[6] Bergius' technique, unlike benzol recovery, was a twentieth-century syn-

[3] Synthetic gas, or "town gas," dates back to the early nineteenth century, when it replaced candles for lighting textile mills with round-the-clock production schedules. By the middle of the nineteenth century most major European and American towns had coal gasworks. The phrase "natural gas" originated to distinguish it from the more widespread man-made product. Coal gasworks were in use until the 1950s. Seattle's "gas park" features brightly painted old coal gas fixtures on which children clamber about.

[4] See Enos 1962 for a history of refining.

[5] Hexner 1946, p. 315; Probstein and Hicks 1982, p. 8. Bergius received the Nobel Prize for his work in 1931.

[6] Pat.Hearings, p. 3704, "Possible Scope of Testimony."

thetic fuel process, and it offered many more industrial possibilities than benzol.

The chemical industry feedstock was coal tar; the refining industry used crude oil. The great chemical similarity between coal and oil meant that developments in oil refining had direct applications in the chemical industry and vice versa. The chemical industry, however, was an integral part of the nineteenth-century industrial base: it bought coal tars from coal gas and steel companies or else burned coal to get the feedstock. The whole industrial complex had a high degree of cross-ownership and cross-investment that had evolved when coal cartels forced steel producers to unite in monopsonies to negotiate better prices.[7] A steel firm bought shares in a coal company to get low coal prices; a chemical firm bought shares in a steel company to get good coal tar prices; the coal company bought shares in the steel company to make sure it would have a loyal large customer; and so on. In the early twentieth century, a European chemical company could not unilaterally replace coal tar feedstocks with oil without affecting the whole network of industries of which it was a part. Since the company directorates were tightly interlocked, such a move would have thrown into disarray the economic structure of the entire coal complex, the social structure of its boards of directors, and the profitability of all the members.

Using a coal-derived feedstock, the 1920s chemical industry made products that today come from oil-derived feedstocks. Figure 7 illustrates this coal-derived family of products, which were and are crucial to industrial activity. Given its capital- and science-intensive nature, the failure of the chemical industry and its industrial allies to become deeply involved in oil refining, and the oil industry, is even more puzzling than the failure of other members of the coal complex to do so. Adapting their technology for refining, the chemical companies might have spearheaded an industrial coalition to sell oil.[8]

Some heavy coal users did try to become involved in the oil industry. Royal Dutch–Shell included the Schneider industrial group and the Peyerhimoff coal syndicate in its ill-fated attempt to monopolize the French oil market. The Schneider steel operation at Creusot was

[7] Newman 1964 describes Germany; ownership patterns in France were much the same.

[8] Some major members of the chemical industry did buy oil companies in the 1980s, when U.S. Steel and DuPont bought Marathon Oil and Conoco, both small oil companies. The two parent firms vertically integrated in a way that had been common in the chemical industry of the coal era. U.S. Steel's (now USX) chemical industry ultimately became more profitable than steel; its heavy coal consumption for blast furnaces produced the by-products that were the basis for chemical manufacture.

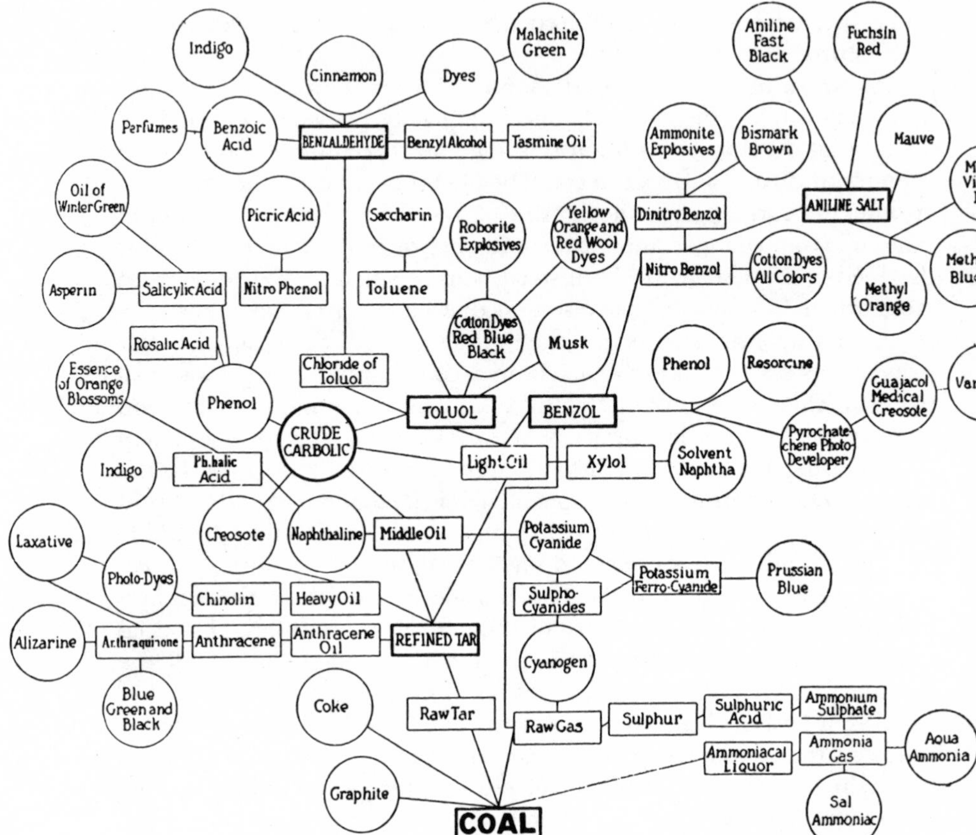

Figure 7. Coal-based chemical industry products, about 1920. *Source:* Oaks 1917.

shared with the de Wendel family; in the late 1920s members of both families sat on the board of directors of the Banque de l'Union Parisienne, which supported the Shell combine's 1919–1920 monopoly attempt but later diverted its oil efforts to become a major stockholder in Pétrofina. The head of the French branch of this company shared a director's seat with the coal syndicate's Henri de Peyerhimoff on the board of Pechelbronn. The Schneider industrial group also manufactured equipment for the chemical industry and, later, oil industry refining: under the right conditions, it could have entered the oil market.[9]

Pechelbronn was an Alsatian oil company that had exploited heavy oil deposits in the east of France for over one hundred years; during

[9] See, for example, the advertisement for refining equipment in the *Revue pétrolifère,* no. 829 (17 May 1939), facing p. 365.

the Prussian occupation of Alsace-Lorraine, German industrial interests worked the deposits. After the First World War the French government handed the oil-bearing area back to the private sector. The area provided only 0.7 percent of sales in the automobile market, but it dominated lubricating oils, where it held between 15 and 18 percent of French sales; its total contribution to France's total oil needs was about 8 to 9 percent.[10] This high-cost oil producer shared the coal industry's worries about large-scale oil imports into France, concerns that dovetailed with the Pétrofina allies' accurate fears that Russian oil imports would be the chief victim of the 1928 import quota.[11]

Pétrofina, oriented toward Russian and Romanian oil, shared Pechelbronn's protectionist objections to low-cost Mesopotamian imports. The coal industry as a group backed the high-cost oil importers and producers, for high-cost oil provided less incentive to convert from coal to oil-burning equipment. The Banque de l'Union Parisienne, as a major stockholder in Pétrofina and a participant in the Schneider group's industrial activities, favored an accommodation between the old coal-based technology and the new oil technology.

The alliance of high-cost oil and coal led to the particular tactics of the French coal complex vis-à-vis the growth of the oil market. The industrial interests with heavy coal investments wanted to adapt their outmoded production to the new era. This meant much more than simple protectionism against all oil imports. It really amounted to a whole program with which to challenge the oil industry; it was envisioned as part of an alliance with the high-cost oil importers. We can outline its phases as follows:

1. Adopt hydrogenation technology to produce chemicals.
2. Once the hydrogenation technology is in place, use it to produce some oil from coal.
3. Refine the oil into gasoline and sell it to independent French distributors.
4. Use the by-products from the previous two steps to make salable chemicals.
5. Branch into related uses of the technology; e.g., synthetic rubber.
6. Enhance refining capacity to use oil imports as well as oil from coal.
7. Produce gasoline from oil or coal depending on market conditions, level of protection, etc. Use oil as well as coal as a feedstock to make chemicals. Over the long term, phase out reliance on coal and

[10] Uhry 1927, p. 312.

[11] Pechelbronn did send survey teams to Mosul, Madagascar, Romania, Argentina, and Morocco. It never became a significant oil discoverer or producer, but through its loose association with Pétrofina and its stockholding in the Compagnie Française des Pétroles, it was affiliated with larger undertakings (Uhry 1927, p. 313).

increase reliance on oil, thus making the transition from the obsolete to the new technology.

8. Purchase own sources of oil supply; establish distribution network (or purchase a retail oil firm).

The idea, in short, was to use the chemical market and other product markets to finance expansion into the transportation market, thus solving the problem of how to manage a late entry into the oil business and also providing a transitional strategy from the coal age to the oil age. In the first, crucial period the chemical firms would not have to worry about the oil industry's monopolistic control over the world's major oil fields: their feedstock would be coal, in which a century of cross-investment had given them substantial protective strength. They could either look for new oil sources or buy from independents, some of whom were invited to testify during the 1928 oil law hearings.

The adaptive strategy hinged on the fact that chemical companies could not only market synthetic oil from coal, but also refine crude petroleum: the economics would thus not hinge on a straight barrel of crude versus equivalent barrel from coal calculation but would average the cost of the oil from coal and the crude oil, and further reduce costs on the whole through the cross-generation of profits from the sale of by-products in the chemical market. In addition, the same technology promised revolutionary possibilities in still unexplored markets, such as synthetic rubber.

The contrast of this ambition with a "pure" oil company is striking: a company like Standard Oil had no concern whatever with slowly phasing out investments in the coal industry, for it had none. Perpetuating the coal anachronisms in the burgeoning oil market could only slow the growth of outlets for its oil. And though the chemical market was "interesting," it was not "essential," as it was to the members of the coal complex. Standard Oil could and did sacrifice entry into chemical production in order to stave off the entry of the world chemical industry into oil refining and gasoline production, whether from oil, coal, or both.

This was the shared nightmare of the world oil companies: that the chemical industry would, through synthetic oil and refining, metamorphose into the oil business. The partners of the chemical companies would be the traditional members of the coal complex. Steel and coal mine owners, and their banks, would form a well-capitalized and politically almost invincible coalition that could easily manipulate market agreements such as the law of 1928 to favor themselves rather than the members of the "as is" agreements. The bitter struggle over hydrogenation patents was the last battle, largely unchronicled, of the whole nineteenth-century industrial complex against that of the twentieth century.

The political language of the struggle was an autarchic fuel program, a recurring feature of French energy debates in the 1920s. Henri de Peyerhimoff, president of the Comité des Houillères, was a major expert witness at the oil hearings led by Charles Baron in 1927 and 1928. Peyerhimoff testified that the coal industry already made a number of chemical and energy products and that it would be "normal" to go on to make petroleum and gasoline. He wanted oil from coal to command 50 percent of the French gasoline market. Liquid fuel sales would insulate the coal industry from the business cycles associated with supplying heavy industry. He foresaw employing 70,000 French workers in the treatment of "by-products," meaning chemicals. For the vast capital outlays required, Peyerhimoff explained that he had already entered into preliminary negotiations with chemical companies, oil importers, and distributors. He claimed to have reached an "agreement in principle" with the "chemical industry." Although he did not name the firms, he did speak of specific negotiations over patent rights for synthetic oil processes with Germany's I. G. Farben.[12] Competition with Mesopotamian production was a clear threat to Peyerhimoff's plans, for the domestic synfuel production he envisioned would need protection and state backing to compete against the international oil majors. In his words,

> We need to talk about centrally rationalized production. We must all recognize right away that an enterprise of this kind cannot be envisioned without the support, and to some extent the control, of the state. You can easily imagine the conversations that will have to take place, when the time comes, with the great trusts of natural oil production as well as a certain neighbor, a powerful producer of synthetic oil.[13] . . . When the day comes for this synthetic oil to play a major role in the national economy, there is no doubt that, regarding prices, there will have to be between this industry and the government either de jure or de facto contacts, which in any event are a habitual practice among our large industries. De facto contacts are, in any event, every bit as useful as de jure relationships.[14]

The coal industry became a major player backing the oil import quota of 1928, seeing a regulated market as essential for its plans. It

[12] Testimony of Peyerhimoff, "L'Enquête parlementaire," *Revue pétrolifère,* no. 259 (3 March 1928): 283, 285.

[13] An allusion to I. G. Farben.

[14] Testimony of Peyerhimoff, "L'Enquête parlementaire," *Revue pétrolifère,* no. 259 (3 March 1928): 285.

was a delicate time for the "as is" companies. They had to use regulation to cartelize the market, *and* they had to keep this regulation from being wrested from their hands by the Soviet oil faction and the coal complex. Louis Loucheur, who played an oversight role in the Office National des Combustibles Liquides, noted in his personal documents that oil market regulation was necessary so that the oil majors "would not be able to lower prices temporarily so as to kill the synthetic oil industry at its first appearance."[15]

Even if only a portion of Peyerhimoff's ambitious 50 percent goal were realized, its impact would have been significant for a whole generation of French technology. The required industry-government cooperation was envisaged relatively early: the French state signed over Bergius patent rights to the Etablissements Schneider in 1924: the plan was to construct hydrogenation plants for synthetic oil.[16] The project stalled, however. Germany's I. G. Farben enjoyed a considerable lead in the development of Bergius-based processes. The free attribution of the 1914 and 1921 Bergius patents, won as war booty from the Versailles peace treaty, was not enough for French firms to compete with the German combine.

In response to the German lead, the French firms had successfully experimented with a process for manufacturing ammonia that had applications for nonoil synthetic fuels, especially methyl alcohol (methanol). Methyl alcohol as a fuel promised to extricate the French chemical industry from competing in the synthetic oil areas where I. G. Farben dominated; but the hydrogenation advances of I. G. Farben gave the firm the crucial patents even in methanol manufacturing. Although the technical feasibility of large-scale methanol production was never in doubt, the French government never backed this alternative.[17] Loucheur testified in 1927 that methyl alcohol had "been in practical use in automobiles, and that a mixture of methyl alcohol, benzol, and ethyl alcohol had been very satisfactory without requiring any modification of carburetors or causing any harm to the engines."[18] In 1927 Loucheur even entered into discussions with both I. G. Farben and Bergius, who held the methanol patents. But the Office National des Combustibles Liquides, which Loucheur tried to push in this direction, had a board of directors so topheavy with major oil interests that even some independent oil companies were suspicious of its composi-

[15] Hoover Inst. Loucheur, box 3A, folder 13, "Monopole des Pétroles," 1926, prob. December. Also cited in Nowell 1983, p. 262.

[16] Hoover Inst. Loucheur, box 3A, folder 13, contract dated 18 March 1924.

[17] Hoover Inst. Loucheur, box 3A, folder 13, letter dated 4 January 1927, signed Daniel Berthélot.

[18] Ass.Nat. Procès Verbaux de la Commission des Pétroles, 22 June 1927.

tion. For the members of the coal and chemical industries, the Office National could serve as an arena to discuss dividing up the energy market under the quota law as passed in 1928, but it was not a place to share sensitive information about synthetic oil or alcohol technologies under development.[19] The methanol alternative ultimately became moot because after 1928 I. G. Farben cartelized this technology in cooperation with Standard Oil, along with synthetic oil, rubber, and other applications.[20]

There was also a domestic economic disadvantage: German applications of the Bergius process went untaxed, whereas the French firms were taxed; the French firms saw themselves seriously disadvantaged vis-à-vis their German rivals, quite apart from technology and patents.[21]

Loucheur's papers do not say why, but he clearly did not get along well with the coal interests represented by Schneider and Peyerhimoff. The coal group did not oppose state equity in a synthetic fuels program, but it opposed Loucheur's conditions on profits and control over patents.[22] The Compagnie Française des Pétroles could get the tax break it wanted, and it had to "endure" state participation only when its major shareholders wanted the help; the coal industry got no such favors. The government that was putting great energy into acquiring a share of Mesopotamia lacked equivalent interest in a strong synthetic fuels program; in any case, the coal complex would have seen no possibility for seriously marketing synthetic fuel products without the protection of the 1928 import quotas, which in fact galvanized its interest.

But as always, the coal complex missed a beat. The same process of world cartelization that caused the law of 1928 also meant that the synthetic fuels program, after 1928, would be under the auspices of the oil companies, which intended to prove the program unworkable. The oil companies "read the play" of the coal complex correctly and mapped out their counterstrategy at the same time that they divided up world oil markets. As a result, synthetic oil programs all over the world, except in Germany, were only technological curiosities. The derailing of coal's pursuit of synthetic oil kept the chemical companies in the chemical industry, redefining their road to modernization. The

[19] Hoover Inst., Loucheur box 3A, folder 13, "Note," n.d., prob. 1927; same file, "Liste des Membres du Conseil d'Administration de l'Office National des Combustibles Liquides."

[20] Pat.Hearings, part 8, pp. 4610–4659.

[21] Hoover Inst. Loucheur, box 3A, folder 13, ". . . visite de M. Aubrun," 2 February 1927.

[22] Hoover Inst. Loucheur, box 3A, folder 13, ". . . visite de M. Aubrun," 2 February 1927; same file, Loucheur to Bokanowski, 6 January 1927; same file, Peyerhimoff to Loucheur, 10 January 1927.

only synthetic fuel that would be produced in any quantity was benzol, which was linked not to the new technology of the twentieth century, but to the dying technology of the nineteenth.

"As Is" and the World Patent Pool

The historiography of the "as is" agreements of 1928 has been dominated by the 1952 Federal Trade Commission Report, widely considered the single most celebrated and authoritative oil source document in the twentieth century. But the report, for reasons that can be known only to its authors, left completely untouched the earlier Senate Patent Hearings that detail the elaborate precautions taken by the same "as is" interests to create an oligopolistic lock on crucial hydrocarbon technologies in world markets.[23] In addition, the Federal Trade Commission report was censored; and even when it was ostensibly declassified and printed in 1975 by the Church committee, dozens of pages remained unprinted.[24] The many analyses of the 1928 agreements that make no reference to the patent hearings are limited to only half of the operations of the world oil cartel, perhaps not even the more important half.[25]

By 1924 the convergence of the oil and chemical industries was clear to the technological leaders in each. I. G. Farben officials visited Standard Oil's refineries, and in 1925 they invited a Standard Oil representative to visit their plants in Germany. They had discovered catalysts that would bring down the hitherto prohibitive costs of the Bergius process

[23] *Hearings before the Committee on Patents,* 77th Cong., 2d sess., part 7, 31 July–4 August 1942, on S. 2303 and S. 2491. Documents subpoenaed by the committee are reproduced in the report, making it a rich and easily consulted archival source. Senator Bone explained that "historically this should be interesting to students who may seek to discover the extent of our 'invisible government' of big business" (p. 3365). Cf. Schattschneider 1935, p. 287: "Influence is the possession of those who have established their supremacy in invisible empires outside of what is ordinarily known as government."

[24] See citations FTC1, FTC2, and Pat.Hearings in "published primary sources" in the bibliography.

[25] Borkin 1978 cites this source; as does Hexner's elegant *International Cartels* (1946). Engler 1976, pp. 96–115, is an important exception among the oil histories. Vietor 1984, p. 28, cites the international patent pool in the context of the "as is" and Red Line agreements; but his conclusion that "rapid technological developments by others either circumvented the pool or forced expansion of its membership to the point of meaninglessness" repeats a common error. The political muscle with which the oil companies discouraged new technologies significantly shaped the distribution of technological abilities at the outbreak of World War II. It was not a "meaningless" international hydrocarbon agreement that provoked the Senate Patent Committee hearings, that evicted coal interests from synthetic oil refining in France, or that formed the basis for business relations between Standard Oil, Royal Dutch–Shell, and I. G. Farben in Germany, to give but three examples.

but would have fundamental applications to oil refining as well. After these visits the companies remained in "continuous contact" and later agreed to collaborate: Standard Oil took the lead on applications of Bergius processes for refining, research for which was centered in the United States, while applications for the chemical industry were carried out by I. G. Farben in Germany.[26] By 1928 the companies knew that 90 percent of the hydrogenation process could be applied to coal and oil and that only minor adaptations were needed to specialize in one or the other, or both.[27]

In the 1920s the International Bergin Company ended up under the control of I. G. Farben, with Royal Dutch–Shell as one of the more important minority stockholders.[28] I. G. Farben's control of the company brought the Bergius patents into the orbit of the ongoing collaboration between I. G. Farben and Standard Oil. This had repercussions in France: the International Bergin Corporation had a contract with "a group of coal, steel, and chemical interests through an organization called Sicol," to which one finds few references in French documents, but which was undoubtedly the Peyerhimoff-Schneider group discussed above.[29] As I. G. Farben's plans to collaborate with Standard Oil and the world oil cartel neared completion in 1927, the German firm cut off agreements for patent sharing with the French group. I. G. Farben's president, Carl Bosch, told the French "of the proposed deal with Standard and that if it went through they would have to deal with the Standard in the future."[30]

I. G. Farben's and Standard Oil's negotiations proceeded concurrently with the "as is" oil negotiations. The two sets of negotiations were inseparable: the Germans could not pursue hydrogenation without guarantees against competition in their domestic market; the international majors could not cartelize world markets if oil companies and chemical companies were poised, with the new technology, to enter the business.[31] An agreement signed by Royal Dutch–Shell, Standard Oil, and I. G. Farben specifically refers to the "as is" agreements as a criterion for determining who would be eligible for licenses of the new

[26] Pat.Hearings, "Possible Scope of Testimony," p. 3704.

[27] Pat.Hearings, "Memorandum of Meeting–July 12th, 1928," p. 3437.

[28] Hexner 1946, p. 315; Pat.Hearings, "International Corporation," 11 August 1927, p. 3652. On p. 3653 is a diagram of company shareholdings, which were held "20% by Shell, 30% I.G., 50% Makot. Makot stock is held 40% I.G., 40% Bergius, 20% English group [Imperial Chemical Industries]."

[29] Pat.Hearings, "International Corporation," 11 August 1927, p. 3652.

[30] Pat.Hearings, "Memorandum of Meeting, March 21st, 1929," pp. 3442–3443.

[31] See letter on I. G. Farben issues from Henry Deterding to Walter Teagle (Pat.Hearings, Deterding to Teagle, 18 October 1929, pp. 3659–3660).

process.[32] The "as is" and hydrogenation agreements competed for time on the same agenda[33]; referring to cartelizing hydrogenation technology in September 1928, Jean Baptist August Kessler of Royal Dutch–Shell wrote: "I feel sure it is the only way to avoid nonoil interests to spoil all the good work that has been done lately."[34]

The Standard–I. G. Farben interests formed a joint holding company called SIG, after the initials of the two parent companies. SIG formed the Hydro Patents Corporation, which in the United States licensed all processes jointly held by the two parent companies. Standard Oil was responsible for licensing all American oil companies, which took shares in Hydro Patents in proportion to the size of their crude oil refining. Payments for stock and licenses were made to the holding company, SIG, and of these American payments Standard Oil pocketed 80 percent. If subscribing companies did not pay the fees to join Hydro Patents at the beginning, they ended up paying even more for patent royalties as nonmember licensees; hydrogenation was so crucial to the future of refining that eighteen major American oil companies succumbed to the industry shakedown.[35]

There were restrictions of great importance to members of the coal complex. The patent licenses applied to coal hydrogenation as well as to hydrogenation required for oil refining; yet no coal companies, steel companies, or chemical companies were invited into the Hydro Patents Corporation, which excluded them from the technology that would have helped them enter the oil business. Member oil companies could not apply their "refining" patents to production for the chemical industry: they too were channeled into the path desired by I. G. Farben and Standard Oil.

The Hydro Patents Corporation took even more unprecedented steps. Refineries erected by licensees were subject to complete review by a separate Standard Oil–I. G. Farben subsidiary, the Hydro Engineering and Chemical Company. For the privilege of this inspection, licensees paid 4 percent of the total capital cost of the plant under construction. Licensees were furthermore required to cross-register all their patents on their own refining discoveries with Hydro Engineering, giving Standard Oil and I. G. Farben a free ride on other

[32] Pat.Hearings, "Hydrogenation Agreement between International Company and Bataafsche Petroleum Maatschappij [Royal Dutch–Shell]," p. 3544.

[33] Cf. question in 1930 by a Standard Oil official "whether we should open the I.G. negotiations with the Shell before or after the discussion on the 'as is' agreement" (Pat. Hearings, Frank Howard to E. M. Clark, 19 March 1930, p. 3669).

[34] Pat.Hearings, Kessler to von Riedemann, 6 September 1928, p. 3662. This Kessler is not to be confused with his father, who ran Royal Dutch in the nineteenth century.

[35] Pat.Hearings, pp. 3341–3342.

companies' technological advances. If a licensee discovered its own catalytic process, it could not take out a patent, but had to turn it over to Hydro Engineering or else use the process without a patent.[36]

The American pattern was repeated worldwide in the International Hydro Patents Corporation, whose International Hydro Engineering subsidiary inspected refineries just as the American Hydro Engineering had done. Royal Dutch–Shell was excluded from exercising patent privileges in the American market, but it became a full partner with Standard Oil in international markets and exercised the licensing privileges enjoyed by Standard Oil for most non-American oil companies around the world. Standard Oil and I. G. Farben included Royal Dutch–Shell because of its great size and because it had independently done significant work in chemicals, leading one legal investigator to call it "an incident of cartelized commercial truce."[37]

The cartel dealt out special privileges to some members. Britain's Imperial Chemical Industries was permitted to make synthetic fuel, on condition that its oil be marketed by Standard Oil and Royal Dutch–Shell; Imperial Chemical Industries earned this privilege from its early foresight in buying a share of the International Bergin Corporation, whose functions were entirely absorbed by the International Hydro Patents Corporation.[38] Anglo-Persian (after 1932 Anglo-Iranian) got special treatment in the world patent pool because of its alliance with Royal Dutch–Shell after their 1927 joint oil marketing agreements. Wherever Anglo-Iranian had "as is" marketing agreements with Royal Dutch–Shell and Standard Oil, it could participate in the International Hydro Patents Corporation.[39] This agreement thwarted the coal industry's support for synthetic oil from coal.[40]

Standard Oil, Royal Dutch–Shell, and I. G. Farben agreed that only they would have the right to use hydrogenation on coal, unless a licensee agreed to market its products through them.[41] This proviso took the crucial technology out of the hands of the world's chemical companies. Though excluded oil and chemical companies tried to develop alternative technologies, the accords remained effective until unilaterally dissolved by United States consent decree in 1942. The excluded oil and chemical companies were naturally the primary constituents backing the patent hearings and the consent decree.

[36] Pat.Hearings, p. 3343.

[37] Pat.Hearings, testimony of Patrick Gibson, pp. 3347–3350.

[38] Pat.Hearings, testimony of Patrick Gibson, p. 3350.

[39] Pat.Hearings, "Outline of Proposal for Standard-Shell Agreement on Hydrogenation," 15 April 1930, p. 3666.

[40] See *Report of the Royal Commission on the Coal Industry*, 1926.

[41] Pat.Hearings, "Outline of Proposal for Standard-Shell Agreement on Hydrogenation," 15 April 1930, p. 3666.

The world's oil companies were kept out of the chemical industry in the same way. I. G. Farben licensed the chemical applications of the patents, whose conditions were as onerous as those imposed on the oil industry. The chemical industry could not use the patents to refine petroleum. Standard Oil agreed not to go into the chemical industry except as a partner of I. G. Farben, and I. G. Farben agreed not to go into the oil business except as a partner of Standard Oil. In Germany the synthetic fuel production of I. G. Farben was marketed through the normal retail outlets of Royal Dutch–Shell and Standard Oil.[42]

The "as is" agreements for oil and the world patent pool created an intimate basis for hydrocarbon control. Dr. Carl Bosch, president of I. G. Farben, observed that "the two companies were *married* for a long period and that each must respect the other's interest; that . . . the I. G. interest and the S.O. interest were the same, and that whatever Standard Oil did would react to the interest of the I.G. as well."[43]

The world hydrocarbon cartel members pledged themselves to be above ordinary national considerations in their work. As an American investigator summarized it: "In the language of high diplomacy and with the suggestion of a reliance upon a private law above the law, this agreement embodies mutual assurance by the parties to each other that they would voluntarily readjust their contracts to carry out their purposes in the event any obstacles were presented by existing or future law, act of governmental authority, or circumstance. This is only the barest effort to state the gist of the agreement, whose actual terms are next to impossible either to summarize or to characterize. That is best left to the text itself."[44]

If the world oil market division accords of 1928 achieve grandeur because of their scale and world vision, they nonetheless remain extensions of cartelistic business practices developed at the national level. The parallel world patent agreements go beyond mere business practicality; they reach for a transnational ideology of world business. Dr. Bosch's reference to the "marriage" of I. G. Farben and Standard Oil was but one example. The hydrocarbon cartel's royalty payments and enforcement provisions continued well after the outbreak of hostilities in World War II. Anglo-Iranian, as a participant in the world patent pool, protested the necessity of making royalty payments to Britain's Nazi enemy. An officer of Standard Oil dismissed the puny nationalist objections; he declared that "technology has to carry on—war or no

[42] Pat.Hearings, "Memorandum of Meeting, Friday Afternoon, 2 P.M., Aug. 31st, 1928," p. 3435; and testimony of Patrick Gibson, p. 3336.

[43] Pat.Hearings, testimony of Patrick Gibson, p. 3326. Italics added.

[44] Pat.Hearings, testimony of Patrick Gibson, p. 3336.

war—so we must find some solution to these last problems." The "problem" was that the "interested parties included Americans, British, Dutch, Germans, and war introduced quite a number of complications. How we are going to make these belligerent parties lie down in the same bed isn't quite clear as yet."[45] The reference to a "bed" is an interesting reification of Dr. Bosch's "marriage," but it shows how these partners envisioned a long-term relationship.

That these agreements could function two years after the outbreak of World War II astonished Senator Homer T. Bone of the investigating committee: "It presents an astounding picture." His witness replied, "I confess that I am at a loss to comment on the international implications of the arrangement."[46] Senator Bone colorfully derided the implicated companies as "cartel addicts."[47] His committee had investigated a case of transnational sovereignty, where power was measured not in military strength among states, but by the arcane criterion of who controlled the technology that the nations of the world and their armies would employ.

Some opined that the hydrocarbon cartel benefited the Nazi regime and even was controlled by it, because by the end of the 1930s patent accords in which German companies participated had to be approved by the Nazi government.[48] But in Germany the government's coordination and cooperation with the cartels predated Nazism and cannot be considered a unique feature of it.[49] By the late 1930s, I. G. Farben officials held posts in the Nazi government; distinguishing "corporate" from "national" purposes is difficult. This book cannot look in detail at the Nazi regime. The world hydrocarbon cartel originated well before the Nazis, however, at least as early as 1924; the collaboration of I. G. Farben and Standard Oil had solidified by 1928. The world patent pool was also part of cartel policies governing world oil markets, of which I. G. Farben was not a member. The ensemble cannot be explained as a function of Nazi strategic interests, which were but one part of an international system whose members, measured on a power-based state-competitive axis, were antagonistic but who were allied in international hydrocarbons. This is a feature of transnational structuring, in which states may "think" their interests are being served while business interests pursue orthogonal goals that may or may not "suit" the needs of states.

[45] Pat.Hearings, Patrick Gibson citing a memorandum dated 14 November 1939, p. 3410.

[46] Pat.Hearings, testimony of Patrick Gibson, p. 3414.

[47] Pat.Hearings, testimony of Patrick Gibson, p. 3411.

[48] Pat.Hearings, testimony of Patrick Gibson, pp. 3358, 3363.

[49] See Neumann 1963, p. 237.

Competition among the international corporations occurred in the recondite medium of patents, market shares, and distribution agreements. These agreements dealt out the technological cards with which nations played their particular power games and structured capabilities in the "politics among nations" that went unperceived until wartime crisis. By that time the assertion of state sovereignty over the world hydrocarbon cartel could alter the distribution of capabilities only with diffficulty, delays, and the commitment of extra resources. The only way to stop the German synthetic fuels program, which the agreements had fostered, was to bomb it. At war's outbreak American capabilities in some chemical manufacturing, such as synthetic rubber, were significantly retarded and had no installed industrial base—a result of the agreements. France and Japan were similarly left with purely symbolic synthetic fuel programs. Hydrogenation patents were used as bait to try to get the Japanese to change their "petroleum control law." Although some documents reveal I. G. Farben's velleity to help Japan get hydrogenation, a British survey after the war, based on interrogations of I. G. Farben officials who worked with the Japanese, indicates that the collaboration between the Japanese and Germans on hydrogenation was marked by interservice jurisdictional rivalry in Japan and by corporate rivalry in Germany.[50] I. G. Farben kept its information transfers to the Japanese to a bare minimum, exactly as it should have done under the terms of the world hydrocarbon cartel.[51]

The world hydrocarbon cartel impeded synthetic fuel development in France and elsewhere as part of its worldwide strategy. The strategy had two phases. In the first, which lasted until about 1936, the lock on technology was so tight that very little progress was made anywhere. In the second phase, the development of a rival technological process forced the world patent pool members into damage-control operations that led them to the business equivalents of "backfire" operations. In forest-fire control small, manageable fires are used to deplete available wood and prevent larger conflagrations; in the world patent pool, limited synthetic fuel programs were encouraged to tie up capital, encourage international respect of patent rights, and exert enough control over the synthetic fuel process to guarantee it would be an unprofitable alternative to oil.

A German steel firm with large coal consumption and commercial ties to Ruhrchemie developed the new synthetic fuel process that forced the transition to the second phase in 1936. The Fischer-Tropsch process was less efficient than the I. G. Farben–Bergius process, requir-

[50] On Japan, see Pat.Hearings, pp. 3349, 3354, 3680–3681, 3691, 3697, 3711–3715.
[51] Peck and Jones 1945.

ing seven tons of coal to produce one ton of oil as opposed to four to five tons of coal for the rival Bergius process. But Fischer-Tropsch was easily adaptable to coal gas production. Developed in 1934, by 1936 the process's advances led the controlling members of the International Hydro Patents group to conclude that they had lost their "complete control" over synthetic fuels.[52]

Negotiations began almost immediately between the International Hydro Patents members and the Ruhrchemie group. By October 1938 I. G. Farben and Ruhrchemie had a new accord that subjected the Fischer-Tropsch process to the same rigid covenants that governed the Bergius patents.[53] The new principal members were Ruhrchemie, I. G. Farben, Standard Oil, Royal Dutch–Shell, and M. W. Kellogg, an American company.[54] A diagram of the international patent pool as it stood after the new agreements of 1938 is shown in figure 8.

What of government and other industry pressure to break up the pool? "Outsider" firms around the world pressured their governments to break the iron lock on the international patent pool. Patent pool members had foreseen this; as early as 1926 I. G. Farben's Carl Bosch observed, "The fact that patents can be expropriated for national defense must be kept in mind, and surely such action would be taken if nationals were ignored in the formation of companies."[55] Yet no government expropriated the patents until well after the outbreak of World War II. Dr. Bosch's strategy of recruiting "nationals" to form a local constituency to respect the patents worked throughout the 1930s.

France furnishes an example of Bosch's strategy in action. The Schneider steel and chemical group wanted a major French synthetic fuels program. Intimidated by the formidable array of Standard Oil, Royal Dutch–Shell, and I. G. Farben, it did not want to fund the project alone. This reluctance was logical in a national market where cheap oil was expected from Mesopotamia, and in an international environment where key patent processes were tied up beyond the reach of ordinary firms. The synthetic oil program needed firm government support for patents and for setting acceptable prices for eventual production. The Schneider group backed the government of Pierre-Etienne Flandin,

[52] Pat.Hearings, Teagle to Howard, 12 February 1936, with accompanying "Memorandum on New Synthetic Oil Processes," pp. 3698–3701.

[53] Pat.Hearings, "Hydrocarbon Synthesis Agreements," 7 October 1938, pp. 3739–3743.

[54] The Kellogg Company developed processes that could break the international patent pool and then sold them to the pool. It became the specialized research and development corporation for the patent pool companies, helping them maintain their technological lead in hydrocarbons.

[55] Pat.Hearings, "Memorandum of Discussion with I. G. at Plaza Hotel, November 6th, 1926," p. 3654.

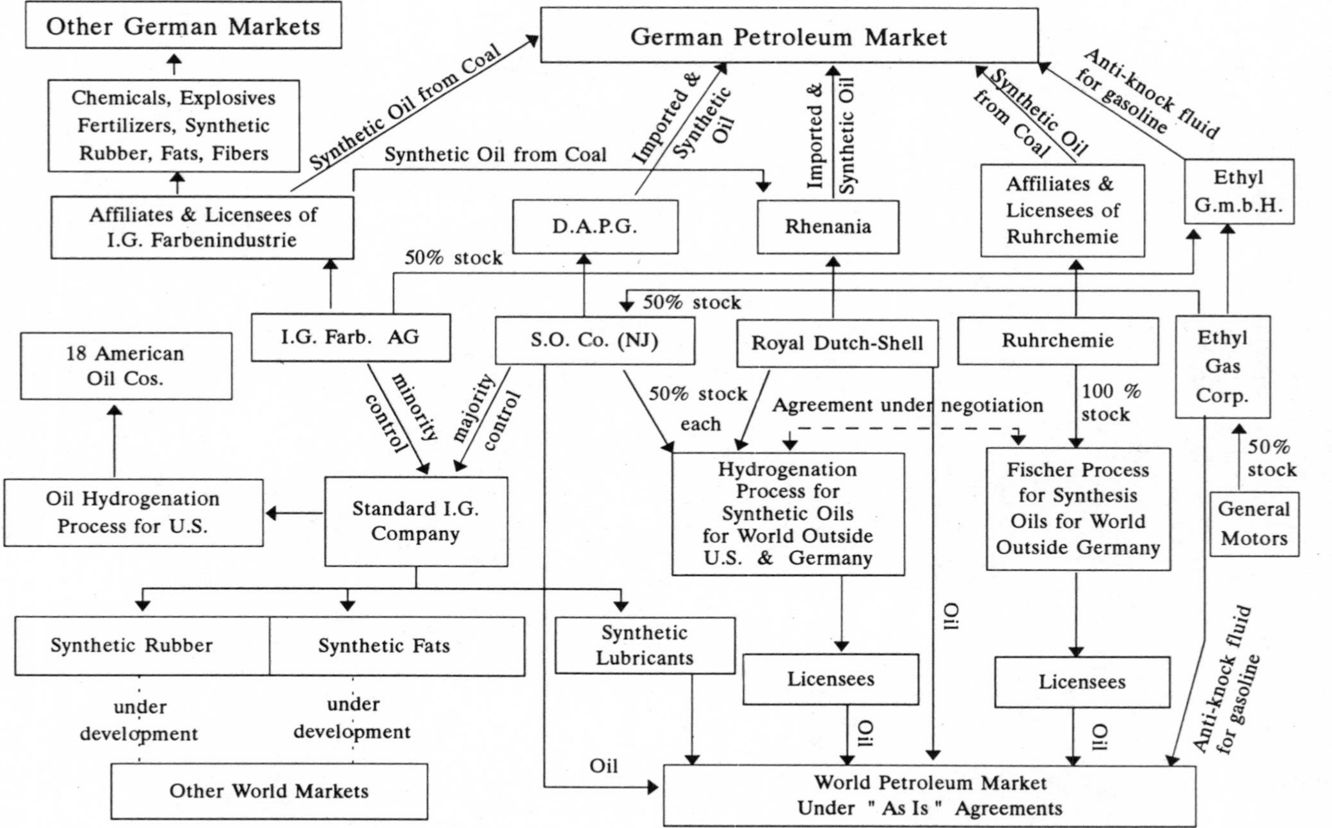

Figure 8. International patent pool, 1938. *Source:* U.S. Government Printing Office, *Investigations of the National Defense Program*, Senate, 77th Cong., 1st sess., March and April 1942, part 11, p. 4724.

and in Standard Oil's assessment it nearly succeeded in getting a synthetic fuel program adopted:

> The situation in France is quite serious. Prior to the fall of the Flandin Government, the prospects for the success of the Schneider project seemed bright. With the fall of that government, the uncertain political situation in France, *and the activities of Pineau,* there seems to be at present very little hope that Schneider will receive the financial support which they require in order to go ahead with this project. They have told us in no uncertain terms that they are not prepared to put up the money themselves. The project, if undertaken, must be subsidized in some way by the French Government. In this connection, I enclose copy of the Minutes of a meeting held on July 6th at the O.N.C.L, which Mr. Boris-Finaly sent me. . . . I would appreciate your impressing upon our people that these communications from him are highly confidential.[56]

This shows how significant an alliance with "nationals" was. Boris-Finaly was one of Standard Oil's connections with the Banque de Paris et des Pays Bas, and it provided the American company with confidential government papers on oil policy. The head of the agency ostensibly in charge of helping France to become energy independent, Louis Pineau of the Office National des Combustibles Liquides, was opposed to sponsoring a synthetic fuel program of significant scope. Pineau's motives are not revealed, but deliveries of Mesopotamian oil to France had begun in late 1934, and he had more than enough problems allocating the French market among existing concerns without having to factor in additional production from synthetic fuel sources. The Saint Gobain case, already discussed, was but one example of the tight competition for market share among existing and aspiring participants in the oil business. Pineau had an oil constituency to satisfy, and his interest coincided with the interest of the international oil companies. The International Hydro Patents group thus used Standard Oil's connections with the Banque de Paris et des Pays Bas and the Office National to keep France from developing the potent new technology for commercial purposes.

France's "etatist" energy czar is revealed by the document to be a conspicuous hypocrite. In 1934 Pineau had reviewed his country's—and his own—oil policy, citing the need for independent French policies, especially in synthetic oil. "During the course of some conflict, submarine warfare or a possible refusal of some countries to supply

[56] Pat.Hearings, letter from W. R. Carlisle to Frank Howard, dated 19 July 1935, pp. 3686–3687. The Flandin government fell in May 1935. ONCL is the acronym of the Office National des Combustibles Liquides. Italics added.

oil could dry up our imports, and the supply of the army would be compromised. *All efforts must therefore be made to bring about, in the best industrial and technical conditions, but in the briefest time,* the manufacture of quantities, which shall be determined, of fuel that will not depend for its raw material on foreign sources."[57] This came from the man who was credited by the Standard Oil Company with having helped kill an ambitious synthetic oil program sponsored by the coal complex in France.

Still, the impulse to break up the patent pool did not disappear. In France a cosmetic was needed to silence the opposition of powerful groups like the Schneider industries. Especially in the three years before the Fischer-Tropsch process came under the same controls that governed the Bergius process, opponents of the patent pool tried to open a wedge in the controls of the world hydrocarbon cartel. The Etablissements Schneider's efforts to launch an independent synthetic fuels program coincided with the moment of opportunity fleetingly presented by the Fischer-Tropsch process, before it was joined to the world hydrocarbon cartel. Efforts such as these, in France and around the world, led the International Hydro Patents group into its second phase. The strategy in the passage below, suggested by Standard Oil's chief officer in charge of hydrogenation, refers specifically to Japan and Italy but also applied to France:

> Although I did not express any conclusion to Mr. von Riedemann,[58] my own idea is that our interest would probably best be served by having the Italians use our hydrogenation process and thus become a minimum factor in the oil business; that is, their crude requirement would be least, their cost of gasoline highest and most artificial, and they would not be producers of any fuel oil. I have all along had the feeling, which I have expressed several times in connection with the Japanese situation, that in some respects it is highly desirable to have foreign governments interested in the oil business go the hydrogenation route. This ties up the most capital and tends to put them all together on a noncommercial plane in which their expansion is bound to be limited and in which they are not capable of becoming a disturbing factor outside of their own bailiwick, one of the most important factors always to be considered in connection with our policy toward governments that are anxious to stick a finger in the petroleum pie.[59]

The patent pool members licensed the hydrogenation process for

[57] Ass.Nat. Commission des Pétroles, *Rapport Général de M. Louis Pineau,* 2 July 1934, pp. 92–93. Italics underlined in original.

[58] Riedemann headed Standard Oil's operations in Germany.

[59] Pat.Hearings, Frank Howard to W. C. Teagle, 9 April 1936, p. 3698.

synthetic fuels to governments because they limited the license to permit only the manufacture of a few fuels. Without those conditions, an independent oil refinery or chemical company could use the hydrogenation technology for a much broader range of applications. Had the government made synthetic fuel and then sold the by-products of that fuel as commercial chemicals, the operation as an ensemble could have broken even or perhaps turned a profit. The sales from the chemical industry would have subsidized the fuel prices.

The world hydrocarbon cartel licensed patents to governments, but the licenses excluded the right to sell hydrogenation by-products along with the fuels. With access to the lucrative chemical end of the hydrogenation product spectrum denied, corporations and governments would deem hydrogenation unprofitable and would limit their production of synthetic fuel. Synthetic fuels under these constraints became by definition the "noncommercial plane," and state and private corporations found the process too expensive to expand production. In this manner the International Hydro Patents group limited governments' ambitions to "stick a finger in the petroleum pie."

These costly and ineffective development programs stifled outsiders' attempts to break up the patent pool. So long as the hydrogenation plants were owned by the patent pool members, whatever synthetic production did come on stream was sold to the advantage of the patent pool members rather than potential competitors. The objective of the second phase was thus to stifle, contain, and maintain control. In the words of one of Standard Oil's top European officers, Heinrich Riedemann:

> You realize the real position as being that if we should decline to make any further investment in France, which I think we all agreed is in itself the soundest thing to do, we run the risk that with the French Government having reached the decision of going ahead with the hydrogenation of the coal problem the coal people may succeed in persuading the government to build a really large plant, in which event our outlet to France would be cut down in proportion to the new indigenous production. In other words, if we can keep the necessary investment down to a comparatively small figure, we should have, on the one hand, the advantage of an increased outlet, while, on the other, if we should refuse to go ahead of these lines, we run the risk of a loss in outlet the amount of which cannot be foretold.[60]

A German industrialist working for an American company set the

[60] Pat.Hearings, Riedemann to Teagle, 18 March 1936, p. 3695. Riedemann was a native German speaker and his syntax in this passage is odd, but the meaning is clear.

boundary to French strategic capabilities in oil. France did establish two small pilot plants for the manufacture of synthetic fuels, but the corporate sponsorship of this effort was limited entirely to oil companies, and among these the leading members of the international patent pool, Standard Oil and Royal Dutch–Shell, were preeminent. The synthetic fuel company, called the Comptoir des Essences Synthétiques, was authorized by government decree on 10 July 1936.[61] Manufacturing only gasoline in accordance with the goals of the world patent pool members, the company built two factories, one in Béthune and the other at Liévin, and from 1937 to 1939 produced a combined output of about 25,000 tons a year, a miniscule fraction (below 1 percent) of total French needs.[62] The coal interests that would have pushed the French program to be as efficacious as its German counterpart (up to 28 percent of wartime oil consumption), such as the Schneider group and the coal companies under Peyerhimoff, were conspicuous by their absence.[63] The French synthetic fuel program that ought to have been but never was, like Sherlock Holmes' dog that failed to bark in the night, is thus of especial interest precisely because it was insignificant: the strategic advantages of autarchic production and the political backing of the large coal and steel lobby *ought* to have produced a program roughly comparable to Germany's; but instead the tightly coordinated action of the international oil companies choked off the endeavor.

The members of the diminutive French synthetic fuels program are listed in table 7. Effective control was exercised by companies directly or indirectly associated with the world patent pool. The second largest bloc of companies were significant shareholders in the Compagnie Française des Pétroles; their interest in coal hydrogenation was tempered by their access to Mesopotamian crude. The third largest block of stock, held by independent chemical and refining concerns, did not have control of the company; an additional 135 miscellaneous members had individual shareholdings too small to act as a unified voting bloc, if they voted at all.

The French synthetic fuel program demonstrated the political strat-

[61] Arch.Nat. Desmarais Frères 130AQ82, file 1, "Comptoir des Essences Synthétiques—Liste des actionnaires au 15 juin 1938."

[62] There were board-level discussions at the Compagnie Française des Pétroles about synthetic fuel production, but no significant program had been established by the time E. Zédet took over fuel policy after the collapse of the Third Republic. See CFP 11SGA15, Comité de Direction, 1937–1941.

[63] There is a brief record of a "Comité Française des Essences Synthétiques" (Arch.Nat. 130AQ83, folder 2, n.d.), whose membership was all coal personalities, and some French national companies, such as Desmarais Frères. I found no records of development activity by this group; the absence of representation by the international majors doomed it to technological irrelevance.

Table 7. Stockholders of the Comptoir des Essences Synthétiques

Company	Number of shares	Percentage of stock
World patent pool members and allied companies		
Sté Générale des Huiles des Pétroles (Anglo-Iranian)	199	6.8
Sté des Pétroles Jupiter (R.D.S.)	304	10.3
Standard Française des Pétroles (S.O.)	434	14.7
Steaua Française	21	0.7
Sté Française des Carburants	43	1.5
TOTAL	1,001	34
French oil firms, members of Compagnie Française des Pétroles		
Desmarais Frères	280	9.5
Sté Fse des Pétroles, Essences, et Naphtes	82	2.8
Cie Industrielle des Pétroles	157	5.3
Lille Bonnière et Colombes	123	4.2
Raffinerie du Pétrole du Nord (Pétrofina)	81	2.7
TOTAL	723	24.5
"Independent" chemical/refining		
Raffineries de la Gironde	135	4.6
Cie des Produits Chimiques et Raff. de Berre (St. Gobain)	134	4.5
TOTAL	269	9.1
Passive shareholders		
135 other small retail and distribution companies averaging seven shares each	955	32.4

Source: Arch.Nat., Desmarais Frères, 130AQ82, folder 1, "Comptoir des Essences Synthétiques," 15 June 1938.

egy of the world patent pool: the French government had a symbolic right to use the patents, but in practice their use and functional installed capacity remained insignificant. The limited amounts of synthetic gasoline were marketed through the retail facilities of Standard Oil and Royal Dutch–Shell.[64]

The world patent pool and "as is" agreements for the control of world oil distribution were such an integral part of the world industry, during such a critical developmental period, that it is difficult to summarize what might have emerged in their absence. The agreements left their imprint on the world hydrocarbon industry long after the formal contracts and cartel accords had been dissolved or rendered irrelevant. In France coal was limited to competing in its traditional

[64] Pat.Hearings, Teagle to Howard, 26 February 1936, p. 3731.

markets. That is to say, coal producers had their traditional status as suppliers of a solid fuel for transportation, industry, and heating. They could not modernize, as Peyerhimoff wanted, through the development of synthetic fuels. As a result of the world hydrocarbon cartel, this avenue was closed off. The coal industry was left, for the liquid fuel market, with its traditional product, benzol.

Benzol marketing developed rapidly in the 1930s. Some coal gas companies found benzol essential to maintaining profits during the Great Depression. Benzol also gave them a new product to market in the face of rising competition from electrification. The very name of the Société d'Eclairage, de Chauffage, et de la Force Motrice,[65] shows the only viable commercial strategy left for coal, once oil from coal had been eliminated. Company records state: "Our equipment for refining coal tars and removing benzol from gas continues to function well. Thanks to it we have made important profits . . . the balance 1932 shows a profit equivalent to the previous fiscal year." In the following year, "refining of coal tars and benzol recovery have maintained our company in a healthy position from an accounting point of view."[66]

Since benzol was often mixed both with alcohol and with gasoline, benzol producers became active importers of gasoline. With synthetic fuels tied up by the world hydrocarbon cartel, this became the avenue for the coal gas industry to increase its position in the liquid fuel market, but it tied the companies to a recovery technology that would disappear once electrification swept away the principal market for coal gas. Still, coal gas companies' requests for licenses to import gasoline to mix with benzol became a durable feature of oil import policies in the 1930s. The proliferation of benzol-related gasoline imports for a time caused the "as is" companies' French market share to decline in the middle 1930s, although their sales in absolute terms increased.[67] The coal gas industry also allied itself with the agricultural lobby, which in 1923, in a display of powerful lobbying influence, had passed a requirement that all companies importing oil in France buy a percentage of alcohol to mix with their fuel. The coal-agriculture lobby's power renamed the Office National des Combustibles Liquides (National Bureau of Combustible Liquids), from the proposed Direction des Essences et Pétroles (Directorate of Gasolines and Oils).[68] The alcohol legislation was a pure income transfer to the agricultural sector, though it was touted as a strategic policy.[69]

[65] Company for Lighting, Heating, and Motor Fuel.
[66] Arch.Nat. 136AQ40, entries for 30 May 1933 and 28 June 1934.
[67] FTC1, p. 324.
[68] Thomas 1934, p. 68.
[69] See Rooy 1925.

The gasohol law remained in effect until 1957. There were also, as mentioned above, some efforts to join agricultural ethanol production to the production of methyl alcohol (methanol).[70] The world hydrocarbon cartel began to develop technical controls over the dispersion of methyl alcohol production patents, however, subsuming this product into its practices on synthetic oil from coal.[71]

When the French coal industry was nationalized after World War II, the high degree of cross-ownership among coal mines and other industrial sectors presented the legal problem of cross-nationalization. When it took over various mines, the government also nationalized by extension the stocks those mines owned in other enterprises. The government thus gained indirect participation in companies that had not been the object of the original nationalization. A special commission determined which cross-nationalized holdings would remain under government control and which would be returned to the private sector.[72]

The commission's work documented an aspect of the coal companies' activities that might otherwise have passed into oblivion: there were coal-owned companies for research on synthetic fuels, which had not had government support. Coal interests had also tried to develop their own conventional oil production in sites in Syria. The fundamental obstacle to these efforts was the monopolization of world oil resources by the members of the "as is" agreement, conjugated with the international patent pool. With hydrogenation and significant oil exploration efforts kept out of reach, the coal industry had only pathetic alternatives: marginal production areas in Syria, benzol marketing, and a failed alliance with the agricultural producers of distilled alcohol fuel. France had the industrial impetus and know-how to enter the modern age of hydrocarbons without the international oil majors; but through patent controls these companies forced French hydrocarbon development down the road they offered, and no other.

The Marginally Independent National Strategy

The independence of France and other countries in the interwar period was compromised by the tight control of world hydrocarbon resources. Still, France marginally *did* enhance its strategic posture.

[70] See Nowell 1985, pp. 110–131.

[71] Pat.Hearings, part 8, 24 May 1932, pp. 4622–4623.

[72] Fin. B9808, "Procès verbaux et comptes rendus de la commission interministérielle de l'article 17 de la loi du 17 mai 1946."

Because transnational structuring works orthogonally and not in opposition to state goals, not all of France's policies responded *solely* to the needs and compromises of the principal contending business groups.

Let us assume that France could not reasonably have resisted the oligopolization of world hydrocarbon resources by companies belonging to other countries. What might have improved the country's strategic posture? Realist theory cannot explain the specifics and timing of French hydrocarbon regulation, but the French state's actions may have had some overall benefit from a realist-mercantilist point of view. There were four arenas of oil policy in which the French might have enhanced their strategic posture without intruding, at least in the near term, upon the overarching objectives of the world hydrocarbon cartel. These were the national reserve requirement; oil exploration independent of the international majors; the development of refining; and transportation.

The National Reserve Requirement

The failure to develop a commercial synthetic fuels program meant reliance on overseas fuel supplies. Mesopotamia was one such source, but French control was sharply limited. France had only one vote out of four in the Iraq Petroleum Company; in wartime the sphere of influence was British. France remained as dependent on overseas supplies as it had been before and during World War I. The notion that the Compagnie Française des Pétroles had a particular *strategic* significance is exaggerated. The company had access only to whatever the British permitted it. The political elites who shaped France's energy policy knew this all too well. It may have been in France's national interest, as a matter of revenue, to have a French company earning money in foreign oil, but the debates over the Compagnie Française des Pétroles' equity structure and oil import quotas had little to do with the national defense. During the ratification debate in 1931 Léon Blum, the socialist leader who led the opposition to the government's proposal for the Compagnie Française des Pétroles, attacked the argument directly: "Let me address the argument . . . about the national defense. It is not a nice subject to have to think about, but let me tell you that any speculation on this subject is just nonsense that is bound to confuse. For if you are ever in a war situation, from the point of view of your petroleum supplies, there is only one question to ask and none other: that is the question of knowing if you will have on your side or not the alliance and the active sympathy of England and America. That is the only question there is."[73]

Louis Loucheur, who had helped in the negotiations of the original bylaws for the Compagnie Française des Pétroles in 1924, who was one of the 1928 oil import law's principal authors, and who was for a time board chairman of the Office National des Combustibles Liquides (and the nominal boss of its director, Louis Pineau), answered Blum's criticism the next day: "In 1924 . . . having discussed the question with M. Poincaré, I was preoccupied, like him, not with wartime supplies—because, on this point, I am as skeptical as you, M. Blum—but in normal times, in time of peace, to reestablish the balance between offers from the Soviets and from the trusts, and it was thus that I thought the Compagnie Française des Pétroles should have as a goal the procurement of crude oil."[74] Loucheur's response confirms the readings of archival sources in this chapter and the previous ones. He has never been quoted in any of the previous English or French monographs on French oil, even though on the whole these have tended to portray French oil legislation as oriented toward furthering the "national defense" and the "national interest." Commercial pressures and competition meant more than any supposed strategic crises or other considerations. His explanation deserves a closer look.

Loucheur's reference to Soviet oil is an opaque allusion to the complex issues regarding Soviet oil. It is not clear what he means by reestablishing a "balance" of Soviet oil in 1924; he probably meant restoring the levels reached before World War I. Loucheur is referring to Mercier's failed efforts to get the Compagnie Française des Pétroles to import Soviet oil, efforts that were bypassed by the Office National des Combustibles Liquides' and Pétrofina's separate arrangements. In other words, Loucheur hoped the Compagnie Française would unite the Soviet and Mesopotamian oil factions; in spite of his best efforts the two split, the Soviet faction pursuing its goals through the Office National, the Mesopotamian faction dominating the Compagnie Française. The second part of Loucheur's statement is precisely the point of this book. Poincaré and he were worried about "normal arrangements," not wartime supplies. The major oil "crisis" France faced was not the ephemeral wartime crisis of the last two weeks of 1917, which indeed, had it been serious, would have cast the issue of oil regulation in a strategic light, but rather the protracted battle for control of the French market that caused Standard Oil to embargo deliveries from November 1919 to April 1920.

[73] Déb.Ch., 20 March 1931, p. 2107.
[74] Déb.Ch., 21 March 1931, p. 2155.

Political elites concluded that reconciling powerful conflicting commercial interests in peacetime was a higher priority than doing so in wartime. As Blum suggested and as Loucheur agreed (claiming Poincaré's concurrence), the wartime oil import game was lost without the help of the world's two major maritime, oil-producing powers. Peacetime relations among companies, on the other hand, were complex. The size and diversity of the industrial and financial groups interested in state policy put the state under extreme pressure to find a modus vivendi among groups whose antagonisms were fundamental. The Compagnie Française des Pétroles, far from being a national champion, was stripped down to the bare function of procuring crude oil, which it did only after the major trusts had taken large shareholdings, and then primarily from a source whose output was controlled by those trusts—the Iraq Petroleum Company.

Loucheur's remarks are a confession of state weakness from a realist-mercantilist point of view. The French state had no articulated oil policy during World War I, and in 1931 one of the leading participants in postwar oil policy averred that he sought to ensure oil procurement in peace only. There was no strategic oil policy during the war and no strategic oil policy afterwards, period. There were various "national goods" that may have been pursued, but once the dominant rhetorical element of national defense policy is removed, as by Loucheur in the passage above, the interpretation of French oil policy from a statist point of view becomes decidedly nonsensical.

Without the "self-help" concept of the national defense dictating appropriate economic policy, realism-mercantilism offers little to determine what goal should be pursued. The regulatory toilings of the French state were designed—for what? To favor the commercial interests that profited from them? To reduce the oil deficit? To freeze market shares of the national companies or their international rivals? To cartelize the market and raise prices? To give major financial institutions a chance to buy into an oil company with state guarantees? To deny synthetic fuel programs to the French coal industry? But if policymakers had concluded that they were helpless to enact any strategically meaningful oil policy, they were free to optimize the rent-seeking activities of groups that were willing to support them: the Banque de Paris et des Pays Bas and its allies in Poincaré's case; Saint Gobain and groups favorable to Russian oil in the case of the Left. Window dressing about the national interest was exactly that.

This brings us to the national reserve oil stocks. The 1928 oil law required all companies importing into France to maintain a three-month supply of oil, using their annual sales from the previous year as the base period. Since investment in Mesopotamian oil was primarily a strategy for commercial compromise in peace, this was one of the few truly strategic elements left in French policy: how much oil would physically be in France in the event of a crisis. The three-month reserve requirement can be analyzed as it evolved and as it worked, as shown by documents from immediately before and after the outbreak of World War II.

Strategic considerations played a minor role in the reserve requirement's adoption. As we saw in chapter 2, a three-month naval reserve was funded by the French parliament before World War I, even though the navy had requested a nine-month reserve; the latter was deemed too expensive. During World War I, as covered in chapter 4, reserve requirements for the industry were put at 22,000 tons and raised to about 100,000 tons towards 1918. In 1921, when the wartime state monopoly was officially dismantled as a result of pressures from Standard Oil, the three-month requirement surfaced again. This time the reserve was not for naval requirements, but for three months' *civilian* consumption. This considerably lowered the quantities needed; it is unclear whether the stocks were rebuilt after the 1919–1920 embargo, but the switch to a "civilian" criterion released sizable stocks, purchased at high prices, that were sold off at a loss to the private sector when the monopoly was ended. The three-month "requirement" for the private sector really served as an excuse for a sell-off of reserves that had grown closer to the nine-month ideal of the navy; after the sell-off, the oil held in strategic reserve declined.[75]

Some kind of speculative maneuver occurred here. Because of Bérenger's 1918 "monopoly" policies, the government "owned" reserves but stored them in the facilities of the private sector; reducing the government reserves opened up additional commercial capacity for the importing companies without forcing them to build more storage. Even though the country had passed through a seven-month oil embargo a year earlier, the 1921 law diminished the size of strategic stocks and turned the responsibility for maintaining them over to the private sector. The "good sense" of the 1921 three-month requirement masks the reality of reducing real reserves only one year after a crisis.

In 1925 the three-month reserve, still in effect, was rewritten into law with the authorization of the Office National des Combustibles Liquides. The same clause was rewritten into the 1928 oil import law, but under radically different commercial conditions. The reserve was a light burden to major companies that were building refineries, but

[75] Doc.Ch., annexe 1782, 9 December 1922; Déb.Sén., 30 June 1921, p. 1471; Fin. B32314, "Le Monopole d'importation en France pendant la guerre et sa liquidation en 1921," n.d., prob. 1931.

when applied to the small importers of refined products it became an instrument of market discrimination. The import licenses were negotiable and had a value independent of the company that held them. They became a means to promote market concentration. In one case a company sold out to another: only one-third of the price was for the capital equipment; two-thirds was for the 38,700 ton import license.[76] The reserve requirement helped companies with storage facilities pry loose import licenses from those that could not afford to build storage.

France's military desired a much larger reserve than the required three months. A report prepared and circulated in 1922–1923 for the prime minister by the Conseil Supérieur de la Défense National, a French body roughly equivalent to the Joint Chiefs of Staff, is interesting to examine in light of what French policy eventually became.[77] The report likened oil's strategic importance to that of coal, iron, cotton, wool, and wheat.[78] The report forecast military needs for a reserve of 1,600,000 tons of various oil products, plus an additional 750,000 for civilian uses: this wartime "minimum" exceeded one year's commercial consumption for the entire country, yet the reserve requirement that prevailed, and that has continued in force until the present, was for only three months. In contrast to the French situation, the report evaluated British reserves at nearly a year's supply of commercial consumption and projected them to increase to well over that.

The report favored the construction of a refining industry in France, but with a six-month, not a three-month, reserve requirement. Japan and Italy were each reported as seeking total reserves of over one million tons—all of which underscores the inadequacy, as measured by the policies of "world peers," of the three-month requirement. The report observed: "The general policy of nations must tend, to the greatest extent possible, to liberate them from foreign supplies of the products necessary to their economic existence in wartime; in spite of this there will unfortunately be some products for which this will prove impossible."[79] The report outlined a series of measures concerning synthetic and substitute fuel production, including shale oil, as well as stockpiling and conservation, none of which alone could create independence for the country, but that together would have done much

[76] Déb.Ch., 25 March 1931, p. 2183.

[77] Fin. B32313, "Rapport au sujet de la politique du pétrole en France," 24 May 1923. All material cited as "the report" comes from this source unless otherwise indicated.

[78] This is an additional argument against the notion that strategic considerations guided state policy in the formation of Compagnie Française des Pétroles. One would expect equivalent "Compagnies Françaises" in cotton, iron, wool, etc. The constellation of private interests in the oil market thrust state policy in the direction of a state company, but there was no equivalent pressure for the other strategic commodities.

[79] Fin. B32313, "Rapport au sujet de la politique du pétrole en France," 24 May 1923.

more than the country actually achieved. The French market was large enough to interest all major oil suppliers, and Poincaré and the government used this interest to incite cooperation for the peacetime regulation schemes.

The report warned, in true realist fashion, that today's friends could become tomorrow's enemies. A policy of total reliance on the international companies was inadvisable. "Everyone remembers," the report said, "how in 1920 Standard Oil refused to contract with the French state under the oil import monopoly regime."[80] The British cutoff of supplies from the Caucasus in 1919, when France was scrambling to make do without Standard Oil deliveries, was also fresh in Louis Pineau's mind in 1928 shortly after the oil import quota law had been enacted.[81] In the face of this perfect bureaucratic memory of earlier supply failures, it is difficult to justify not having more significant oil reserves than the three-month supply required in the 1928 law.

By 1939 the situation had not changed. The 1928 oil legislation was eleven years old, the Compagnie Française des Pétroles was producing oil and making profits, but the government had yet to make the oil industry hold significant oil reserves. The situation looked deplorable even to an industry representative, Léon Wenger of Pétrofina, who in May 1939 completed a survey for the French government of available oil sources and observed that "the question of war stocks . . . is still not resolved, in spite of all the efforts in that direction." What these efforts might have been, Wenger does not say; but they could not have included large oil purchases, for he was still talking about war stocks in the conditional mood: "I have always said that the creation of war stocks would permit the reabsorption, for example, of the 150 million francs in commercial back payments and 300 million francs in longer-term debt that Romania owes France. In addition, these stocks should be protected, in underground storage."[82]

Even France's adversaries could not believe the pitiful state of its oil stocks. After France's capitulation in June 1940 the Italian admiralty sought, during armistice negotiations, to lay its hands on the vaunted French stocks:

> Admiral Raineri Biscia took the floor and pushed hard concerning the interest the Italian navy had in getting a favorable response to its requests

[80] Fin. B32313, "Rapport au sujet de la politique du pétrole en France," 24 May 1923.

[81] Pineau refers to the British control of the pipeline from the Caucasus in Fin. B32310, folder 4, "Reunion chez M. de Saint Quentin," 16 April 1928.

[82] Fin. B32313, Wenger to Coulondre, 15 May 1939. The linkage of oil imports to debt, now focused on Romanian payments, remained even in 1939 characteristic of Pétrofina's concerns as a "bankers' oil company."

for fuel oil. He was astonished at the refusal of the French minister of the navy. He said that he knew about the fine organization of this navy and the high quality of its officers, especially Admiral Darlan, whom he saw once at a conference in London; and he thinks he can infer that the French naval reserves must have been quite large. He made an especial mention of Bizerte [in Tunisia], where there ought to be, he said, veritable "underground lakes" of oil. He added that warships near Toulon must have, in their holds, at least 300,000 tons of fuel oil.

Our representatives answered that, as the Italian delegation knew, French naval needs remained important; *that the guarantee of the freedom of the seas had allowed us not to accumulate stocks as large as Admiral Biscia claimed to know about;* and that a large part of the stocks had been destroyed during the German advance.[83]

French representatives had every reason not to tell the truth to their Italian counterparts in 1940, and it would have been to their credit if, in this situation, they had been deceiving the Italians. One indication that they were is the navy's claim that "naval needs remained important," surely an exaggeration after the British sank the bulk of its fleet off the coast of Africa, at Mers El-Kebir, in July. Still, in the main they were truthful. Their statements are remarkably similar to a much earlier report made to the government in 1939:

It is said that the Anglo-French accords will permit, in wartime, sufficient supplies in gasoline and fuel oil and that the real thing to worry about is the impossibility of finding anywhere sufficient quantities of aviation fuel. *Yet the Germans have synthetic gasolines made in Germany that have the necessary quality (octane rating) to be used for aviation.*

It may be that [current policy] is enough, from the military viewpoint, but I have to say that I can only with difficulty admit that the Anglo-French accords can replace a stock, which would be immediately available on French soil from the moment hostilities began, a stock that could be used for aviation, in case of an unexpected intensification of aerial or submarine warfare.

I have equal difficulty admitting that we could be permanently subordinated in our political and diplomatic activity to the desires of England alone, which could, simply by refusing us fuel, forbid us any freedom in our own European interventions.[84]

Wenger underscores that the French three-month reserve requirement, as of mid-1939, offered no margin of additional security. France's

[83] Arch.Nat. Desmarais Frères 130AQ16, folder 3, "Note," n.d., but cites text as referring to a meeting between French and Italian delegations in Rome, 6 August 1940. Italics added.

[84] Fin. B32313, Wenger to Coulondre, 15 May 1939. Italics are underscored typewriting in original.

oil policy was really reliance on British formal assurances, country to country, and American informal assurances, which were more of the nature of company to country.[85] France's oil policy had not gotten it much independence, nor, it appears, much security either: after two decades of the government's supposedly "strategic" intervention in the oil market, by mid-1939 there was not enough aviation fuel on hand to meet battle conditions.

Why then did the 1928 legislation stipulate a three-month reserve requirement? What was the purpose of such legislation, given its utter inadequacy? We have seen that most aspects of French oil policy during this period, from oil import quotas to the fight over state equity participation in the Compagnie Française des Pétroles, trace their origins to specific interest group conflicts. The three-month reserve requirement was no exception.

Any enterprise maintains stock for commercial purposes. Raw materials wait to be processed into finished products; some materials are in transition from the raw to the finished state; the finished products await shipment. The state required an oil company to keep on hand three months' oil sales as determined by total sales the year before. How much was this, measured against the quantity of oil that the company would keep on hand anyway? Ernest Mercier testified in 1929 that oil refineries normally needed a forty-five-day supply of crude, a quantity of "intermediate products" waiting for further processing equal to about fifteen days of sales, and an additional thirty days' worth of finished products waiting to move to market.[86] That totals ninety days, or three months. The ninety-day "reserve requirement" was no requirement at all, at least not for the large companies that planned to go into the refining business.

If the reserve requirement was not in fact a requirement, what function did it serve? France continued to have, as before World War I, many small importers who typically imported a petroleum wagon, or shipload, of refined products from Romania or elsewhere. This oil petty bourgeoisie competed against the larger firms, whose state-cartel prices made outsiders' attempts to enter the market a certainty. The three-month reserve requirement was particularly onerous for these small firms, for whom it represented not only excessive capital to be invested in maintaining an unnecessary level of stock, but the additional costs of proper storage. Conflicts over enforcement of this requirement, an integral part of the oil-import quota system, were a recurrent element

[85] The French had repeatedly agreed in the naval conferences of the 1920s and 1930s to a smaller navy than other major powers, relying on British and possibly American naval support in the event of hostilities.

[86] Arch.Nat. Desmarais Frères 130AQ1, "Notes pour les experts—Déposition de M. Mercier devant la commission des experts," 6 June 1929.

of the oil market in the mid-1930s.[87] A common quota evasion tactic was for a small importer to rent sufficient oil storage space to qualify for assignment of an import license, and then import without using the tanks. In short, the cost of the three-month stock was onerous. The small importers would have been brushed aside by the oil majors had not many of them sold their refined products to benzol producers: the true conflict over refined imports was between oil and the coal sector.

The scheming of one oil importer illustrates the reserve requirements' effects. To eliminate the "independents" of "lesser importance" than himself, he proposed, "The most effective means would be to buy their products using third parties as buyers . . . to the degree that their reserve requirement will increase [in proportion to their sales], so will they have financial problems; for it is not always easy for these groups to buy fractions of one- or two-thousand-ton cargoes. Their progressive weakening can be hastened if we buy their products. . . . After letting a certain amount of time go by, and letting their required reserve stock go up, it will be easy to eliminate them from the market altogether by offering to buy their entire reserve stock, and thus their right to import."[88]

The elaborate scheme of front men buying on behalf of the larger oil companies never took place, but the passage's anecdotal value lies in the author's perception that the reserve requirement was going to be a significant burden on small companies. Along these lines, another French company, which imported refined products from Romania, testified that the three-month reserve requirement "constitutes an incredible cost for us."[89]

Reliance on the "freedom of the seas" as guaranteed by Britain was repeatedly used as an excuse not to build up large oil supplies. If the First World War had brought about a major supply crisis in 1917, as standard histories say, one cannot explain how French elites could have come around to this position. The interpretation offered here is more consistent: Standard Oil had pulled France through the previous war, and Standard Oil could pull them through the next one. Therefore the best thing to do was to make Standard Oil feel at home. Give Standard Oil a share in the French "national" company; give Standard Oil a generous quota under the 1928 legislation; adhere to the worldwide hydrocarbon patent agreements of Standard Oil and its friends: these things would make the company feel welcome.

[87] Arch.Nat. DF130AQ1, letter from Détinguy to Cayrol, 27 June 1935; "Unification du régime des autorisations," 4 June 1935; "Fédération Française des Carburants," 26 June 1935.

[88] Arch.Nat. Desmarais Frères 130AQ6, "Les Indépendants," 12 August 1928. The mastermind was A. Kahan of the Union Général des Naphtes.

[89] Testimony of Marcel Champin, "L'Enquête parlementaire," *Revue pétrolifère,* no. 261 (17 March 1928):364.

France succeeded in making Standard Oil feel at home: hydrogenation as the backbone of synthetic rubber and fuels progressed no more in France than in the United States. One-hundred-octane aviation gasoline, as Wenger notes above, was not to be had in France; but Standard Oil also refused the United States Army's request to make the fuel available. The fuel might have given the United States "the lead over the military aviation of the rest of the world," but "Standard Oil . . . to comply with the wishes of the United States Army and to so serve the country's defense . . . would have to violate its contracts with I. G.," which the company would not do.[90] France's policy was not "national independence." Nor was it particularly shrewd, since the single major oil crisis the country had faced had come from a dispute with Standard Oil. And when the favors were handed out to Standard Oil every other member of the oil industry also held its hand out, and in the mad rush for market shares strategic considerations disappeared altogether.

A white knight would supply oil to France in a crisis. Strong state intervention had engendered a weak state's policy. Did it have to be *so* weak? Even the Italians, who have not been noted for military prowess in this century, had difficulty believing France's lack of preparation. Perhaps French oil regulation has passed as a "strong state's" market intervention because the truth is, in mercantilist terms, unbelievable.

The strategy of neglecting domestic oil stocks while relying on allies to guarantee freedom of the seas did not win the approval of military strategists; it was the product of civilian control, especially Poincaré's governments. The 1923 military report on oil policy was willing to ask the difficult and pertinent questions: "Can we imagine the possibility of a great struggle without also having, thanks to our alliances, the freedom of the seas? Must we take into consideration the possibility of a more limited struggle where our own fleet would have to guarantee a certain relative freedom of the seas?"[91] For the military, the problem was how to achieve independence from the trusts. "Above all," the 1923 report continued, "France must establish direct relations with other productive countries, such as Mexico, Central America, Venezuela, Colombia and others of the South American states, as well as Romania, Poland, and Russia, which would undoubtedly favor such a policy, which our country could pursue through an accord with the Belgians and the Italians."

These are the realist-mercantilist goals of supply diversification. They assume that the international oil majors were the powers from which France ought to be independent. The military wanted to unite

[90] Pat.Hearings, testimony of Patrick Gibson p. 3353; memorandum, 29 March 1935, pp. 3708–3710.

[91] Fin. B32313, "Rapport au sujet de la politique du pétrole en France," 24 May 1923.

the underdogs against the dominant world oil interests, the opposite of the collaboration with them followed under Poincaré. The recommendations followed naturally from the assumption that the "oil problem" was how to achieve independence from the trusts. Independence connoted autarky, indigenous production, stockpiling procedures, and finding ways to bypass the trusts in world oil markets to make direct ties with the "underdog" undeveloped nations who also, so the report's authors presumed, had a strong desire to bypass the oil internationals. France had "realists," but they were not relevant.

The navy wrote: "Concerning acquiring oil-producing terrain, the minister of the navy points out that it will be necessary to avoid operations in countries too far from France or in countries in which our land and sea links might be cut in time of war. Such would be the case, for example, of the Caucasus and Georgia, if the straits of the Dardanelles could not be neutralized or were under the domination of the Turks, who might again be allied with our adversaries. On the contrary, except for the risks of naval warfare, it will always be possible to bring into France oil from Mexico or South America; thus the purchase of oil-bearing terrain in these countries seems particularly interesting."[92]

Two years later this same French navy, in order to help Pétrofina's oil purchases, began to supply itself almost exclusively from Soviet sources, even though the Dardanelles and the Caucasus had been singled out as unreliable regions. The procurement continued through 1939.[93] Although Mesopotamian oil was more secure than Soviet oil, it too depended on a control of the Mediterranean that the French armed forces considered questionable. Yet French oil policy focused exclusively on this oil source, and one finds few efforts directed at commercial relations with the oil-supplying countries whose Atlantic supply route would have posed fewer strategic problems.[94]

Independent Exploration

The 1923 report of the Conseil Supérieur de la Défense Nationale strongly supported oil prospecting in French territories and colonies. The French mandate of Syria was part of the Red Line agreement; oil

[92] Fin. B32313, "Rapport au sujet de la politique du pétrole en France," 24 May 1923.

[93] Fin. B32313, Wenger to Coulondre, 15 May 1939. Léon Wenger's survey of Eastern European sources of oil commented that "the [Soviet] Pétronaphte still enjoys a kind of monopoly on fuel oil supplies for the navy."

[94] CFP 11SGA15, Comité de Direction, 1937–1941, has some records of board-level discussions by the Compagnie Française des Pétroles to establish a commercial presence in Venezuela and Paraguay. Nothing came of these pre–World War II efforts.

exploration was deliberately crippled by the Iraq Petroleum Company, which included the French "national champion." The procedure was to drill shallow holes in order to avoid finding oil. As the Iraq Petroleum Company's director said, "the Company should drill shallow holes which would constitute technical compliance with the Company's obligations." The idea was to dupe the Syrians into thinking that real efforts were being made. Fake oil exploration, tolerated by the French government, extended into other areas of the world as well. Restraints of trade were encouraged by Ernest Mercier, who himself in September 1930 had negotiated the "pro-ration," or inclusion in the "as is" agreements, of the French national market so long as oil imports from Mesopotamia were allowed.[95]

Not all of Mercier's colleagues viewed these collusive practices with equanimity. Louis Tronchère, for example, was a top lieutenant of Ernest Mercier and a director of the Compagnie Française des Pétroles and the Compagnie Française de Raffinage. In 1930 Tronchère complained about his own company's treatment by the French colonial government in Gabon and Cameroon: "It is too much to ask for the help of the CFP for three colonies when *it is believed* that the chances [of finding oil] are weak and that we get eliminated when *it is believed* that the chances [of finding oil] are strong. [The colonial governor] answered that for this year it was too late, but that next year we would be allowed in. By that time *our participation will come too late for us to have given our own opinion.* We want to be *consulted now* . . . because we are convinced that someone is going to *do something stupid.*"[96] How the exploratory mission described above turned out is not on record, but the end result is known: Gabon did not become an oil producer until much later. Tronchère's letter shows how the Compagnie Française des Pétroles was the last to be called in, even in French territory, when the prospects for oil appeared promising. This is consistent with the pattern of suppressed production in Syria. It is inconsistent with the idea of a "national champion" charged with finding independent sources of French oil, as the French admiralty desired. It fits the expected behavior of a firm that collaborated in the 1930s with world oil interests who wanted to hold back global overproduction.

Significantly, Tronchère lodged his complaint about exploration in Gabon not with his superior, Ernest Mercier, but with Robert Cayrol, head of the French independent Desmarais Frères. The most faithful repository of the "public interest" was the single largest French *private*

[95] FTC2, pp. 149, 168.

[96] Arch.Nat. Desmarais Frères 130AQ32, folder 3, Tronchère to Cayrol, 15 April 1932. Italics are underlined handwriting in original.

interest, which wanted to increase the amount of oil available to the Compagnie Française des Pétroles rather than hold back that production in line with the desires of the international majors. The Compagnie Française des Pétroles' oil exploration record in the pre–World War II oil market remained unimpressive.[97]

Perhaps capitulation to the world hydrocarbon cartel brought France "security through interdependence." The country's collapse in World War II was so rapid that no long-term strategic procurement policy could have been tested. The alliance with Britain and the United States could hardly perform when the national government had been replaced with the Vichy regime. Still, sudden turnabouts of allies and enemies are the very stuff of realist thinking about the international arena. In the eyes of the French military, whose tradition of continental power balancing included centuries of opposition to the English, reliance on the British "guarantees" of "freedom of the seas" must have seemed the height of naïveté. The Conseil Supérieur de la Défense Nationale voiced a pure realist skepticism about the value of France's oil policies: "What does it matter if we have all these accords, all these contracts, if we are not masters of a productive territory? Someone can deny us oil, when the time comes, in spite of all the written promises."[98]

REFINING

The 1928 import quotas fostered peace between the international majors and domestic French oil companies, resulting in the growth of a genuine refining industry on French soil. The refining industry had two important effects. Refining investments increased the stake of the oil multinationals in France and gave an incentive to keep them working and profitable. Standard Oil would think twice about suspending oil deliveries, as it had in 1919–1920, if that meant idling its own refinery. The presence of the refineries on French soil also gave the government something to seize if a national emergency required it. Though the quotas that made the refining industry possible were the artifact of private capital, France's strategic position objectively improved.

[97] Forbin, a "pro–oil company" author, reports (1940, p. 227) that after exploring in Algeria the companies "recognized the uselessness of their efforts and abandoned them *too* discreetly. Contrary to established custom they closed up the bore holes without giving to the Algerian authorities the logs (stratigraphic sequences, daily memoranda kept on the site of the exploration) which would have informed us about the composition of the underground strata. The explanation of abandoned effort because of failure to find oil is not considered true by the Algerian colonists; for them, the trusts indeed discovered oil, but for unknown reasons decided not to develop it."

[98] Fin. B32313, "Rapport au sujet de la politique du pétrole en France," 24 May 1923.

Opportunities to maximize the strategic value of the refining industry were passed up, however. Ordinary market forces would have brought some refining capacity into France even without the quota system. In the nineteenth century free traders had mocked the refining industry in France: "If you cannot refine oil in France, then don't do it. Would you take French wine abroad in order to distill it into cognac? No, you distill it in France, so do the same for oil . . . you don't have oil, too bad for you, your industry is impossible."[99] It was thought that without protection there could be no refining in France even much later. By the late 1920s, refining advances meant that higher proportions of crude oil could be distilled into a marketable product rather than discarded as waste. The old analogy with cognac no longer applied, largely because of the convergence of chemical synthetics and oil-refining technologies. Residues from oil refining could in theory have been processed directly by the refinery into chemical products, but the world patent pool preempted this "natural" development. The second alternative was for a refinery to sell its "residual waste" to a chemical company for processing into chemicals.

Refining near the country of consumption affected shipping costs. An oil company could ship crude from the point of production to the point of consumption, where it made its normal spread of gasoline, fuel oil, and residuals for sale to a chemical firm. All it needed was a tanker equipped for crude oil. Refining at the source required specialized tankers for each kind of product: at a minimum, different holds had to be outfitted to transport each finished product. Since crude oils varied in their product yields depending on their region of origin, a company would have to calculate the optimal proportions of product-carrying ability necessary to build into the holds of the various ships. Since market demands shift, how to constitute an ideal corporate fleet was a problem. There were also differential rates of wear on tankers designed for refined products as opposed to crude oil; refined products were more corrosive and shortened the useful life of tankers.[100]

Thus the advantage shifted, though not completely, in favor of locating refineries near the point of consumption. To this day the world oil market offers examples of refining both near the source of production and near the point of consumption, or at a convenient point of transshipment. Some factors argued in favor of putting a refinery near production, especially when the industrial profile of the market area did not permit the sale of a full range of products. The Abadan refin-

[99] *Annales du Sénat et de la Chambre des Députés,* vol. 5 (20 April–8 May 1880), p. 200.

[100] Testimony of Marcel Champin, "L'Enquête parlementaire," *Revue pétrolifère,* no. 261 (17 March 1928): 364.

ery in Iran could sell its fuel oil to the British navy and furnish kerosene to the Asian markets, which used lamp oil longer than the European market. Shipping Iranian crude to India made no sense, for part of the spectrum of products would have to be shipped back to industrial countries for consumption. On the other hand, Anglo-Iranian did locate a full refinery at Swansea in Britain. Britain could absorb a full product spectrum. As a "late developer," Anglo-Iranian was less committed to refining near oil sources than Royal Dutch–Shell and Standard Oil. It was one of the first companies to locate modern refining plants in France.

The French domestic companies used the national oil producing and refining companies to achieve complete vertical integration. They wanted to insulate themselves from depending on the refined products supplied by the world trusts. Other firms interested in refining included Pétrofina, which wanted to enlarge its presence in French markets, and the Saint Gobain group. For them the import quotas of 1928 meant that the expensive refineries would be protected from the risk of fluctuating market demand. Investment in the refining industry also became a means of promoting one's eligibility for a quota, so that the investment itself became part of the competitive process to obtain a legal claim on market share. Saint Gobain's frustration in this process was one of the underlying causes for the struggle over the size of state equity in the Compagnie Française des Pétroles, as discussed above. On the whole, however, the legislation was accurately summarized in a report to Louis Pineau as having "sheltered from the trusts" Desmarais Frères and other French national companies.[101]

Even within multinational firms, there was no consensus about where to put refineries. Standard Oil confessed before parliament to internal disagreement about the relative merits of refining in France in 1927. A. C. Bedford thought that "refining in a non-oil-producing country was not economically sound" but added, "in my company, not everyone is of the same opinion." He thought the oil import quota system would work if "it allows us to survive and develop our business affairs according to normal commercial practices"; he worried that if the regime was not maintained long enough, investments made on the basis of the law would not be adequately amortized.[102] Privately, Standard Oil wanted the French companies lumped into one group so as to control production in the market as a whole: in Teagle's words to I. G. Farben representatives, "We could let the Compagnie Nationale build a refin-

[101] Arch.Nat. Desmarais Frères 130AQ1, letter to Pineau dated 24 June 1930.

[102] Testimony of Bedford, "L'Enquête parlementaire," *Revue pétrolifère,* no. 269 (12 May 1928): 665.

ery in one place, we in another, Standard to give an outlet to Compagnie Nationale's product."[103]

The syntax "we could let" is extraordinary: Americans negotiating with the Germans "could let" the French build a refinery. And in fact that is what happened: Standard Oil got one refinery; it "let" the French "Compagnie Nationale" have another, it did not become the distributor for the French national company's products, but that was a detail: as the I. G. Farben representative said, "What you want is the selling," to which Teagle replied, "Yes; same percentage as now. What we hope for in France is a unification of the various French interests behind a policy satisfactory to the majority."[104] And that is what Ernest Mercier and Standard Oil put together; or rather, what "they let" France do, with the knowledge, consent, and prior discussion of the matter with I. G. Farben, which sought to prevent the spread of hydrogenation technology.

The state's concern with refining reflected financial and strategic considerations. National refining would reduce foreign exchange losses by keeping the profit on refining within France. In 1927 Poincaré himself spoke of "the connection of oil affairs with general political considerations as well as with finance politics."[105] In 1923 a French admiral argued for bringing the refining industry into France because "any diminution of our financial power acts against our military power."[106]

In theory, refineries in France meant the government could tell them where they had to buy their oil and so could have diversified sources. In practice this option depended on political and technical factors. The political factor was the will to tell companies where they could buy their oil; there is not much evidence that such a will existed. The technical factor was outfitting the refineries with the equipment they needed to switch from one crude source to another. Refineries were designed to handle specific kinds of crude. Different crudes either did not produce optimal yields or actually damaged refineries not able to handle them, sulfur content being a key variable. The government could have forced a "flexibility requirement" on refining companies as a condition of the import quota, requiring them to accept crude from almost anywhere in the world. As with stocking requirements, however, the government

[103] Pat.Hearings, Memorandum of Meeting, 31 August 1928, p. 3430. "Compagnie Nationale" means the Compagnie Française des Pétroles.

[104] Pat.Hearings, Memorandum of Meeting, 31 August 1928, p. 3430.

[105] Fin. B32310, folder 1, Poincaré to Minister of Commerce, 8 November 1927.

[106] Fin. B32313, "Rapport au sujet de la politique du pétrole en France," 24 May 1923. Forbin 1940, p. 222, puts the increase in the number of refineries from two in 1930 to fifteen in 1936.

backed down, and refineries specialized in specific crudes.[107] The ostensible reason was the cost, which Pineau estimated for the entire country at FF200–300 million, or about one-twelfth of total investments in refining made in France during the 1930s.[108] The government made no effort to capture the strategic advantages of having a refining industry; the one advantage it did capture was that of having the industry on its soil, which many of the oil companies themselves either favored or did not strenuously oppose.

The world hydrocarbon cartel kept hydrogenation processes from being adopted in French refineries, which limited the gains in technical expertise and related industrial knowledge that a completely modernized refining industry could bring. In 1934 Standard Oil's brand new one-million-ton French refinery did not have the hydrogenation technology that would have greatly enhanced the plant's productive flexibility.[109] Pineau argued that hydrogenation could not be used in "small" refineries; but this is not backed up by the United States Senate Patent hearings of a decade later. If the process was not profitable in France, it was because the Standard Oil–I. G. Farben accords prevented marketing the full range of by-products that would have rendered it so.

The oil-refining industry's development in France resulted primarily, but not entirely, from the law of 1928. That legislation was strongly backed by French national firms; the international firms had independent reasons to support it. Changes in technology and demand allowed refining operations to move from sources of production to the more developed market areas. Where strategic objectives indicated specifying technologies, the state backed down. Refineries specialized in specific crudes, integrating France ever more with the overall logic of international production organized by the 1928 "as is" agreements. Though some strategic benefits were realized, the whole can scarcely be called the result of realist-mercantilist state policy.

Transportation

During the First World War, the availability of oil had been less of a problem than who would deliver it, with attendant effects on company

[107] Ass.Nat. Commission des Pétroles, *Rapport Général de M. Louis Pineau,* 8 July 1934, pp. 35, 40.

[108] Fin. B32314, "Monopole d'importation et monopole de raffinage: Aspects technique et économique du problème," n.d., prob. 1933.

[109] Ass.Nat. Commission des Pétroles, *Rapport Général de M. Louis Pineau,* 8 July 1934, p. 90.

market share. The small French tanker fleet, then under British registry, was only with difficulty made available to French firms.

In the 1920s French policy brought more tankers under the national registry. There was no great opposition from any party, French or otherwise, importing oil into France. Why bother? During the First World War Standard Oil had pulled most of its fleet under German registry out from German service with complete ease. The Germans might well have had other uses for those tankers, but Standard Oil reassigned them to Asian waters, and the Germans were powerless over any ship not directly in a German port.

At any given time the bulk of an international oil company's tankers are at sea, beyond the reach of any single sovereign power. A company fleet remains de facto independent even if it is not so de jure. Sometimes a nation can get the better of a company, as for example when Britain requisitioned Socony's Hong Kong fleet and assigned it to Royal Dutch–Shell. Such actions guarantee a wartime corporate struggle for political control of industry regulation. In the 1920s the major international companies registered some of their ships under the French flag with no objection; the tonnage of oil tankers under French registry increased throughout the 1920s. French tankers carried only a small fraction of the total deliveries to France in 1919 but delivered about 13 percent by 1928.[110]

Since companies could redirect ships at sea, this gave France a marginal increase in power over shipping compared with World War I. In World War II, after the capitulation, the tankers working for foreign firms under the French flag were pulled from French service, confirming the purely symbolic character of the "control."

Of greater consequence was the development of a purely French tanker fleet. Some of these ships were military: the practice, still used today, was for oil tankers to accompany fleet movements and refuel ships at sea. The tankers so employed were not strictly speaking commercial, and any naval power needed them whether or not it was in the oil business. In the French case, however, a commercial tanker fleet was developed to carry the Iraqi oil of the Compagnie Française des Pétroles. This part of the tanker trade serviced the 25 percent of the French market reserved for the French national firm. The tankers were operated by the Compagnie Française des Pétroles along with a private firm, the Compagnie Auxiliaire de Navigation.

By 1938, however, the Office National des Combustibles Liquides, in conjunction with top management of the Compagnie Française des

[110] Figures are in Doc.Sén., annexe 77, 18 February 1926, pp. 73–74; and in Doc.Ch., annexe 5449, 3 February 1928, p. 377. A table will be found in Nowell 1983, p. 247.

Pétroles, was still of the opinion that the "oil fleet under the French flag was insufficient" and proposed a new mixed company, the Société Française des Transports Pétroliers, to purchase and operate additional tankers.[111] The state was to take a 30 percent shareholding, the Bank of Worms 22 percent, the Louis Dreyfus Bank and the French domestic firm Desmarais Frères 12 percent each, Saint Gobain 7 percent, and a combined shareholding of the Compagnie Auxiliaire de Navigation and the Compagnie Navale des Pétroles 18 percent. This commercial fleet could in theory have meant an additional margin of maneuver for the state, or those interests it wished to favor, in the event of intense rivalry over tanker tonnage under conditions similar to those of World War I. French policy in shipping was neither controversial nor, it seems, exceptionally vigorous; but it was certainly better than nothing.

A greater clash developed over pipeline routes in the Middle East, to run from the Kirkuk fields in Iraq to the Mediterranean. The French favored a pipeline route through their mandate in Syria; the British favored a route through Palestine. The Solomonic compromise was a pipeline that bifurcated in western Iraq, one branch heading to the Mediterranean through Syria to Tripoli (Lebanon) and the other through Palestine to Haifa. The French were suspicious of allowing the British exclusive control over any oil pipeline facility; their experience with Britain's seizure of oil export facilities in the Caucasus in 1919–1921 left a bad taste. So said Pineau in 1928, but the crucial producing end of the pipeline remained under British control in Iraq. The British could easily cut off French oil from Iraqi territory, whether the final port was in Syria or Palestine. Since there would be a need for some storage space at the point of export, however, the facilities on French soil provided a small buffer against a short British cutoff.[112]

The French navy also wanted an oil fueling facility in a "French" port in the eastern Mediterranean; the bifurcated pipeline answered this need.[113] French banking interests favored the increase in economic activity that the pipeline would bring in areas of the Middle East where they had investments. The Banque de Paris et des Pays Bas favored a route through Alexandretta, where it had substantial investments, but agreed to the Tripoli route when it was brought into the financing.[114]

British desires for a pipeline route through the mandate in Palestine

[111] CFP 11SGA15, Comité de Direction, 1937–1941, 12 September 1938.

[112] Fin. B32310, folder 4, "Réunion chez M. de Saint Quentin," 16 April 1928; Fin. B32310, folder 4, "Note pour le président," 25 July 1928.

[113] MAE E-Levant, Mésopotamie-Irak, 1918–1929, vol. 35, "Ministre de la Marine à Monsieur le Ministre des Affaires Etrangères," 26 August 1928.

[114] Fin. B32310, folder 4, "Beyrouth, 5 June 1928."

were much the same. The Admiralty pushed for Haifa over Alexandretta, which was an early candidate over Tripoli, because it was farther from the Turkish frontier and more insulated from a potential foe whose prowess had been demonstrated under Mustafa Kemal.[115] Anglo-Iranian was no more eager to be dependent on French jurisdiction than the French were to trust the British. The French distrusted the British to care for their pipelines, yet relied on them to guarantee "freedom of the seas," an excuse not to build up large oil reserves. The strategic reasoning was inconsistent, but the commercial reasoning was not: everyone wanted control over distributing what would be the least expensive oil in the eastern Mediterranean.[116]

Royal Dutch–Shell collaborated in Asian markets with Anglo-Iranian and echoed its preferences on the pipeline. Some of the Shell combine's management and stockholders sympathized with Zionism and wanted the pipeline to contribute to the economy of the Palestine mandate.[117] Marcus Samuel was a Zionist, but how much his influence continued after his death is hard to tell. With the Rothschilds as major stockholders, it was easy to call the firm pro-Zionist, and an anti-Semitic appeal was addressed to the government of Iraq by an independent British firm that wanted concessions in 1924. Phoenix Oil and Transport combined naked opportunism with a vulgar appeal to the basest motives imputable to the Iraqi government. It vaunted its board of directors, who were "real British interests and also shows an entire absence of that Jewish and Dutch financial control so noticeable on the Directorates of those Companies who apparently had been selected to form the group of Turkish Petroleum Company combine. . . . the Anglo-Saxon Company [Royal Dutch–Shell] have a directorate composed of 50 to 60 percent Jews while [in] others including the Shell and Asiatic the same preponderance of Jewish Directors [is] also noticeable and therefore should the British Government eventually decide to sell the interests they now hold in the A.P.O.C. to the Shell group the whole financial position in Iraq would be absolutely dominated by Jewish Financiers, thus producing the same conditions with its attendant evils as now exist in Palestine."[118] Shell's Henry Deterding, swayed by anti-Sovietism, became increasingly pro-Nazi in the 1930s, a development hard to reconcile with his career, which included thirty years of close collaboration with prominent Jews. Replaced by August Kessler in the 1930s,

[115] Fin. B32310, folder 4, "M. de Fleuriau à . . . M. le Ministre des Affaires Etrangères," 1 December 1928; Fin. B32310, folder 4, "Rapport au président du Conseil," 27 September 1928.

[116] Fin. B32310, folder 4, "Note pour le président," 25 July 1928.

[117] Fin. B32310, folder 4, "Note pour le président," 25 July 1928.

[118] MAE, 1930–1940, Iraq, vol. 67, Cheney to H. E. Jafar Pasha, 21 January 1924.

he died honored by the Nazis and shunned by the company he had made one of the twentieth century's foremost firms. Given the impassioned context of Zionism in the 1930s, the allegation that pro-Zionist considerations dictated pipeline policy must be considered skeptically. The combine would have had to depart from the single-minded devotion to oil profits that had motivated it over several decades. A sudden interest in routing its pipeline out of eleemosynary concerns for Zionism is implausible, unless those concerns matched commercial goals.

These commercial goals were not lacking, and they suffice to explain the company's position. Royal Dutch–Shell and Anglo-Iranian keenly desired to control Iraqi oil in the Mediterranean, allowing Royal Dutch–Shell to recoup the dominance it had enjoyed twenty years earlier as the second largest producer in the Caucasus. For precisely the same reason, Standard Oil opposed the British attempt to control oil exports in the Mediterranean, and the American company weighed in on the side of the Compagnie Française des Pétroles, supporting a bifurcated route instead of the Haifa-only route. "Standard Oil," read one report, "would be happy to align itself with a solution that would avoid the total control of the English authorities and would have an international character."[119]

Decisions of the Iraq Petroleum Company required a majority of the board of directors, which was divided equally between Royal Dutch–Shell, Anglo-Iranian, the Standard Oil companies, and the Compagnie Française des Pétroles. A major policy dispute in which the parties divided evenly was a recipe for trouble.[120] The pipeline debate was not just over routes, but over whether the oil should be developed at all. Given the intense thirty-year struggle for control of the Mesopotamian oil fields, this was ironic. But the world oil market was tending towards overproduction, and to maintain prices Royal Dutch–Shell and Anglo-Iranian slowed down the development of the area to keep its oil from reaching the market. In the words of the Federal Trade Commission: "The Anglo-Persian group was a large producer of oil in Persia (Iran), and if Iraq oil were brought to full development, it would compete with Iranian oil. Consequently, it was to the interest of Anglo-Persian to retard the arrival of Iraq oil on world markets. This interest was shared by the Royal Dutch–Shell and American groups. Royal Dutch wanted to protect its holdings in the Dutch East Indies, and the American group wanted to prevent the competition of Iraq oil with oil supplied by Venezuela and the United

[119] Fin. B32310, folder 4, "Rapport au président du Conseil," 27 September 1928.

[120] The majors had made Gulbenkian a nonvoting member of the company board to prevent his vote from becoming the controlling tiebreaker.

States. Thus three of the major group ... had a general interest in 'curtailing world oil production.'"[121]

The three of the four major parties that had an interest in curtailing Iraqi production did not seek to suppress it totally. Commercial strategy dictated caution in bringing such a large source of oil into world markets; the conflict was over the timetable. According to a Standard Oil memorandum, Sir John Cadman of Anglo-Persian at first pushed for an early settlement of the pipeline route question but gave in later to dilatory pressure from Royal Dutch–Shell:

> Sir John Cadman seemed very much impressed with his visit to Iraq and the prospects the TPC[122] had before them. He also stated that although there was apparently an overproduction of oil in the world today, yet in a few years' time the companies might be in a position of wanting to draw on the Iraq fields, and he was advocating quite strongly the necessity of pushing on the development as quickly as possible, and was anxious to get this question of the route of the pipe line settled, and thereafter, the construction of the line put in hand.
>
> Sir Henry Deterding, who as you know has just returned from New York after attending conferences in connection with the limitation of production, said nothing on this subject, but looked, according to my observation, as if he were not entirely agreeing with Sir John Cadman's enthusiasm for pushing matters on.[123]

Calouste Gulbenkian favored rapid production. In September 1930 he opined that the prolonged dispute over the pipeline route was attributable to "the three oil groups [desire] to prevent any pipeline being built at all." He exaggerated, however, to put pressure on the recalcitrant parties. Standard Oil was willing to go along with the French, but only if the French were willing to adhere to the general principles of the "as is" agreements. The pipeline route dispute led Ernest Mercier to agree to pro-rationing of Iraqi oil in the French market in September 1930, on the condition that such "pro-rationing" not reach 100 percent, a total shutdown of Iraq.[124]

Construction finally started in 1932 and was completed in 1934. The two-year project had been dragged out for five unnecessary years. The first tanker was loaded on 2 August 1934 at Tripoli and on 26 October 1934 at Haifa.[125] The British manager of the Iraq Petroleum Company, John Skliros, used the extra five years to reduce the oil found, restrict

[121] FTC2, p. 145.
[122] The Turkish Petroleum Company, renamed the Iraq Petroleum Company.
[123] FTC2, p. 145, reproducing Standard Oil memorandum dated 26 March 1929.
[124] FTC2, pp. 148–149.
[125] FTC2, p. 157.

the number of wells drilled in Iraq, and cut back the depth of those that were drilled. He tried to push the completion of the pipeline project back to 1938. While he delayed, negotiations continued on how to integrate Iraqi oil into a world market that had been hit hard by the Depression. The Royal Dutch–Shell directors of the Iraq Petroleum Company "considered that an essential preliminary to any decision on the finance of pipeline construction was a full and open discussion between the Major Groups of the whole world oil with a view to the building up of their Iraq business on secure foundations and not on sifting [*sic*] sands."[126]

By November 1934, however, the groups could agree "to make every effort to take their full share of oil, but it was agreed that if the sales coordination committee could not suggest an outlet the quantity not accepted would not be produced." Iraqi production, and with it the fraction of French production, was securely anchored in the overall objectives of the management of the world oil cartel. This led to a continual tug-of-war between the "as is" principal companies and the French company, whose interest in increasing production stemmed from its role as supplier of domestic companies that, unlike the oil majors, had few alternative sources on world markets. The French government and the national companies (especially Robert Cayrol of Desmarais Frères) had similar desires to bring in more "French" oil from Iraq. The conflict between the desire to hold back Iraqi production and the desire to increase it had no clear winners. Iraqi production was shut back by 400,000 tons because of cartel agreements in 1935 but increased to full pipeline capacity from 1936 to 1938. When the French group proposed increasing pipeline capacity in 1939, the three major oil groups were opposed.[127]

As a result of the world cartel, Iraq became a swing producer. The "as is" companies kept Iraqi production down in order to favor production in other areas where the French did not have a stake. The effects of efforts to hold back Iraqi production are perhaps best seen by comparison. In 1936 Iranian production, under the exclusive control of Anglo-Iranian, was double Iraq's, and by 1948 it would be seven times Iraq's, even though the size of Iraqi reserves would have merited, under other conditions, more rapid development. From 1934, the year Iraqi production began, to 1939, Venezuelan production increased 50 percent; during the whole interwar period when the role of Iraqi production in world markets was debated, from 1928 to 1939, Venezuelan

[126] FTC2, p. 150.
[127] FTC2, pp. 159–160, 163–164.

production nearly doubled.[128] Fighting with the French over the route delayed production and was in itself a victory for the oil majors, for whom each year of delay meant additional productive capacity developed elsewhere. This was true regardless of whether the French ultimately "won" by getting their pipeline branch where they wanted it. As the Federal Trade Commission summarized, "Delaying tactics and restrictive policies of IPC have contributed substantially to the lagging position of Iraq relative to other Middle East oil developments."[129]

The refining industry that grew up in the 1930s and the pipeline from Iraq gave significant benefits to the French state and economy, but these benefits do not explain the evolution of regulatory policy. Multinational companies got extra profits from the quota and used it to contain Soviet oil production. The bias of the world oil industry moreover, was, shifting toward refining in the country of consumption rather than in the country of production. The oil import quota allowed the international firms to replicate the "as is" agreements in French domestic policy, solving the problem of cartel enforcement.

Strategic opportunities were missed. The government failed to impose a flexibility requirement on refineries; the refineries that were built did not employ the hydrogenation technology controlled by the world hydrocarbon cartel. France's share of Iraqi oil, which would reach 25 percent of the domestic market, benefited the balance of payments. Without dogged French persistence, the Iraq pipeline would not have been routed through the French mandate, thus forfeiting the advantage of French-controlled refueling facilities in the eastern Mediterranean.

The military disliked depending on Mediterranean oil delivery routes. Control of the seas was not a French prerogative, and so France relied on allies, particularly Britain, who had been unreliable in the past. Léon Blum and Louis Loucheur admitted in public debate that oil deliveries could be ensured only through the proper alliances. Others, especially Louis Pineau at the Office National des Combustibles Liquides, hypocritically spoke for independence while acting to the contrary. His office caved in to the world hydrocarbon cartel on hydrogenation and synthetic fuels and rubber; his inattention to large reserves for military needs was systematic and prolonged.

The military never had control over, nor even great influence on, energy policy. By the late 1920s the army had lost its fight for significant reserve stocks in France. The navy fared better, but only marginally; the three-month reserve requirement was used to stop all discussion about enlarging reserve stocks. Civilian policy and the pow-

[128] Figures in Tugwell 1975, p. 183.
[129] FTC2, p. 171.

erful companies that drove it overwhelmed purely "etatist" strategic considerations, as defined by the people in charge of fighting.[130] There were desultory efforts to increase the reserves, but the private sector's unwillingness to finance these proposals, and the state's unwillingness to make the private sector finance them, kept a serious reserve policy from developing in the 1930s. This was ironic because internal memorandums show that Standard Oil was willing to go much further than the French government ever required. The American oil giant would have accepted a storage consortium in which the state and private interests participated.[131]

The tremendous interest group conflicts of the 1920s shunted aside strategic goals, which were always secondary. This was underscored in 1932. As a result of parliamentary pressures to purchase a refinery for military use, Premier André Tardieu put the Office National des Combustibles Liquides in total control of French oil policy, in both war and peace.[132] Pineau consulted the military from time to time, but primarily he cartelized the French market in oils and hydrocarbons.

Marginalized in national policy, the army pursued quixotic alternatives that now appear quaint. Convinced that autarky was the only reliable answer to oil dependency, the army explored an alternative technology based on the gasification of wood. Burning wood gives off gases that can be collected and run through a carburetor to provide locomotive power. The large, awkward collection devices were fitted onto tanks and automobiles; in spite of Pineau's opposition, army experimentation on these devices continued through the 1930s. Trucks fueled by wood gas supplied the Maginot Line in 1936.[133] This seemingly ludicrous alternative became a primary means of automotive locomotion in France during the Second World War. The army's worst-case scenario about oil supplies had proved too true. It is no tribute to Pineau's foresight or pro-cartel bias that his Vichy successor, Emile Zédet, could write that by 1938 alternative fuels, such as coal synthetics and wood-burning vehicles, "did not correspond to any systematic plan of production."[134]

Conclusion

What were the transnational structuring effects of the world hydrocarbon cartel? Government policies worldwide were designed not to

[130] Nayberg 1983, pp. 519–521.

[131] Pat.Hearings, p. 3693, document dated March 1936.

[132] Fin. B32313, "Instruction sur la production, l'importation, et la repartition des combustibles liquides, carburants et lubrifiants en temps de guerre," signed Tardieu, 14 May 1932.

[133] Nayberg 1983, p. 564.

[134] Hoover Institution, René de Chambrun Collection, Emile Zédet deposition, n.d.

favor independence, but to favor the needs of the world hydrocarbon cartel members, whose "cooperation is emphatically illustrated in their unanimity of policy in the face of political demands of such foreign governments as France, Germany, Italy, Spain, Hungary, and Japan."[135] Other countries such as the United States and Great Britain were no exception.[136] The improved refining, synthetic rubber, and synthetic oil that would have gone with this technology, as well as a closer relationship between the chemical industry and the oil market, were delayed, by international corporate agreement, in every country except Germany.[137] Worldwide the coal industry, trying desperately to adapt to the evolution of the new liquid fuel market, for want of better technology and government help was forced to rely on benzol. After major efforts to get hydrogenation in France were defeated, the principal coal lobbying went into seeking import permits to mix gasoline with benzol.

The coal industry backed the creation of the Office National des Combustibles Liquides, backed quotas, and pushed both in the early 1920s and in the 1930s for a synthetic fuels program. In spite of their commercial importance, the French coal interests failed to get the kind of state cooperation for their synfuel program that oil had gotten for its projects in domestic refining and Mesopotamia. The international oil companies and the financial houses pushed their oil agenda, which included the supposed tools of "French independence," the oil import quota and the Compagnie Française des Pétroles. These same interests opposed the coal agenda for synthetic fuels. They won in French oil policy, and they won in controlling independent technical progress towards autonomy from the world oil market. Their success in France was only one of many victories, for they were successful elsewhere as well.

A market structuring effect was that chemical companies were given a strong bias not to move into refining; there were attempts at this, but the operations of the world hydrocarbon cartel, both in the United States and in France, prevented serious movement. The Saint Gobain failure shows what happened when a major chemical firm tried to buck the tide and expand its operations to include the oil market.

These things occurred in spite of a strong autarchic bias (predicted by realism-mercantilism) in the nation's military. Its bias against imports from the eastern Mediterranean were swept aside, as was its desire for national reserve stocks. Premier Tardieu consolidated power in the hands of Louis Pineau, whom Standard Oil counted as its own,

[135] Pat.Hearings, p. 3349.

[136] Pat.Hearings, testimony of Patrick Gibson, pp. 3350–3351.

[137] Pat.Hearings, "Outline of Proposal for Standard-Shell Agreement on Hydrogenation," 15 April 1930, p. 3666.

and who opposed a major synthetic fuels program. Can anyone doubt that the armed services would have preferred oil from coal? Napoleon had answered the continental blockade with widespread substitution policies: beets replaced sugarcane, chicory replaced coffee. His bold substitutions show the pathos of his Third Republic descendants, whose only effective answer to the world hydrocarbon cartel, once it had got its hands around Pineau and the Office National des Combustibles Liquides, was to bolt wood-burning gasification units onto their tanks and trucks. The irony of history is that even though some preached the necessity of "freedom of the seas" guaranteed by Britain, the armed services and Pineau himself knew full well that Britain had once proved unreliable and might again prove so. That France had vehicles that could move at all during World War II was due to the armed services' wise distrust of their nation's "mercantilist" oil policy.

But the mercantilist view is an illusion created by distorted histories of the oil market and idealized, erroneous conceptions of the state. Ernest Mercier was explicit about the need for French domestic policy to match the goals of the international oil cartel. Louis Loucheur was equally candid: France did not try to satisfy the needs of national defense in war, but attempted to appease the intense commercial conflicts that had erupted in time of peace. The founders of French oil policy themselves said and wrote these things. Negligent historiography and the dominance of realist-mercantilist, state-centric ideology have allowed them to go unremarked.

Transnational structuring is seen even in state victories. The French victory over Anglo-Persian and Royal Dutch–Shell in getting a pipeline route through Syria to the Mediterranean was possible only because of an accommodation with Standard Oil; while this fight dragged on, the world hydrocarbon cartel successfully brought on stream millions of tons of annual production and stalled development in Iraq. This had a durable effect on Iraqi national development and left France further at the mercy of the international market, while the cartel's ally in the Office National staved off the development of synthetic fuels.

The achievements of France came to this: a pipeline through Syria whose source point was controlled by the British; a small increase in tanker capacity that was virtually unopposed by the oil majors; and refining, some of which would have come to the country anyhow, and which the government failed to require to have flexible technology to handle crudes from more than one area. That there were some gains for France, particularly in the movement of the profit margin on refining to the national territory, is not to be disputed; but this was more the by-product of intense commercial competition between domestic and international interests than it was the triumph of a realist-mercantilist policy.

As for describing not what France did from a realist-mercantilist viewpoint but rather *what was in fact done in the world political economy,* the state-centric international relations vocabulary of realism-mercantilism has no terminology, no conception, and therefore no history of these events.

CHAPTER SIX

Conclusion: Transnational Structuring and the World Order

> Who will write this modern *Iliad,* about this gigantic struggle fought with blows of billions between two groups, English and American, who seek to assure themselves the mastery of the oil fields? Who will tell of the strategems employed, of the fraudulent maneuvers attempted by the audacious producers and unscrupulous speculators in the pursuit of an economic victory that, however, has yet to be decided?
>
> —Deputy Charles Baron, 7 March 1928

"It presents an astounding picture," said Senator Bone, commenting on what we may now recognize as transnational structuring. "I confess that I am at a loss to comment on the international implications of the arrangement," agreed his chief witness.[1] The investigators of the world hydrocarbon cartel had reached a true aporia; this book has attempted to go beyond their dilemma. Transnational structuring has many implications for our conception of the state and the international system; and of course there are implications for the more limited domain of what transnational structuring tells us about France. Other questions remain; two of interest are, first, what other cases of transnational structuring might be found, and second, the phenomenon's relationship to technological transitions. This chapter also provides some indications about international relations theory, French statism, and technological transitions as they relate to the historical exposition and theoretical approach of this book.

INTERNATIONAL RELATIONS THEORY

The traditional realist-mercantilist approach to the reasons for state intervention in strategic sectors has been shown to be wholly inade-

[1] Pat.Hearings, testimony of Patrick Gibson, p. 3414.

quate. As a theory, it fails to explain the nature, timing, and kind of economic regulation as it appears. It also leads to bad history. In failing to explain what it purports to explain, it actually prevents our seeing what happened. A revealing example is furnished by Waltz: "The ability to make do in the face of adversity is one of the most impressive qualities of large and advanced economies. The synthetic production of rubber provides the most impressive example. The need was not recognized until the fall of Singapore in 1942. The United States was able to develop an entirely new large-scale industry while fighting a global war on two fronts."[2] This passage is remarkable for its self-fulfilling view of the world; it admits no rebuttal because it does not admit the existence of confounding evidence. The United States did build a synthetic rubber industry during World War II, after breaking the world hydrocarbon cartel's lock on patents. The accomplishment was remarkable, especially since the United States fought a war on two fronts and did other research such as developing the atomic bomb.

A false theory that sees the international order as composed primarily of states leads to the conclusion that states can reorder the economy as they wish. There is no recognition of how states are themselves structured by actors in the economy. "The need was not recognized until the fall of Singapore in 1942." That is not true; reductionist theory leads to bad history. Many firms and government agencies wanted hydrogenation technology: They could not muster the political power to confront the world hydrocarbon cartel until crisis conditions forced a confrontation between the state's wartime needs and the existing distribution of technological capabilities as established by the cartel.[3] The price paid was time: synthetic rubber was not produced in significant quantities in the United States till 1944, and other countries such as France had to do without it altogether.

The United States was fortunate to develop hydrogenation-related industries during the war (refining was affected as well as rubber). To do so, it allocated resources that might have been used elsewhere. Rubber recycling drives consumed manpower. Hurriedly developing technology required brainpower that might have been better used in weapons design or for increasing factory production. That the United States won the war does nothing to vindicate all of its policies before and after—and that Germany lost the war does not mean it erred by building up a strong synthetic rubber and hydrogenation capability

[2] Waltz 1970, p. 212.

[3] See, for example, Larson, Knowlton, and Popple 1971, p. 415: "By mid-1940 the rubber issue had been taken up by the Advisory Commission to the Council of National Defense." But hydrogenation had come up as a military matter as early as 1935 (see p. 261).

before the conflict. Whether in Germany, France, Britain, the United States, Italy, or Japan, the fundamental decision to build or not to build hydrogenation technology was not related to realist-mercantilist power drives. That was determined by the world hydrocarbon cartel, which used "national interest" rhetoric to justify whatever course it wished to take.

With a realist-mercantilist theoretical bias, everything becomes self-fulfilling because the contrary evidence is not taken into account. Waltz finds the history of synthetic rubber so "obvious" that he dispenses with it in three lines. There is no trace of the awe that so clearly struck Senator Bone and the Justice Department investigator. Realism simplifies world history down to the primitive level of the theory. The study of international relations becomes the celebratory fetish of the state and the glory of its power. Would realist-mercantilist writings have the same tone if they admitted that the capabilities of the state could be structured, limited, and defined by nonstatal actors who saw the state as nothing more than a resource to pursue their own agenda?

Transnational structuring assumes the operation of business groups in a world of states. Might we not object, therefore, that the states and their "anarchy" are the "basic" components of the system, with transnational structuring confined to the margins?

The question is legitimate. A deeper understanding of the world system requires looking at how several logics, or systems of rules, operate at the same time, in the same space. In the first paragraphs of this book I stated that the world hydrocarbon cartel had "three major effects on the world order: the first at the level of national, strategically oriented regulatory policies, the second at the level of the interrelationships of the members of the oil industry, and the third at the level of the oil industry as a whole and its relation to the coal industry."

We can model this conceptually by imagining a chessboard that is perhaps three or four times the size of a standard board. Let us first arrange, on two sides, the standard complement of white and black pieces. We shall call these pieces the "realist" game and assign to them two players we shall call state leaders. Now rotate the board ninety degrees. Along the other two sides, distribute red and green checkers and assign them to two players we shall call oil companies. Now rotate the board once more. On two corners diagonally placed, imagine two players of the Chinese game of go, with either black or white stones, who are the coal complex members. We now instruct each player: "Win the game to achieve what you define as victory, and each of you play primarily by the basic rules of your own games."

The chess players seek checkmate, which may or may not require eliminating each other's pieces. The checkers players seek to crown their pieces and eliminate each other completely; the go players seek to occupy the board in the manner they deem advantageous. As each player tries to "get at" his or her nominal opponent, the pattern of play affects the other two games. The red checkers player may study the rules of chess and kibitz the black chess player to move a piece which, he says, is to her advantage but in reality may or may not help the chess player; it will block a crucial move by his green checkers opponent, however. The chess players do likewise, and also the go players. All three sets of players "play" simultaneously; sometimes they are winning, sometimes they are losing.

Which of these games is "fundamental" to the "world" of the board? In transnational structuring, which was most fundamental to the world? The national struggle for power among nations? The struggle to divide world oil production and markets among oil companies? Or the longer-term technological succession between the nineteenth-century industrial system, based on coal transportation, and the twentieth-century technological system, based on oil?

Let us look at the three contending logics as they played out in world history, simplified from the material in this book:

Great Britain to Germany (realist logic): We are at war and therefore, to weaken you, I will blockade your ports.

Germany to Great Britain (realist logic): We are at war and I, to weaken you, shall help Turkey blockade the Dardanelles.

Standard Oil (transnational structuring): I cannot stop this war. To me the German blockade means that my oil fleet has been released from the German market. Rather than have it idle, I shall expand into Asian markets.

Anglo-Persian (transnational structuring): This war has crippled my adversary, Royal Dutch–Shell. If I could get the whole of Mesopotamia to myself, I could outcompete Russia, in which I have no oil; I could become a major world producer.

Royal Dutch–Shell (transnational structuring): To me the German blockade means I shall have a powerful competitor in Asia. To me the Turkish blockade means I no longer have a short supply route to Europe; I have few ships and a long route to the Dutch East Indies. I stand to lose a major oil field in Mesopotamia. I shall persuade the French that it is in their national interest to help me win.

Coal complex (technological transition; transnational structuring): Oil use is growing in this war, and it is a threat to us. I shall use the oil blockades to show how unreliable these quarrelsome oil companies are. I shall try to convince the government of (France, Great Britain, Germany, Japan, etc.) that we can do everything oil can do.

International oil industry to coal industry (technological transition; trans-

> national structuring): Our coal complex opponents have mastered a technology that is crucial to us. We must get it for ourselves and keep it from spreading. We must try to divide the members of the industry against themselves. We must seek to persuade governments not to use it, or to use it only on terms that will not threaten us.

We will return to the technological transition from oil to coal in a section below. Let us work a while longer with the chess-checkers-go game as an analogy for transnational structuring.

Through incredible concentration, the chess players ignore as best they can what the other two sets of players are doing, and the black chess player, through many adaptations of strategy, manages to pin the white chess player down and checkmate his king. Now we ask the victorious chess player to write her history of the game. It is clear that she will have a particular slant to what happened on the board and the "meaning" of the various moves. The player may think that at certain points she persuaded the green checkers player to move his pieces in a certain way, and thus "explain" the movement of the checkers pieces in terms of "chess logic." The victorious chess player "sees" the "total picture" in terms of her own particular game and its impact on the board. She will have only a partial idea of what has happened between the go players and the checkers players. At times she may have had to reroute an attack on the white chess pieces owing to a skirmish between the checkers players; but she will minimize this in describing how she achieved victory. Her account will be "accurate" within the confines of her logic and the rules she plays by. It will by no means be an "accurate" description of what happened on the board, nor will it be a particularly "fundamental" description.

That is exactly what Waltz does in his cursory description of synthetic rubber. To him it is one incident around one strategic commodity in a vast realist conflict. He sees this conflict as "fundamental," but in fact his realist "protagonists," states, are playing their "game" with technologies and capabilities that have been developed by other logics. Moreover, *what states can do* in terms of "ordering" such things as their domestic markets and technological research programs is determined by actors who move through the world system following nonstatal logics.[4]

In the game of chess-checkers-go, the best chess and the best checkers and the best go strategies will be played by those who are also good at the other games. It therefore makes sense that each player will try to incorporate the logic of the "other games" into his or her own game,

[4] The idea of contending logics, though not the game metaphor, owes much to Alker 1981, 1987.

and that each will to some extent seek control over how the other game players move. Waltz's capsule summary of synthetic rubber shows his profound indifference to the "other games" and illustrates how a poor understanding of the other aspects of international power not only fails to describe them on their own terms, but leads to bad realism. A *good* realist, truly concerned about the distribution of power and capabilities, would have to take an extreme interest in how multinational corporations were not only allocating technologies but affecting the articulation of policy and goals in his own state. Accepting the simplistic realist worldview of Waltz would set a good realist up for a sucker punch from other players with a more sophisticated understanding of the multidimensional game.

Though the need for hydrogenation technology and synthetic rubber was recognized as early as 1935, "the state" did not actually swat Standard Oil until 1942; synthetic rubber production did not hit high gear until 1944. That is a significant cost in time: the world hydrocarbon cartel controlled crucial technology for fourteen years plus the two it took to build new plant capacity. Countries like France had to muddle through with no hydrogenation at all. Nor was this the only thing achieved: the basing-point system, it will be recalled, bled states of revenue, a key component of their ability to fight, right through the entire world war. True, "the state," using coercion, can get corporations to go along with its logic; but corporations also have ways to affect the behavior of the state that regulates them.

This brings us to the question of the state's "relative autonomy" from interest groups—its ability, as it were, to ignore them or its permeability to influence. In chess-checkers-go, it is meaningless to say that the chess player is "relatively autonomous" from the checkers player. Both are constitutive elements of the "total game" and have many methods of persuasion to use on one another. That the state has a monopoly on the use of legitimate force does not make it supreme; it only means that those who seek to influence the state will resort to more subtle methods of control. The checkers player, trying to influence the chess player to move a knight to one square rather than another, does not violate the rules of checkers or get the chess player to violate the laws of chess; each operates according to the nature of what he is. A chess player kibitzed by a checkers player on the same board may occasionally, though not always, make moves that "make no sense" from a chess point of view, just as in the world a nation might pass up hydrogenation technology without quite "understanding" what it was doing. What is interesting about the French case, however, is that many key leaders knew perfectly well what they were doing.

Corporations mold the world through cash, information, and technology because that is what they have the most of; governments coerce because force is their endowment. But governments manipulate information, and corporations on occasion help groups purchase weapons and so influence the use of force. If we are worried about the "state's relative autonomy" or its "instrumentality," we are revealing more about a particular fetish for characterizing the state than about understanding how the world works. Too great a focus on seeing the world through state-centric logic chains us to an exiguous theoretical perspective.

Realism-mercantilism is not so much *wrong* as *partial.* The play of alliances, the balance of power, and the importance of military conflict are all rightly stressed. Yet the theory does not account for economic regulation and is prone to error as a consequence. This is especially true of the three principles identified in the introduction of this book:

1. States are the primary actors in the world system; all other entities are ultimately subordinate to them. Because firms hold both wealth and know-how, states use them to pursue power.
2. International competition among states creates an objective need for access to strategic resources. Interventionist policies to secure supplies develop where the "free market" is unreliable. Public discourse about "national security," the "public welfare," and "the national interest" largely reflects objective threats and efforts to deal with them.
3. The "free market" is "natural," and "strong interventionism in the market" is an artifact of state power. State power is therefore measurable by the "degree of intervention."

States are only one set of actors. Firms use states as resources, on a global scale. National interest rhetoric, though reflecting genuine crisis, also shields the more obscure objectives of private interests. Free markets and strong state power are not respectively "natural" or an artifact of the "national will." They may reflect the contending forces of transnational structuring, which sometimes build state regulatory strength to achieve certain ends and sometimes tear it down. In short, the three principles of transnational structuring constitute a contending logic in the world order and explain more, better, than realism-mercantilism:

1. Transnational firms view individual states as resources to provide conditions amenable to business. Large firms may pursue an agenda for rationalizing world markets across many states at a time. Tactics in each state will vary. Deregulation (weakening the state) and regulation (enhancing state power) are possible tools in a larger strategy. Of the many ways to influence state policies, the most effective is to become one of the players in that state's domestic politics.
2. States routinely engage in wars and other struggles for advantage with other states. These struggles provide good opportunities to

gain special privileges, dressed in the rhetorical garb of supporting the "national interest."

3. Like cartel agreements, the coalitions of influence that weaken or enhance state power are ephemeral. State authorities, once created, may take on an institutional life of their own or be captured by other interest groups.

More goes on in the international system, of a "fundamental" nature, than the realist theory admits; transnational structuring affects how states fight for power. The chess-checkers-go game may look anarchic; in fact its highly structured operating principles are extremely difficult to comprehend. Structured interplay that is at the limit of comprehension is not the same as anarchy: the stars at night may look like a random scattering, but getting to the point of understanding galaxies and our place in one of them took considerable human ingenuity over centuries.

A worldwide transnational structure determined whose trucks had synthetic rubber tires in 1942 and who could draw upon coal reserves to power planes, tanks, and submarines. Nations could play "catch up" and develop new industries, but only by paying a penalty in time and resources. By the time a nation has to "notice" that transnational structuring has denied it a capability, it is because transnational structuring has already fundamentally altered the flow and rhythm of power upon the board. Nations fought World War II within a structure of hydrocarbon technology and rules for its use that had been developed by extrastatal actors.

With regard to theories of international relations that are more receptive to considering the actions of economic actors, transnational structuring still points to aspects of the world system that current theories do not describe. Mercantile theory in the realist view would have states use their economic actors to maximize power; in world-systems theory, states can become the instrument of coalitions of economic interests that seek to use state power for their own ends. Most typically, "American capital" seeks to push "American imperialism," "British capital" pushes "British imperialism," and so on. Nothing in this book confirms or confutes such a view. The economically oriented international relations theories do not have much to say about how Standard Oil and I. G. Farben might reach an agreement on the way France's government should regulate its domestic hydrocarbon production and markets.

Transnational structuring operates through many kinds of states. The oil imperialism that tied up Iraq's development as a state and as an oil producer had an impact on Iraq's entire society. So far no one has really thought through the implications of the Federal Trade Commission's discovery that the world cartel agreements "proved applicable in all kinds of markets, in large countries and small, in industrialized economies and in agricultural and even 'undeveloped' economies."[5]

We therefore may distinguish transnational structuring from Lenin's and Hobson's brands of "imperialism." Wallerstein's world-systems approach is amenable to world cartels, but in spite of his detailed description of the interdependence between the "core" and the "periphery," he has not really shown us anything of the character or scale described in this book.

In the current state of theory transnational structuring is largely sui generis. The realist-mercantilist theory has been the primary obstacle to noticing how far it can occur. Speculation about the larger meaning of transnational structuring is best left to the theorists of the respective schools who care to undertake the task. Some may find that, as with the many fairly accessible primary materials presented here, it is easiest to deal with counterfactual arguments and evidence by ignoring them altogether.

TRANSNATIONAL STRUCTURING AND FRENCH ETATISM

This book began in 1981 as a short, "simple" test of the etatist thesis. The idea was to show that the French state had the power to overcome the resistance of the private sector and implement a national program of oil independence. Had the primary sources upheld that thesis, the book would not have grown to its current dimensions. The primary sources not only failed to uphold the statist thesis, they failed to show that the opposite argument—that France was a "weak" state—was of any greater utility. Arguing that France is strong or weak is important only if France is somehow unique or special. If France became "independent" it would be proved "strong"; otherwise it would be "weak." The importance of the adjectives stems from the literature that holds that France is an archetypal "strong state."

None of the descriptive vocabulary commonly used about France (or for that matter international relations) fits the event. This conclusion was not easy to reach. What happened in France was of a piece with what happened everywhere else. There is no current cultural or institutionalist theory of French politics that links the enhancement of oil regulatory authority in places such as Texas and France with production controls in Iraq and patent cartel agreements between American and German corporations.

[5] FTC1, p. 346.

This is an interesting intellectual problem. For proponents of the "etatist tradition" that goes back to Jean-Baptiste Colbert's mercantilism, it seems clear that there is a "pattern" of statist intervention in the economy into which the pattern of oil regulation fits.[6] This book, on the contrary, finds that the etatist tradition matters little because it explains only one country, whereas we must interpret the world hydrocarbon cartel across many countries.

To support the statist tradition, one might defend French oil policy against the considerable body of evidence presented here; alternatively, one might label the oil case an anomaly and retain the basic premises of traditional analyses. The first course requires falsifying this work; the second requires little more than sitting tight upon the conventional wisdom and hoping that more studies like this one do not happen. A more interesting challenge is to develop transnational structuring. This book does not furnish enough material to dismiss the etatist thesis, which is a generalization about the state's regulatory proclivities across many areas of the economy. To attack the whole tradition would require many works like this one. No one person could write such a body of literature. If several studies like this one were to be completed, one can imagine several outcomes. If we found a number of cases of deregulation due to transnational structuring as well as regulation, then the etatist thesis would be seriously challenged. If we found that, on the whole, France is "statist" because when transnational structuring occurs there it results in a regulated market more often than in other countries, then we would have to ask why this one country, and perhaps a few others like it, typically responded to transnational structuring in this way. It could lead us right back to the traditionalist or institutionalist arguments about the state, albeit with a new twist. The new twist would be the diminution of nationalist or public interest rationales for state behavior.

Transnational structuring does not replace all other theories of economic intervention, nor does it replace realism by explaining the causes of war. One can think of examples of government economic intervention, such as the interwar proliferation of fuel alcohol programs, whose *origins* conform to the convergence models described in chapter 1. The post–World War II creation of a worldwide "free market," which removed alcohol subsidy programs and gave oil more room to grow, was certainly the result of the combined desires of the international oil and automobile industries: transnational structuring worked to deregulate rather than regulate, and replaced a worldwide regulatory

[6] Examples of this tradition have already been cited in chapter 1, p. 37.

pattern that had developed through convergence.[7] Other cases of regulation may certainly be local—milk standards, for example. Still other cases that appear local may not be: the state's takeover of national railroads may be a matter of local politics in countries such as France and Germany but is certainly not a matter of purely local politics in countries whose railroads were built by foreign interests; and even in a country like France or Germany one might find that if railroad bonds are traded in international markets, international financial interests could play a major role in pushing for the nationalization of insolvent firms.

Perhaps transnational structuring is characteristic of the twentieth century and is a "new development" in the long history of etatism. Mercantilism has always been a public policy designed under pressure from private interests, however. "By a strange absurdity," says Adam Smith, business interests "regard the character of the sovereign as but an appendix to that of the merchant, as something which ought to be made subservient to it."[8] This argument is the basis of Smith's polemic against state interventionism in his *Wealth of Nations;* it is repeated by Ekelund and Tollison (1981). One might think a biographer of Colbert would endorse the realist-mercantilist thesis, but Cole often leans towards a neo-Smithian interpretation. English mercantilism, he writes, "was in good part the result of pressure from business groups with specific ends in view," but in France business interests "were made to behave in a manner which . . . the royal government . . . believed to be for the best interests of all concerned." This may seem to be state-centric analysis, but Cole then qualifies his distinction between French and British mercantilism, saying that the "difference may be more apparent than real. The English bourgeoisie had to bring its influence to bear on a public institution, the Parliament, in a semipublic manner. The French bourgeoisie could gain its ends by backstairs influence, private meetings with officials, and all the subtlety and intrigue that goes with the court of an absolute monarch."[9]

The process of rent seeking affects the evidence it generates. It is easier to write the statist version of French mercantilism, in the seventeenth century as well as the twentieth, because it is easier to find the "public story" than the more complicated backstairs influence, subtlety, and intrigue that were a part of Colbert's time as well as part and parcel of the world hydrocarbon cartel.

Does the influence of economic interest groups—the French

[7] Nowell 1985, pp. 118–131.
[8] Smith 1976, vol. 2, bk. 4, chap. vii, pt. iii, p. 154.
[9] Cole 1939, 2:553.

bourgeoisie—necessarily indicate transnational rent seeking? No: taken alone it merely calls into question the public, self-justifying rhetoric of the realist-mercantilist thesis. Confined to one nation, this looks more like a seventeenth-century version of the work of business-oriented pluralists such as Schattschneider (1935), Beard and Smith (1934), or Key (1967) than like transnational structuring.

Cole does furnish an early example of transnational structuring, however. The Compagnie de la Nacelle de Saint Pierre Fleurdelysée proposed in 1627 would have brought together French, Flemish, and Dutch interests into an extraordinary company that would have monopolized a wide variety of trade, from shipbuilding to such items as cheese, butter, crystal, and porcelain. Cardinal Richelieu approved the plan, and it received royal letters patent, but it was torpedoed by the Paris Parlement. Cole observes that "officials, merchants, and financiers who saw their interests threatened undoubtedly fought the proposals by every means in their power."[10] The political conflict bears striking similarities to the fight for control of the French oil market, with the difference that in the twentieth century the international interests succeeded. But transnational structuring does not have to result in successful regulation in order to exist as a process of political struggle. State goals and capabilities, and the perceptions of other interest groups, are influenced by the very fact of having had a struggle, no matter what the outcome.

We might say that transnational structuring diminished in world politics with the end of the pre–World War II cartels. This does not seem likely: The American foreign tax credit inaugurated by the American oil industry in 1950 was quietly written into French law in 1965.[11] The celebrated oil depletion allowance made its way into French law in 1953.[12] As the world market became more competitive, the cartelistic powers of the French government increased: state control over distribution points (gasoline stations) was established in 1964, enhancing the oil import quota authority of 1928 and providing an alternative "domestic" quota system that would endure assaults made in the name of free trade and the Common Market.[13]

It would be unwise to generalize about the frequency of transnational structuring, now or in the past. The documented cases are too

[10] Cole 1939, 1:170–173.

[11] Schvartz 1974, p. 32. In French the foreign tax credit is called the *régime du bénéfice consolidé;* see also Blair 1976, pp. 193–203.

[12] Schvartz 1974, p. 20. The French equivalent of "oil depletion allowance" is the *provision pour reconstitution de gisement,* or PRG. The probable motive was the development of limited oil reserves in Aquitaine and the search for oil in Algeria, then part of "France."

[13] Schvartz 1974, p. 209.

few, probably more through inattention than through lack of examples. The evidence of this book is that it has been, and is, an important part of politics among nations. Government economic intervention is a multicausal process. All capitalist countries share certain functional approaches to the economy, and the tools of intervention are the same everywhere. What varies is how frequently they are applied. Even the "liberal" United States has the "socialist" Tennessee Valley Authority and municipal utilities; all capitalist states have used tariffs; one finds production controls and market quotas in many places at many times. Through these tools, governments attempt to control economies; through these tools, interest groups influence governments' attempts to control economies.

I do not argue that all the etatist literature about France is wrong. A more useful approach is to integrate transnational structuring into the concept of etatism. This would radically reorder the whole idea in some ways, but in other ways it would not disturb a body of literature that is too big to ignore. As one example, Kuisel's thesis (1967,1981) about the "new generation" of technically trained and highly professional businessmen and civil servants as a major influence on the development of the French economy falls into the etatist tradition but is not incompatible with transnational structuring. We would have to reconsider what it was about Ernest Mercier and his generation's collaboration with the world oil cartel that was different from the family capitalism of an earlier era. Possibly the integration of France into the world hydrocarbon cartel required the technical competency and leadership of someone like Mercier, as opposed to the limited horizons of the French domestic oil companies.

Kuisel's thesis that with the twentieth century an identifiably new kind of business leader developed in France, and that this had an impact on the country's development, may be altogether true. Transnational structuring, however, alters our view of "what was done" by this class. In oil it promoted the development of a French capitalism that harmonized with the desiderata of major international interest groups. This required a different kind of business thinker than the cartelized, parochial family enterprises, called "Malthusians," of the late nineteenth-century French business class. The national cartel of French domestic firms was replaced by an international cartel of multinational firms; and much more than just an oil market was divided up.

Transnational structuring is a *process* that can deregulate or regulate. Once interest group conflict has led to the creation of a new regulatory agency, that agency may be able, given time, to alter the scope of its powers and change its relation to the regulated sector of the economy. Or it may be "captured" by yet another group of interests. I am describing how the game is played, not predicting its outcome, in the same manner that chess games remain full of surprises even though the rules are known. This book has shown how a fuller picture of the rules provides a better accounting of the facts.

Transnational structuring asks us to consider that both regulated and unregulated markets can be the artifacts of multiple-level international and domestic interest group struggles. I do not say statism does not exist; I attribute its existence to different reasons.

France's apparently homegrown etatism may in fact result from international processes that work through states and are difficult to see. Realism-mercantilism, and traditional etatism, have always acted on the supposition that "outside the logic of states" there really are no other fundamental forces. The oil case presented here is analogous to the Michelson-Morley experiment about the existence of ether. The physicists failed to find ether, but we have not: there are transnational forces that condition and shape the form of behavior of states.

What to do with this information? Opinions will differ. Theories such as realism and traditional approaches such as French etatism are wrong only to the extent that they spuriously explain specific instances of economic intervention or nonintervention. The more interesting question is how to reduce these theories to partial logics of the world system, in which multiple power struggles, following multiple rules, are going on at once.

Transnational Structuring as a Research Program

How is transnational structuring likely to fare as a research program? The answer is extremely doubtful and raises questions about likely cases, primary source availability, investigative resources, and even the individual abilities of the researchers. I will take up these issues one at a time.

On likely cases, we can distinguish between complex and simple transnational structuring. I have not introduced this distinction until now. A simple case of transnational structuring would be when one firm influences regulatory politics in its sector in one other country. Business arrangements between Canada and the United States are likely to be a fertile field of investigation,[14] and France and Belgium another.[15] The "Koreagate" scandal in the United States in the 1970s is highly suggestive. Possibly the simple cases of transnational structuring are

[14] See Johnson 1972.
[15] See Nouschi 1975.

more common and collectively exercise a greater impact on world politics than spectacular, complex cases like the one in this book.

Complex transnational structuring, which involves many countries and many firms, such as the hydrocarbon case, may be truly exceptional. Ervin Hexner's *International Cartels* and the 1942 Senate Hearings before the Committee on Patents suggest other major commodities that are possible cases of complex transnational structuring.[16] Today the international grain trade, the computer industry, and the aviation industry are likely arenas for transnational structuring.

Transnational structuring may depend to some extent on the number of players. There are now hundreds of oil companies operating worldwide instead of the half-dozen of consequence before World War II. Following Mancur Olson's *Logic of Collective Action,* we might conclude that as the companies become more numerous it becomes more difficult for them to reach a consensus, weakening their ability to push their collective agenda. Testing this in the case of oil poses formidable problems. Much of the regulatory structure that has allowed the oil companies to multiply into the hundreds was created before World War II as a consequence of transnational structuring. The oil companies are running wild in an international market environment created by their forebears. They flourish because they are in an environment designed by those like themselves: getting the permission of the Direction des Carburants to sell oil in France, even if one is an outsider, is an incremental adjustment compared with setting up the authority in the first place. A world hydrocarbon cartel in the style of that of 1928–1942 is a superfluity because the social need for the oil industry is now physically embedded in the structure of cities, the roads, the means to distribute goods, the cars owned by hundreds of millions.

A true test of oil's ability to push an international agenda for regulation or deregulation would occur only if the worldwide oil-consuming society began the transition to some other fuel, as it once did from coal to oil. In California, where alternative fuels have been on the political agenda for some years, the indications are mixed. Some of the oil companies appear to lack the capital, the know-how, or the leadership to produce the so-called clean gasolines that might make the emissions benefits of alternatives look less desirable and thus derail their use. There have been disagreements in strategy among major oil firms. Collectively, however, oil's resistance has remained formidable.[17]

[16] Hexner 1946; Pat.Hearings; see also Mirow and Maurer 1982.

[17] The likely appearance in Japan of alternative fuel regulation mimicking California's is probably a new permutation of transnational structuring. Japanese car manufacturers may be replicating domestically the demand for cars that they know they will need to

On *primary source availability,* we note that it is difficult to research contemporary cases because documents are not made available. The oil industry has been annoyed often enough by investigations such as that by the 1952 Federal Trade Commission that it now routinely destroys documents after a few years. Moreover, many corporations destroy documents after commissioning an official history, which is invariably written in a hagiographic tone and thus is of little real use except as a monument to the company's self-image, the modern equivalent of the tomb of the Medici. Unless documents have found their way into someone else's files, such as a politician or an agency, or have been leaked, access is nearly impossible. Even if one is lucky enough to have found a treasure of documents, a recent revision of copyright law may make it impossible to use them in their most potent form: the direct quotation. The bias is against finding information, and when it is found it must be used subject to restrictions and clearances.

Investigative resources are another sharp limitation on research into transnational structuring. More than one language is usually required. The research is time consuming and requires grants; many of the "best" documents cited in this book were not found until my second year of work abroad, and many scholars are not lucky enough to have two years of funding. Moreover, as will readily be seen from the citations in this book, much good material came from government investigations that were backed by subpoena power. The best material in the Federal Trade Commission Report was not declassified until the 1980s—notwithstanding that the Church committee *thought* it had declassified the study in 1975.

Not every sector of the international economy is going to provoke government investigation. Ironically, to investigate transnational structuring we need narrowly state-centric politicians and bureaucrats. From time to time they are presented with such incongruities—such as not having a synthetic rubber industry—that they are forced to bring to bear the legal staffs, consultants, and subpoena power that generate much more information, and more quickly, than the neurotic labors of a poor scholar wasting away in the dusty tedium of the archives. Wherever these kinds of government investigations are lacking—and they usually are—available evidence will most likely favor a state-centric interpretation of policy.

The abilities of researchers to understand what they have in front of them merits a brief discussion. Chess players, to return to our analogy, know exactly where they stand in relation to one another because of

make for the California market, thereby expanding the market base and lowering total costs ("MITI to Set Up Electric Power and Methanol Stations in Japan" 1992).

the international rating system for players. We could ask two city champions to analyze a world match between two brilliant international grand masters. They would probably do a fair job. But even with full knowledge of the rules and all the moves listed in front of them, their description and "understanding" of what was going on would not be equivalent to the "understanding" of a half dozen or so truly world-class contenders.

By extension, it is fair to say that a political scientist or historian analyzing a "realist" event such as a war, even with "knowledge of all the realist rules" *and* full evidence "under his nose" might still miss much of "what happened." How much more problematic, then, is transnational structuring. Not all the rules are known, and not all the evidence is available. And were the rules known and the evidence available, there is still the disturbing question of what it means to "put it all together" into an interpretation. Unless transnational structuring is seriously considered along with other interpretive approaches to political economic events, it will not develop as an approach.

Without an awareness of alternative rule structures, logical but fundamentally distorted analysis will result. Consider the two city chess champions. We hand them the recorded moves of black's victory over white in the game of chess-checkers-go. We omit any record of the checkers game and the go game that were played on the same board. We ask the two "to explain" why the chess pieces were moved in the way they were. They will "make sense" of the game according to the rules they know. Because chess-checkers-go uses an expanded board, some of the notations will be for squares that do not exist in standard chess. The two will throw out this information as an "error" and try to reconstruct "what must have happened" from within their known body of rules and strategies. They will see, nonetheless, some truly mystifying moves, some of which they call strokes of genius and others inexplicable idiocy. But they will create an intelligible order to the game, even if it means discarding or ignoring factual evidence. Their definition of an intelligible order is based on a rule structure shared with other chess players; if we ask them to accept a different set of interpretive rules, they must perforce leave the community of accepted belief that is the whole basis of their understanding of what they do, and what others do.

So it has been with the many standard histories of the world oil industry in France and elsewhere. Much in them is right; they all make sense; and they are wrong. Of all the reasons they are wrong, the principal one is this: the dominance of realist-mercantilist theory, not just in international relations theory but as an ideology in the common discourse of politics and history. Transnational structuring requires that we see a game that claims to be total, realism, as only partial.

In addition, scholarship requires productivity, and productivity requires richness of raw material. Since the raw material for the study of transnational structuring is anything but abundant, productive scholarship focuses on what is at hand: the national interest views promulgated by self-interested actors. There is a systemic bias to reproduce what is known rather than delve into what is unknown.

Transnational Structuring and Technological Transitions

It is one thing to divide up a market among firms in a cartel-like fashion; it is another thing to divide up the technology that will shape the evolution of world industry. The confrontation between coal and oil as related in this book *is* part of transnational structuring, but it has a different order of significance than the confrontation among rival oil interests. The confrontation between coal and oil looks like transnational structuring because firms use the same basic tools: governments are influenced with money or information, the public level of discourse is focused on some version of the "national interest," and the more important issues remain in back corridors. The tools of government intervention are the same: tariffs, taxes, equity participations, quotas, and so on.

The stakes in a technological transition differ from the competition that leads to ordinary cartels. If an oil firm goes under, as many have, other oil companies take up the slack. The struggle for market power goes on, but there is very little difference in the evolution of technique, in the development and spread of the technology that is the physical basis of society. In a technological transition, whole classes of people may face unemployment; the geographical resource that is the basis of a whole pattern of development may alter.[18] The face of society begins to change as the technology changes; urbanization; the ways of making war—all can be transformed.

In short, the stakes are much higher. The evolution of whole new patterns of technological development must be a political process, as opposed to a purely economic one. If that is so, then the struggle between coal and oil described in the previous chapter is a model of future conflicts that will be of the same magnitude, or greater.

The reason has to do with the nature of technological evolution. There is a tendency for each new technology to be more capital inten-

[18] Weber 1971.

sive than the one that preceded it. The problem of exiting from an existing investment will therefore increase with each new generation of productive technology; as each new system becomes more expensive, the loss inherent in "shutting it down" increases.

Hilferding thought this tendency in capitalism could actually lead to the end of competition.[19] Banks, worried about repayment of their loan capital and also, in the German system, about the viability of their shareholdings, protected their investments from the anarchy of market competition by totally cartelizing the capital-intensive productive sectors. The control of technological evolution, and of production, would eventually center in the banks. Neumann shows that Hilferding's argument is false: the cartelized economy generates surplus revenues for the cartelized industries.[20] Much of the revenue is retained by industry, allowing it to shake off the controlling influence of the banks. A competitive dynamic is restored *as a result of* the period of domination by "finance capital."

Still, Hilferding was pursuing an important idea. His general principle is that the owners of capital-intensive investments try to stabilize their markets by manipulating supply, prices, and demand. As time goes on, production becomes more capital intensive, and the need for this stabilization increases. But the stabilizing policies can come from the political arena as well as from a tight oligopoly of banks; and even where there is a tight banking oligopoly, government regulation may still be the preferred instrument of control. In Japan today, for example, automobile inspections force vehicles off the road after six years, an obvious demand management device that helps to keep the factories humming.[21] In the United States the automobile firms are independent of the banks and even run their own credit operations; they are still able to get "voluntary import quotas" to protect their interests. The basic idea is the same: political power is used to shore up capital-intensive investments.

The kinds of relationships between government and industry that allow political power to smooth out the effects of competition and the business cycle are also the kinds of relationships that would be used by a "defending" industrial sector against the encroachment of a new one. New industry, to make any headway at all, must therefore compete against the political power of its precedessor; the more capital intensive the predecessor, the greater will be its tenacity in the face of a challenge. As each new successively more capital-intensive technological transition

[19] Hilferding 1970, p. 328.
[20] Neumann 1963.
[21] Womack, Jones, and Roos 1990, pp. 67, 247.

occurs, it will become a progressively more politicized process. The transition from coal to oil therefore was a more intense regulatory struggle than the transition from wind power to coal-powered ocean-going technology,[22] and the transition from oil to something else will easily outdistance in intensity the transition from coal to oil.

Political struggles over technological transitions should also become more extensive in space as well as more intensive in terms of the resources dedicated to the political struggle. The more capital intensive production becomes, the more extensive are the markets needed to generate a sufficient return on the investment. On the one hand, capital-intensive production must seek to globalize, and to the extent that access to many markets is needed, it becomes "free trade" oriented or, to be precise, oriented to smashing down the protective barriers of smaller firms and older production processes. On the other hand, the same capital intensiveness increases the need to regulate and to protect the investment from competitive effects. So industry will become increasingly global and seek to be increasingly regulated: complex transnational structuring should, over time, become more frequent rather than less so.

Part of this process will include cycles of deregulation as coalitions determine, temporarily, that they could fare better with "no" regulation than with active discrimination against them, just as Standard Oil preferred an "unregulated" market in France in 1920. Such shifts in preferences are not ideological and are not culturally based, but are due to specific circumstances. Consequently we will see the seeming contradiction that the increased power of multinational corporations will also lead to increased regulatory power of states; far from having their "sovereignty at bay," states will find their regulatory powers greatly strengthened in some dimensions, exactly as happened in France.

Since corporations are *not* in total control, however, we will find that the logic of state power continues to apply in some dimensions. The game of world politics will be to determine the limits to that power, which cannot be deduced from theoretical principles and which in any case will be in constant flux. The increasing politicization of technological development may represent an opportunity: environmental challenges seem to present compelling cases for "state logic" to intervene over "transnational structuring logic." That is, if Standard Oil's agreements with I. G. Farben could be broken, even belatedly, by government intervention during World War II, then it is theoretically possible that the overwhelming propensity of the business sector to try to control

[22] France subsidized sailing vessels in the late nineteenth century (Shonfield 1965, pp. 74–75).

politics for its own ends can be broken for some other compelling need. The regulatory attempt to forge a transition from oil to environmentally preferable alternative fuels in California is therefore an interesting contemporary test case of a fundamental, deeply rooted element of world politics.

* * *

The evolution of the world oil market and its hydrocarbon processing technologies expressed itself through rivalries among states, rivalries among oil firms, and rivalries among sectors such as coal, oil, and finance. Ordinary theories of state behavior describe the process partially and, because of that, falsely. I have therefore had to rework some traditional approaches to the French state and to the system of states of which France is a part. In a world of transnational structuring, no state develops in isolation. The idea that a history of the oil industry could be written without reference to the worldwide universe of affected interests is the first step into mythmaking.

Regulation nearly enacted before World War I was part of a pan-European effort to make room for Mesopotamian oil development. During the Great War, the tardy efforts of the French state to regulate oil responded primarily to the needs of Royal Dutch–Shell; in the immediate aftermath of that war, French policy was reversed by the aggressive action of Standard Oil. The development of French policy in the 1920s was linked to the response of French domestic firms to multinational penetration, and also to their efforts to increase imports of Soviet Oil. The Mesopotamian oil faction led by the multinationals (principally Standard Oil and the Banque de Paris et des Pays Bas) was the primary influence leading to the creation of the Compagnie Française des Pétroles; the Belgian Pétrofina and French firms were the primary backers of the Office National des Combustibles Liquides. The oil import law of 1928 responded to the "as is" desiderata of the world oil cartel and worked to restrict Russian oil imports; the world hydrocarbon cartel outflanked the autarchic impulses of French chemical and coal interests who sought to pursue synthetic fuels. These events were not peculiar to France but occurred in many other places around the world.

The "mercantile state" of France did little more in its policies than gyrate back and forth among the conflicting pressures of a complex array of interest groups. The "national defense" pretensions of French oil regulation were nonsense and were recognized as such by government leaders. This surely seems to be a weak state: yet it was precisely during this period that the government's legal and practical authority over the oil market developed and grew as it had at no other time.

Such was the paradox of the mercantile illusion: a state seeming to stand strong over a host of rival interest groups—its powers in reality created by them—and yet gaining real powers nonetheless. The oil market was most certainly regulated, though to what kind of "national interest" it is hard to say. So it was all over the world, whose separate dominions became, for the sultans of hydrocarbons, the building blocks of an invisible empire.

"Influence," wrote Schattschneider in 1935, "is the possession of those who have established their supremacy in the invisible empires outside of what is ordinarily known as government."[23] He ended his study of rent seeking and the tariff with a strong appeal for more openness in government and greater democratization of the processes by which economic decisions are made. In the absence of such democratization, "the function of pressure politics is to reconcile formal political democracy and economic autocracy."[24] Recurrent circumstances produce recurrent concerns: nearly forty years later, the French deputy Schvartz would write, "We have to recognize that a number of decisions concerning oil in our country have been made without . . . consultation of the parliament, the press, or public opinion."[25] That such similar appeals should come from such diverse cultures and times indicates the problematic nature of business influence for the pragmatic concerns of the state and theories of governance. It also points to the major obstacles in the way of studying transnational structuring: the frequent invisibility of evidence and the belief that the practice is somehow "wrong" or "exceptional" and therefore not a routine part of the affairs of nations.

These moral sentiments have little to do with the study of the world hydrocarbon cartel, whose actions violated a number of the most sacrosanct officially promulgated norms of state and society: the liberal principles of free competition; the answerability of governments to their publics; the sovereign power of nations to maximize their strategic capabilities. These norms were, and are, merely common deceits. The world is far too complex to be encompassed within these comforting but empirically inadequate concepts. The agents of transnational structuring do not refer to them in defining their worldwide priorities, save to employ them as hypocritical masks to conceal their own manipulations. Ironically, it is through this very statist hypocrisy that, throughout the world, both the reality and the illusion of sovereignty are scattered.

[23] Schattschneider 1935, p. 287.

[24] Schattschneider 1935, p. 287; cf. Weber 1968, pp. 283–284, where the "big capitalistic interests" tend to favor "monocracy" because it "is, from their point of view, more 'discreet.'"

[25] Schvartz 1974, p. 233.

Bibliography

Archival and Other Primary Sources

Archives Nationales, Paris

(Arch.Nat. in notes)

324AP77, André Tardieu files on Compagnie Française des Pétroles

Desmarais Frères, series 130AQ

130AQ1
130AQ3
130AQ4
130AQ6
130AQ7
130AQ8
130AQ9
130AQ10
130AQ11
130AQ16
130AQ20
130AQ24
130AQ32
130AQ35
130AQ41
130AQ82
130AQ83
130AQ94

Series on Sociétés Filiales de Gaz de France, 1941–1963

Materials on World War I period and nineteenth century.

F^{7}13978
F^{7}7662
F^{12}6848^{c}
F^{23}81
F^{23}84
F^{12}7662
F^{12}7715
F^{12}7716
F^{30}1506

Ministère de l'Industrie: (files provided courtesy of Philippe Mullerfeuga). (Min.Ind. in notes)

B245
12892

Archives of the Assemblée Nationale, Château de Versailles

(Ass.Nat. in notes)

Procès Verbaux de la Commission des Pétroles, 1927–1928

Procès Verbaux de la Commission des Pétroles, 1933 projet de monopole
Procès Verbaux de la Commission des Pétroles, Rapport Pineau, 1934
Procès Verbaux de la Commission des Mines et de la Force Motrice, 1929–1931
Procès Verbaux de la Commission des Affaires Etrangères, 1929–1931, vol. 1
See also parliamentary hearings partially reprinted in the *Revue pétrolifère,* under published primary sources.

Archives of the Compagnie Française des Pétroles, Paris

(CFP in notes)

81.1/41 (Courtesy of Philippe Mullerfeuga) 81.1/59 81.1/62 81.1/63 81.1/81 81.1/87 82.7/1 11SGA15

Archives of the Ministère des Affaires Etrangères, Paris

(MAE in notes)

Levant, Turquie, 1918–1940, vols. 429, 430, 431. Also microfilm reel, vols. 657–662
Levant, Syrie-Liban, 1918–1929, no. 344
E-Levant, Mésopotamie-Irak, 1918–1929, vols. 32, 33, 34, 35, Irak, 1930–1940, vols. 66, 67
Europe-Russie, 1918–1929, vols. 526, 527, 528, 529, 530, Etats-Unis, 1918–1929, vols. 250, 251
Papiers d'Agents, André Tardieu, vol. 56, "dossier Tardieu," and vol. 57, sous-série "Commissariat Général aux Essences et Combustibles"
Papiers d'Agents, Paul Cambon, vol. 4, dossier 9

Ministère des Finances, Paris

(Fin. in notes)

B9808 B32310 B32311 B32312 B32313 B32314 B32866 B34034 B27305 B31828 B32023 B32308 B32309

Hoover Institution Archives, Stanford, California

(Hoover Inst. in notes)

Collection title: Louis Loucheur
Box 3A, folder 13
Box 2, folder 11
Box 9, folder 2

Box 11, folder M
Collection title: Renée de Chambrun; Deposition of E. Zédet

Harry S. Truman Library, Independence, Missouri

(FTC2 in notes)

Report of the Federal Trade Commission on Petroleum Cartels, Confidential File, Truman Papers. Forty-two pages of deleted material from "Report to the FTC by Its Staff on the International Petroleum Cartel" (1952) listed below under published primary sources and abbreviated FTC1. Although the 1975 Church committee (see below) did publish part of the formerly classified portions of FTC1, they missed this material.

Rockefeller Archive Center, Tarrytown, New York

(Rock.Arch. in notes)

RG 1 John D. Rockefeller Business Correspondence, 1879–1897, box 65, folder 483
JDR Jr. Business Interests, box 126, folder SO Indiana, 1919–1923 (87.1.S91)
JDR Jr. Business Interests, box 137, folder SONY; folder Vacuum Oil (87.1.S91)

Published Primary Sources

French Parliamentary Documents, Nineteenth Century

Nineteenth-century forms, which vary considerably, are cited in full; e.g., *Annales du Sénat et de la Chambre des Députés.*

Annales de l'Assemblée Nationale, vol 3, 27 May 1871, pp. 146–147
Annales de l'Assemblée Nationale, vol. 28, 29 December 1873, pp. 614–616
Annales de l'Assemblée Nationale, vol. 28, annexes, 17 December 1873, pp. 270–271
Annales de l'Assemblée Nationale, vol. 15, annexes, 5 February 1873, pp. 287–288
Annales du Sénat et de la Chambre des Députés, vol. 5, 29 April 1880, pp. 190–222; vol. 5, 30 April 1880, pp. 233–255
Annales de la Chambre des Députés, Débats, 29 June 1893, pp. 830–851
Annales de la Chambre des Députés, Débats, Session Extraordinaire, 19 December 1893, pp. 460–461

French Parliamentary Documents, Twentieth Century

See also bibliography in Nowell 1983.

All documents, debates, and reports, Senate and Chamber of Deputies, are separate series of the *Journal Officiel.* Some research libraries in the United States have the less compendious *Annales* series of parliamentary transcripts. If the word *Annales* appears on the title page of the published volume, the pagination will be different and materials found by date only, if at all.

Débats Parlementaires de la Chambre des Députés (Déb.Ch.)
Débats Parlementaires du Sénat (Déb.Sén.)
Documents Parlementaires de la Chambre des Députés (Doc.Ch.)
Documents Parlementaires du Sénat (Doc.Sén)
Déb.Ch., 6 March 1902, pp. 1193–1205.
Déb.Ch., 28 February 1903, pp. 923–936
Doc.Ch., annexe 988, 11 June 1903, pp. 624–630
Doc.Ch., annexe 3382, 15 January 1914, pp. 18–19
Doc.Ch., annexe 227, 3 July 1914, pp. 1935–1938
Déb.Ch., 26 April 1920, pp. 1367–1385
Déb.Ch., 10 July 1923, pp. 3313–3321
Déb.Sén., 30 June 1921, p. 1471
Doc.Sén., annexe 77, 18 February 1926, pp. 73–74
Doc.Ch., annexe 5170, 6 December 1927, pp. 355–372
Déb.Ch., 6 March 1928, pp. 1210–1211, 1218–1224
Déb.Ch., 7 March 1928, pp. 1249–1278
Doc.Ch., annexe 3110, 28 March 1930, p. 368
Doc.Ch., annexe 3173, vol. 1, 28 March 1930, pp. 492–495
Doc.Ch., annexe 3366, vol. 1, 3 June 1930, pp. 829–831
Doc.Ch., annexe 3366, vol. 1, 4 July 1930, pp. 1094–1096
Déb.Ch., 8 July 1930, pp. 2966–2984
Doc.Ch., annexe 3729, vol. 1, 8 July 1930, pp. 1108–1109
Doc.Ch., annexe 3787, vol. 2, 10 July 1930, pp. 1375–1377
Déb.Ch., 11 February 1931, pp. 570–583
Déb.Ch., 12 February 1931, pp. 626–636
Doc.Ch., annexe 4764, 12 March 1931, pp. 587–590
Déb.Ch., 24 March 1931, pp. 2151–2167
Déb.Ch., 29 March 1931, pp. 2181–2192
Déb.Ch., 29 March 1931, pp. 2096–2113
Doc.Ch., annexe 1960, 31 May 1933, pp. 1222–1223
Doc.Ch., annexe 5059, 17 January 1939, pp. 42–56

American Congressional Documents

(FTC1 in notes)

International Petroleum Cartel, Staff Report to the Federal Trade Commission, Select Committee on Small Business, U.S. Senate, 82d Cong., 2d sess., 22 August 1952 (Washington, D.C.: USGPO, 1952).

Extensive portions of this report were classified. Some were published in the Multinational Corporations report (1975) below. The bulk of the forty-two declassified pages is in the Harry S. Truman Archives, cited above.

U.S. Senate, Senate Foreign Relations Committee, Subcommittee on Multinational Corporations, Multinational Corporations and United States Foreign Policy, 93d Cong., 2d sess. (Washington, D.C.: USGPO, 1975), part 8, app. 5, pp. 529–531.

(Pat.Hearings in notes)

Hearings before the Committee on Patents, S. 2303 and S. 2491, part 7 (31 July, 3–4 August 1942) and part 8 (4, 5, 12 August 1942) 77th Cong. 2d sess. (Washington, D.C.: USGPO, 1942).

La Revue Pétrolifère

Reprints of testimony before the Commission des Pétroles under the general title, "L'Enquête parlementaire."
3 March 1928, no. 259, pp. 281–286.
10 March 1928, no. 260, pp. 329–335.
17 March 1928, no. 261, pp. 361–364.
24 March 1928, no. 262, pp. 393–399.
31 March 1928, no. 263, pp. 435–439.
21 April 1928, no. 266, pp. 545–552.
28 April 1928, no. 267, pp. 581–587.
5 May 1928, no. 268, pp. 623–631.
12 May 1928, no. 269, pp. 661–667.
19 May 1928, no. 270, pp. 691–702.

SECONDARY SOURCES

Adelman, Morris A. 1972. *The World Petroleum Market.* Baltimore: Johns Hopkins University Press.

Alker, Hayward R., Jr. 1981. "The Dialectical Foundations of Global Disparities." *International Studies Quarterly* 29 (March): 69–98.

———. 1987. "Fairy Tales, Tragedies, and World Histories: Towards Interpretive Story Grammars as Possibilist World Models." *Behaviormetrika* 21:1–28.

Almond, Gabriel, and G. B. Powell, Jr. 1978. *Comparative Politics: System, Process, and Policy.* 2d ed. Boston: Little, Brown.

Althusser, Louis. 1970. *Reading Capital.* Trans. Ben Brewster. London: New Left Books.

American Committee for the Independence of Armenia. 1925. *The Senate Should Reject the Turkish Treaty.* New York: The Committee.

Amin, Samir. 1974. *Accumulation on a World Scale: A Critique of the Theory of Underdevelopment.* Trans. Brian Pearce: New York: Monthly Review Press.

André, Robert. 1910. *L'Industrie du commerce et du pétrole en France.* Paris: Arthur Rousseau.

Antonius, George. 1939. *The Arab Awakening: The Story of the Arab National Movement.* Philadelphia: J. B. Lippincott.

Apostol, Paul, and Alexandre Michelson. 1922. *La Lutte pour le pétrole et la Russie.* Paris: Payot.

Arendt, Hannah. 1966. *The Origins of Totalitarianism.* New York: Harcourt, Brace, and World. Originally published 1951.

Art, Robert J. 1973. "Influence of Foreign Policy on Seapower: New Weapons and Weltpolitik in Wilhelmian Germany." In *Sage Professional Papers: International Studies Series,* ed. Vincent Davis and Maurice A. East, vol. 2, ser. 02-019. Beverley Hills, Calif.: Sage.

Ashley, Richard K. 1984. "The Poverty of Neo-Realism." *International Organization* 38 (spring): 225–286.

Augé-Laribé, Michel. 1950. *La Politique agricole de la France de 1880 à 1940.* Paris: Presses Universitaires.

Baldacci, A. 1926. "Les Pétroles albanais dans les négociations anglo-italiennes." *Revue économique internationale* 1 (February): 303–327.

Baron, Charles. 1933. "Rapport No. 1960." In *Documents Parlementaires de la Chambre des Députés,* 31 May, pp. 1222–1223.

Bates, J. Leonard. 1963. *The Origins of Teapot Dome: Progressives, Parties, and Petroleum, 1909–1921.* Urbana: University of Illinois Press.

Beard, Charles A., and G. H. E. Smith. 1934. *The Idea of National Interest: An Analytical Study in American Foreign Policy.* New York: Macmillan.

Bérenger, Henry. 1920. *Le Pétrole et la France.* Paris: E. Flammarion.

———. 1926. *France and Her Capacity to Pay.* Paris: Reference Service on International Affairs of the American Library in Paris.

Bergier, Jacques, and Bernard Thomas. 1968. *La Guerre secrète du pétrole.* Paris: Editions Denoël.

Bhagwati, Jagdish. 1982. "Directly Unproductive, Profit-Seeking (DUP) Activities." *Journal of Political Economy* 90, no. 5: 988–1002.

Biersteker, Thomas J. 1988. *Multinationals, the State and Control of the Economy: The Political Economy of Indigenization in Nigeria.* New Haven: Yale University Press.

Birnbaum, Pierre. 1982. *The Heights of Power: An Essay on the Power Elite in France.* Trans. Arthur Goldhammer. Chicago: University of Chicago Press.

Blair, John M. 1976. *The Control of Oil.* New York: Random House.

Block, Fred. 1984. "The Ruling Class Does Not Rule: Notes on a Marxist Theory of the State." In *Political Economy: Readings in the Politics and Economics of American Public Policy,* ed. Thomas Ferguson and Joel Rogers, pp. 32–46. Armonk, N.Y.: M. E. Sharpe.

Borkin, Joseph. 1978. *The Crime and Punishment of I. G. Farben.* New York: Macmillan.

Brewer, Anthony. 1980. *Marxist Theories of Imperialism: A Critical Survey.* London: Routledge and Kegan Paul.

Bronson, H. E. 1972. "Continentalism and Canadian Agriculture." In *Capitalism and the National Question in Canada,* ed. Gary Teeple, pp. 122–140. Toronto: University of Toronto Press.

Brunner, Christopher. 1930. *The Problem of Oil.* London: Ernest Benn.

Buchanan, James, Charles Rowley, and Robert Tollison. 1987. *Deficits*. Oxford: Basil Blackwell.

Buck, Philip W. 1964. *The Politics of Mercantilism*. New York: Octagon. Originally published 1942.

Buell, Raymond Leslie. 1925. *International Relations*. New York: Henry Holt.

Bull, Hedley. 1977. *The Anarchical Society: A Study of Order in World Politics*. New York: Columbia University Press.

Calfas, P. 1921. "Le Chemin de fer de Bagdad." *Le Génie civil* 78 (8 January): 25–29.

Cardoso, Henrique Fernando, and Enzo Faletto. 1979. *Dependency and Development in Latin America*. Trans. M. M. Urquidi. Berkeley and Los Angeles: University of California Press.

Carr, Edward Hallet. 1964. *The Twenty Years' Crisis, 1919–1939*. New York: Harper and Row.

Chauveau, C. 1916. *La France agricole et la guerre*. Paris: Librairie Baillière.

Chazel, A., and H. Poyet. 1963. *L'Economie mixte*. Paris: Presses Universitaires de France.

Colander, David C., ed. 1984. *Neoclassical Political Economy: The Analysis of Rent-Seeking and DUP Activities*. Cambridge, Mass.: Ballinger.

Cole, Charles Woolsey. 1931. *French Mercantilist Doctrines before Colbert*. New York: Richard R. Smith.

——. 1939. *Colbert and a Century of French Mercantilism*. 2 vols. New York: Columbia University Press.

——. 1943. *French Mercantilism: 1683–1700*. New York: Columbia University Press.

Corley, T. A. B. 1983a. "Strategic Factors in the Growth of a Multinational Enterprise: the Burmah Oil Company, 1886–1928." In *Growth of International Business*, ed. Mark Casson, pp. 214–235. London: George Allen and Unwin.

——. 1983b. *The History of the Burmah Oil Company*. Vol. 1, *1886–1924*. London: Heinemann.

——. 1988. *The History of the Burmah Oil Company*. Vol. 2, *1924–1966*. London: Heinemann.

Coston, Henry. 1975. *Dictionnaire des dynasties bourgeoises et du monde des affaires*. Paris: Editions Alain Moreau.

Darby, H. C., and Harold Fullard, eds. 1970. *The New Cambridge Modern History Atlas*. Cambridge: Cambridge University Press.

Dégouy, Contre-Amiral. 1920. "Le Pétrole et la marine." *Revue des deux mondes*, 6th ser., 56 (April): 661–686.

Délégation de la Republique Arménienne. 1922. *L'Arménie au point de vue économique*. Paris: Presses Universitaires de France.

DeNovo, John. 1956. "Movement for an Aggressive American Oil Policy Abroad, 1918–1920." *American Historical Review* 61 (July): 854–876.

Duvall, Raymond, and Alexander Wendt. 1987. "The International Capital Regime and the Internationalization of the State." Paper prepared for the German-American Conference on International Relations Theory, Bad Homburg, Federal Republic of Germany, 31 May–4 June.

Earle, E. M. 1925. "The Turkish Petroleum Company." *Political Science Quarterly* 39 (June): 265–279.

Ekelund, Robert B., Jr., and Robert D. Tollison. 1981. *Mercantilism as a Rent-Seeking Society: Economic Regulation in Historical Perspective.* College Station: Texas A&M Press.

Encarnation, Dennis J., and Louis T. Wells, Jr. 1985. "Sovereignty en Garde: Negotiating with Foreign Investors." *International Organization* 39 (winter): 46–78.

Engler, Robert. 1976. *The Politics of Oil: A Study of Private Power and Democratic Directions.* 2d ed. Chicago: University of Chicago Press.

Enos, John L. 1962. *Petroleum Progress and Profits: A History of Process Innovation.* Cambridge: MIT Press.

Erard, N. R. 1934. *World Petroleum Directory.* New York: Russell Palmer.

Evans, Laurence. 1965. *United States Policy and the Partition of Turkey, 1914–1924.* Baltimore: Johns Hopkins University Press.

Evans, Peter. 1979. *Dependent Development: The Alliance of Multinational, State, and Local Capital in Brazil.* Princeton: Princeton University Press.

Fanning, Leonard M., ed. 1945. *Our Oil Resources.* New York: McGraw-Hill.

Faure, Edgar. 1939. *Le Pétrole dans la paix et dans la guerre.* Paris: Editions de la Nouvelle Revue Critique.

Feigenbaum, Harvey B. 1985. *The Politics of Public Enterprise: Oil and the French State.* Princeton: Princeton University Press.

Feis, Herbert 1964. *Europe: The World's Banker, 1870–1914.* New York: Augustus M. Kelley. Originally published 1930.

Ferrier, R. W. 1982. *The History of the British Petroleum Company.* Vol. 1, *The Developing Years, 1901–1932.* Cambridge: Cambridge University Press.

Fichte, Johann Gottlieb. 1980. *L'état commercial fermé.* Trans. from German by Daniel Schulthess. Lausanne: Editions de l'Age d'Homme. Originally published 1800.

Fischer, Fritz. 1967. *Germany's Aims in the First World War.* Paper ed. New York: W. W. Norton.

Fischer, Louis. 1926. *Oil Imperialism: The International Struggle for Petroleum.* New York: International.

Fontaine, Pierre. 1960. *Le Pétrole du Moyen Orient et les trusts.* Paris: Sept Couleurs.

———. 1961. *Alerte au Pétrole Franco-Saharien.* Paris: Sept Couleurs.

———. 1967. *L'Aventure du pétrole français.* Paris: Sept Couleurs.

Forbin, Victor. 1940. *Le Pétrole dans le monde.* Paris: Payot.

Frank, André Gunder. 1982. "Crisis of Ideology and Ideology of Crisis." In *Dynamics of Global Crisis,* ed. Samir Amin, pp. 109–166. New York: Monthly Review Press.

Frankel, P. H. 1969. *The Essentials of Petroleum: A Key to Oil Economics.* New York: Augustus M. Kelley. Originally published 1946.

Gargas, Sigisimund. 1927. "Politique pétrolière dans les Pays-Bas." *Revue économique internationale* 1 (January): 71–114.

Gerretson, F. C. 1953–1957. *The History of the Royal Dutch.* 4 vols.; vol. 1: 1953; vol. 2: 1955; vols. 3 and 4: 1957. Leiden: E. J. Brill.

Gibb, George Sweet, and Evelyn K. Knowlton. 1956. *The History of the Standard Oil Company (New Jersey): The Resurgent Years, 1911–1927.* New York: Harper and Row.

Gibbs, David N. 1991. *The Political Economy of Third World Intervention: Mines, Money, and U.S. Policy in the Congo Crisis.* Chicago: University of Chicago Press.

Gilpin, Robert. 1975. *U.S. Power and the Multinational Corporation: The Political Economy of Foreign Direct Investment.* New York: Basic Books.

———. 1981. *War and Change in International Politics.* Cambridge: Cambridge University Press.

———. 1987. *The Political Economy of International Relations.* Princeton, N.J.: Princeton University Press.

Golob, Eugene Owen. 1944. *The Méline Tariff: French Agriculture and Nationalist Economic Policy.* New York: Columbia University Press.

Goodspeed, D. J. 1977. *The German Wars: 1914–1945.* Boston: Houghton Mifflin.

Gordon, Richard. 1970. *The Evolution of Energy Policy in Western Europe: The Reluctant Retreat from Coal.* New York: Praeger.

Gouldner, Alvin W. 1970. *The Coming Crisis of Western Sociology.* New York: Basic Books.

Goulévitch, Alexis de. 1924. "Le Pétrole russe: La Défense des intérêts privés engagés dans l'industrie pétrolière russe." *Revue économique internationale* 4 (December): 320–333.

———. 1927. "Le Pétrole russe." *Revue économique internationale* 4 (November): 232–246.

Gourevitch, Peter. 1978. "The Second Image Reversed: The International Sources of Domestic Politics." *International Organization* 32 (autumn): 881–912.

Green, Diana. 1983. "Strategic Management and the State: France." In *Industrial Crisis: A Comparative Study of the State and Industry,* ed. Kenneth Dyson and Stephen Wilks. New York: St. Martin's Press.

Grunberg, Isabelle. 1990. "Exploring the 'Myth' of Hegemonic Stability." *International Organization* 44 (autumn): 431–477.

Guide pratique de la réglementation pétrolière. 1984. Paris: Comité Professionel du Pétrole.

Gulbenkian, Calouste S. 1945. "Memoirs of Calouste Sarkis Gulbenkian with Particular Relation to the Origins and Foundation of the Iraq Petroleum Company Limited" (16 September). Washington, D.C., U.S. State Department Library.

Haas, Ernest. 1964. *Beyond the Nation-State: Functionalism and International Organization.* Stanford, Calif.: Stanford University Press.

Hall, Peter A. 1984. "Patterns of Economic Policy: An Organizational Approach." In *The State in Capitalist Europe,* ed. Stephen Bornstein, David Held, and Joel Krieger, pp. 21–53. London: George Allen and Unwin.

———. 1986. *Governing the Economy: The Politics of State Intervention in Britain and France.* New York: Oxford University Press.

Hall, W. A. 1919. "Oil Prospects in France." *Petroleum World* 16 (January): 16.

Hamilton, Alexander. 1968. *The Industrial and Commercial Correspondence of Alexander Hamilton Anticipating the Report on Manufactures.* Ed. Arthur H. Cole. New York: Augustus M. Kelley. Originally published 1792.

Heckscher, Eli F. 1935. *Mercantilism.* Trans. Mendel Shapiro. 2 vols. New York: Macmillan.

Henriques, Robert. 1960. *Bearsted: A Biography of Marcus Samuel, First Viscount Bearsted and Founder of "Shell" Transport and Trading Company.* New York: Viking Press.

——. 1966. *Sir Robert Waley Cohen, 1877–1952.* London: Secker and Warburg.

Hewins, Ralph. 1957. *Mr. Five Per Cent: The Biography of Calouste Gulbenkian.* London: Hutchinson.

Hexner, Ervin. 1946. *International Cartels.* London: Sir Isaac Pitman and Sons.

Hidy, Muriel E., and Ralph W. Hidy. 1955. *The History of the Standard Oil Company (New Jersey): Pioneering in Big Business, 1882–1911.* New York: Harper and Brothers.

Hilferding, Rudolf. 1970. *Le Capital financier: Etude sur le développement récent du capitalisme.* Trans. from German by Yvon Bourdet. Paris: Editions de Minuit. Originally published 1910.

Hirschman, Albert O. 1980. *National Power and the Structure of Foreign Trade.* Berkeley and Los Angeles: University of California Press. Originally published 1945.

Hobson, J. A. 1902. *Imperialism: A Study.* London: James Nisbet.

——. 1966. *International Trade: An Application of Economic Theory.* New York: Augustus M. Kelley. Originally published 1904.

"Home Sources of Oil Supply." 1919. *Petroleum World.* 16 (May): 214.

Hovannisian, Richard G. 1967. *Armenia on the Road to Independence, 1918.* Berkeley and Los Angeles: University of California Press.

Hubbard, Gustave. 1903. "Annexe No. 988." In *Documents Parlémentaires de la Chambres des deputés,* 11 June, pp. 624–630.

Huntington, Samuel. 1968. *Political Order in Changing Societies.* New Haven: Yale University Press.

———. 1973. "Transnational Organizations in World Politics." *World Politics* 25 (April): 333–368.

———. 1984. "Congressional Responses to the Twentieth Century." In *Political Economy: Readings in the Politics and Economics of American Public Policy,* ed. Thomas Ferguson and Joel Rogers, pp. 180–202. Armonk, N.Y.: M. E. Sharpe.

Hymer, S. A. 1972. "Internationalization of Capital." In *La Croissance et la grande firme multinationale,* pp. 581–601. Colloque International 549, 28–30 September, Centre National de la Recherche Scientifique. Paris: Editions du CNRS.

Jeanneny, Jean-Noel. 1981. *L'Argent câché: Milieux d'affaires et pouvoirs politiques dans la France du vingtième siècle.* Paris: Librairie Fayard.

Johnson, Arthur Menzies. 1956. *The Development of American Petroleum Pipelines:*

A Study in Private Enterprise and Public Policy, 1862–1906. Ithaca: Cornell University Press.

Johnson, Leo A. 1972. "The Development of Class in Canada in the Twentieth Century." In *Capitalism and the National Question in Canada,* ed. Gary Teeple, pp. 122–183. Toronto: Toronto University Press.

Jolly, Jean, ed. 1962. *Dictionnaire des parlementaires français: Notices biographiques sur les ministres, députés, et sénateurs français de 1889 à 1940.* Paris: Presses Universitaires de France.

Katz, Friedrich. 1981. *The Secret War in Mexico: Europe, the United States, and the Mexican Revolution.* Chicago: University of Chicago Press.

Kautsky, Karl. 1970. "Ultra-imperialism." *New Left Review* 50 (January–February): 41–46. Originally published 1914.

Kazemzadeh, Firuz. 1968. *Russia and Britain in Persia, 1864–1914: A Study in Imperialism.* New Haven: Yale University Press.

Kent, Marian. 1976. *Oil and Empire: British Policy and Mesopotamian Oil, 1900–1920.* New York: Harper and Row.

Keohane, Robert O. 1982a. "Hegemonic Leadership and U.S. Foreign Economic Policy in the 'Long Decade' of the 1950s." In *America in a Changing World Political Economy,* ed. David P. Rapkin and William P. Avery, pp. 49–76. New York: Longman.

——. 1982b. "State Power and Industry Influence: American Foreign Oil Policy in the 1940s." *International Organizations* 36, no. 1:165–183.

——. 1984. *After Hegemony: Cooperation and Discord in the World Political Economy.* Princeton: Princeton University Press.

Keohane, Robert, and Joseph Nye. 1977. *Power and Interdependence: World Politics in Transition.* Boston: Little, Brown.

Key, V. O. 1967. *Politics, Parties, and Pressure Groups.* New York: Thomas Y. Crowell. Originally published 1942.

Keynes, John M. 1964. *The General Theory of Employment, Interest, and Money.* New York: Harcourt Brace Jovanovich. Originally published 1936.

Kindleberger, Charles. 1973. *The World in Depression, 1929–1939.* Berkeley and Los Angeles: University of California Press.

Krasner, Stephen D. 1978. *Defending the National Interest: Raw Materials Investments and U.S. Foreign Policy.* Princeton: Princeton University Press.

——. 1983a. "Structural Causes and Regime Consequences: Regimes as Intervening Variables." In *International Regimes,* ed. Stephen D. Krasner, pp. 1–21. Ithaca: Cornell University Press.

——. 1983b. "Regimes and the Limits of Realism: Regimes as Autonomous Variables." In *International Regimes,* ed. Stephen D. Krasner, pp. 355–368. Ithaca: Cornell University Press.

Krueger, Anne O. 1974. "The Political Economy of the Rent-Seeking Society." *American Economic Review* 64 (June): 291–303.

Kuisel, Richard F. 1967. *Ernest Mercier: French Technocrat.* Berkeley and Los Angeles: University of California Press.

——. 1981. *Capitalism and the State in Modern France: Renovation and Economic Management in the Twentieth Century.* New York: Cambridge University Press.

Labarrière,Guillaume de. 1932. *Un Instrument de politique économique nationale: Les sociétés de pétrole à participation de l'état dans divers pays: La Compagnie française des pétroles.* Brest: Imprimerie Commerciale et Administrative.

Larson, Henrietta M., Evelyn H. Knowlton, and Charles S. Popple. 1971. *History of the Standard Oil Company (New Jersey), 1927–1950: New Horizons.* New York: Harper and Row.

Lee, Dwight. 1987. "Deficits, Political Myopia, and the Asymmetric Dynamics of Taxing and Spending." In *Deficits,* ed. James Buchanan, Charles Rowley, and Robert Tollison, pp. 289–309. Oxford: Basil Blackwell.

Lender, Wolf. 1934. *Le Pétrole en Pologne.* Paris: Editions Internationales.

Lenin, V. I. 1977. *Imperialism, the Highest Stage of Capitalism: A Popular Outline.* In *Selected Works,* 1:633–731. Moscow: Progress Publishers. Originally published 1917.

Leo, Peter M. 1978. "The Emergence of Strategic Resources in World Politics: A Study of Petroleum and the International Order." Ph.D. diss., Columbia University.

Leone, Robert. 1986. *Who Profits: Winners, Losers, and Government Regulation.* New York: Basic Books.

Lepoutre, Jean. 1923. "L'Industrie pétrolifère Belge: Son champ d'action, ses 'possibilités.'" *Revue économique internationale* 4 (October): 80–114.

L'Espagnol de la Tramerye, Pierre. 1921. *La Lutte mondiale pour le pétrole.* Paris: Editions de la Vie Universitaire.

Levi, Margaret. 1988. *Of Rule and Revenue.* Berkeley and Los Angeles: University of California Press.

Lipson, Charles. 1981. "International Organization of Third World Debt." *International Organization* 35 (autumn): 603–632.

———. 1982. "The Transformation of Trade: The Sources and Effects of Regime Change," *International Organization* 36 (spring): 417–455.

List, Friedrich. 1966. *The National System of Political Economy.* New York: Augustus M. Kelley. Originally published 1837.

Luxemburg, Rosa. 1980. *L'Accumulazione del capitale: Contributo all spiegazione economica dell'imperialismo.* Trans. from German by Bruno Maffi. Turin: Einaudi. Originally published 1913.

Machlup, Fritz. 1952. *The Political Economy of Monopoly: Business, Labor, and Government Policies.* Baltimore: Johns Hopkins University Press.

Marchand, Jacques. 1937. *La Renaissance du mercantilisme à l'époque contemporaine.* Paris: Librairie Economique et Technique.

Martin, Bradford G. 1959. *German-Persian Diplomatic Relations: 1873–1912.* The Hague: Mouton.

Martin, Katherine. 1985. *Multinationals, Technology, and Industrialization: Implications and Impact in Third World Countries.* Lexington, Mass.: Lexington Books.

McKay, John P. 1983. "Entrepreneurship and the Emergence of the Russian Petroleum Industry, 1813–1883." *Research in Economic History,* vol. 8, Greenwich, Conn.: JAI Press.

Melby, Erik D. K. 1981. *Oil and the International System: The Case of France, 1918–1969*. New York: Arno.

Menjaud, Henri, and Marguerite Dion. 1961. *Desmarais Frères: Un siècle de l'industrie française du pétrole, 1861–1961*. Paris: Draeger Frères. Desmarais Frères centennial commemorative edition.

Mény, Yves, and Vincent Wright. 1987. "The State and Steel in Western Europe." In *Politics of Steel: Western Europe and the Steel Industry in the Crisis Years (1974–1984)*, ed. Yves Mény and Vincent Wright. Berlin: Walter de Gruyter.

Mertens, Jean. 1926. "Vers une politique belge du pétrole." *Revue économique internationale* 2 (May): 245–260.

Meyer, Lorenzo. 1977. *Mexico and the United States in the Oil Controversy: 1917–1942*. Trans. Muriel Vasconcellos. Austin: University of Texas Press.

Mirow, Kurt Rudolf, and Harry Maurer. 1982. *Webs of Power: International Cartels and the World Economy*. Boston: Houghton Mifflin.

"MITI to Set up Electric Power and Methanol Stations in Japan." 1992. *Oxy-Fuel News: A Weekly Update on Reformulated Gasoline, Oxy-Fuels, and Alternative Fuels Worldwide* 4 (8 June): 1–2.

Montgelas, Count Max. 1925. *The Case for the Central Powers: An Impeachment of the Versailles Verdict*. Trans. Constance Vesey. New York: Alfred A. Knopf.

"Morality in Business: Standard Oil Chairman Touches on Troublesome Topic." 1918. *Petroleum World* 15 (May): 182.

Morgenthau, Hans. 1973. *Politics among Nations: The Struggle for Power and Peace*. New York: Alfred A. Knopf. Originally published 1948.

Morse, Edward L. 1976. *Modernization and the Transformation of International Relations*. New York: Macmillan.

Muffelmann, Leo. 1907. "Un Trust européen du pétrole." *Revue économique internationale* 1 (January): 135–146.

——. 1908. "La Situation nouvelle du marché européen du pétrole." *Revue économique internationale* 2 (April): 55–96.

——. 1914. "Le Monopole d'état en matière de pétrole en allemagne." *Revue économique internationale* 2 (April): 7–33.

Murat, Daniel. 1969. *L'Intervention de l'état dans le secteur pétrolier en France*. Paris: Editions Technip.

Nash, Gerald D. 1978. *United States Oil Policy, 1890–1964*. Westport, Conn.: Greenwood.

Nayberg, Roberto. 1983. "La Question pétrolière en France, du point de vue de la défense nationale, de 1914 à 1928." 3 vols., doctoral thesis, directed by Guy Pedroncini, Université de Paris I (Panthéon-Sorbonne).

Neumann, Franz. 1963. *Behemoth: The Structure and Practice of National Socialism, 1933–1944*. New York: Octagon Books. Originally published 1942.

Neumann, Robert. 1935. *Sir Basil Zaharoff: Le Roi des armes*. Trans. from German by Denise Van Moppes. Paris: Editions Bernard Grasset.

Newman, Philip C. 1964. *Cartel and Combine: Essays in Monopoly Problems*. Ridgewood, N.J.: Foreign Studies Institute.

"News from Producing Fields: Over-production and Over-drilling." 1915. *Petroleum World* 12 (May): 236–237.

Nouschi, André. 1970. *Luttes pétrolières au Proche-Orient.* Paris: Flammarion.

——. 1975. "Les relations franco-belges de 1830 à 1934." Acte du Colloque de Metz 15–16 November 1974. Metz: Centre de Recherches Relations Internationales de l'Université de Metz.

——. 1979. "Pipe-lines et politique au Proche-Orient dans les années 1930." *Relations internationales,* no. 19 (autumn): 279–294.

Nowell, Gregory P. 1983. "French State and the Developing World Oil Market: Domestic, International, and Environmental Constraints, 1864–1928." In *Research in Political Economy,* vol. 6, ed. Paul Zarembka. Greenwich, Conn.: JAI Press.

——. 1985. "International Relations Theories and Technological Development: The Oil Market and the Case for Nondeterminate Outcomes." In *Research in Political Economy,* vol. 8, ed. Paul Zarembka and Thomas Ferguson, Greenwich, Conn.: JAI Press.

——. 1988. "Realpolitik vs. Transnational Rent-Seeking: French Mercantilism and the Development of the World Oil Cartel, 1860–1939." Ph.D. diss., Massachusetts Institute of Technology.

——. 1991. "Blood, Oil, and the Balance of Power: The Kurds and Nation-State Competition in the Persian Gulf." Paper presented at American Political Science Association Meeting, Washington, D.C., 28 August.

Oaks, George M. 1917. "Coal Tar and Its Products." *Petroleum World* 14 (2 July): 302.

"Oil for Germany up the Danube." 1916. *Petroleum World* 13 (July): 319.

"Oil Industry of France and Alsace." 1919. *Petroleum World* 16 (September): 370–374.

Olson, Mancur. 1965. *The Logic of Collective Action: Public Goods and the Theory of Groups.* Cambridge: Harvard University Press.

Péan, Pierre, and Jean-Pierre Séréni. 1982. *Les Emirs de la République: L'aventure du pétrole tricolore.* Paris: Seuil.

Pearton, Maurice. 1971. *Oil and the Romanian State: 1895–1948.* Oxford: Clarendon Press.

Peck, E. B., and Irvin H. Jones. 1945. *Technical Assistance on Synthetic Oils Rendered the Japanese by the I. G. Farbenindustrie A.G.* Combined Intelligence Objectives Sub-committee. London: HMSO.

Penrose, Edith, and E. F. Penrose. 1978. *Iraq: International Relations and National Development.* London: Ernest Benn.

"Pétroles de Mésopotamie." 1928. *Revue pétrolifère* 262 (24 March): 385–386.

Peyret, Henry. 1952. "Les Pétroles." In *Le Monde des affaires en France de 1830 à nos jours,* ed. Jacques Boudet. Paris: Société d'Editions de Dictionnaires et Encyclopédies.

Phillips, Frank. 1992. "Smoking-Tax Backers Are Bracing for Blitz." *Boston Globe,* 21 September, pp. 17–18.

Polanyi, Karl. 1957. *Great Transformation: The Political and Economic Origins of Our Time.* Boston: Beacon Press. Originally published 1944.

Posner, Richard A. 1974. "Theories of Economic Regulation." *Bell Journal of Economics and Management Science* 5, no. 2: 335–358.

Poulantzas, Nicos. 1978. *State, Power, Socialism.* Trans. Patrick Camiller. London: New Deft Books.

Probstein, Ronald F., and R. Edwin Hicks. 1982. *Synthetic Fuels.* New York: McGraw-Hill.

Przeworski, Adam, 1990. *The State and the Economy under Capitalism.* New York: Harwood Academic.

Report of the Royal Commission on the Coal Industry. 1926. London: HMSO.

Report on the Petroleum and Synthetic Oil Industry of Germany. 1947. London: Ministry of Fuel and Power, HMSO.

Robert, Paul. 1981. *Dictionnaire alphabétique et analogique de la langue française.* Paris: Société du Nouveau Littré, Dictionnaire Le Robert.

Roberts, Glyn. 1938. *The Most Powerful Man in the World: The Life of Sir Henri Deterding.* New York: Covici, Friede.

Roncaglia, Alessandro. 1983. *L'economia del petrolio.* Bari: Editori Laterza.

Rondot, Jean. 1977. *La Compagnie française des pétroles: Du franc-or au pétrole franc.* New York: Arno Press. Originally published 1962.

Rooy, Marcel. 1925. *Le Carburant national: Etude sur le régime de l'alcool en France.* Paris: Presses Universitaires de France.

Ropp, Theodore. 1937. "Development of a Modern Navy: French Naval Policy, 1871–1904." Ph.D. diss., Harvard University.

Rowley, Charles K. 1987. "The Legacy of Keynes: From the General Theory to Generalized Budget Deficits." In *Deficits,* ed. James Buchanan, Charles Rowley, and Robert Tollison, pp. 143–172. Oxford: Basil Blackwell.

Ruggie, John G. 1975. "International Responses to Technology: Concepts and Trends." *International Organization* 29 (summer): 557–584.

Sampson, Anthony. 1975. *The Seven Sisters: The Great Oil Companies and the World They Shaped.* New York: Bantam Books.

Samuels, Richard J. 1987. *The Business of the Japanese State: Energy Markets in Comparative and Historical Perspective.* Ithaca: Cornell University Press.

Schattschneider, R. E. 1935. *Politics, Pressures, and the Tariff.* New York: Prentice-Hall.

Schvartz, Julien. 1974. *Sur les sociétés petrolières opérant en France. . . : Rapport de la commission d'enquête parlementaire.* Paris: Union Générale d'Editions.

Shapiro, Michael J. 1989. "Textualizing Global Politics." In *International/Intertextual Relations: Postmodern Readings of World Politics,* ed. James Der Derian and Michael J. Shapiro, pp. 11–22. Lexington, Mass: Lexington Books.

Shonfield, Andrew. 1965. *Modern Capitalism: The Changing Balance of Public and Private Power.* New York: Oxford University Press.

Simmons, Beth, and Stephan Haggard. 1987. "Theories of International Regimes." *International Organization* 41 (summer): 491–517.

Simonnot, Philippe. 1978. *Le Complot pétrolier.* Paris: Alain Moreau.

Skocpol, Theda. 1979. *States and Social Revolutions: A Comparative Analysis of France, Russia, and China.* Cambridge: Cambridge University Press.

———. 1985. "Bringing the State Back In: Strategies of Analysis in Current Research." In *Bringing the State Back In,* ed. Peter Evans, Dietrich Rueschemeyer,

meyer, and Theda Skocpol, pp. 3–43. Cambridge: Cambridge University Press.

Smith, Adam. 1976. *An Inquiry into the Nature and Causes of the Wealth of Nations.* Ed. Edwin Cannan. Chicago: University of Chicago Press. Originally published 1776.

Smith, Michael Stephen. 1980. *Tariff Reform in France, 1860–1900: The Politics of Economic Interest.* Ithaca: Cornell University Press.

Solberg, Carl. 1976. *Oil Power.* New York: New American Library.

Stigler, George J. 1975. "The Theory of Economic Regulation." In *Citizens and the State: Essays on Regulation,* pp. 114–141. Chicago: University of Chicago Press.

Strange, Susan. 1983. "*Cave! Hic Dragones:* A Critique of Regime Analysis." In *International Regimes,* ed. Stephen Krasner, pp. 337–354. Ithaca: Cornell University Press.

——. 1985. "Protectionism and World Politics." *International Organization* 39 (spring): 233–259.

——. 1986. *Casino Capitalism.* Oxford: Basil Blackwell.

Streeck, Wolfgang, and Philippe C. Schmitter. 1991. "From National Corporatism to Transnational Pluralism: Organized Interests in the Single European Market." *Politics and Society* 19, no. 2: 133–164.

Tarbell, Ida. 1904. *History of the Standard Oil Company.* 2 vols. New York: McClure, Phillips.

Thomas, Gregory H. 1934. *Le Régime juridique du pétrole en France: L'Office National des Combustibles Liquides.* Paris: Librairie du Recueil Sirey.

Thompson, E. P. 1978. *Poverty of Theory and Other Essays.* New York: Monthly Review Press.

Tolf, Robert W. 1976. *Russian Rockefellers: The Saga of the Nobel Family and the Russian Oil Industry.* Stanford, Calif.: Hoover Institution Press.

Touret, Denis. 1968. *Le Régime français d'importation du pétrole et la Communauté Economique Européenne.* Paris: R. Pichon and R. Durand-Auzias.

Tugwell, Franklin. 1975. *Politics of Oil in Venezuela.* Stanford, Calif.: Stanford University Press.

Tullock, Gordon. 1989. *Economics of Special Privilege and Rent Seeking.* Boston: Kluwer Academic.

Uhry, Alfred. 1927. "L'Alsace pétrolière étudiée dans ses rapports avec la vie économique internationale." *Revue économique internationale* 2, no. 2: 292–331.

Ullman, Richard. 1968. *Anglo-Soviet Relations, 1917–1921: Britain and the Russian Civil War.* Princeton: Princeton University Press.

———. 1989. "The Covert French Connection." *Foreign Policy,* no. 75 (summer): 3–33.

Vernon, Raymond. 1971. *Sovereignty at Bay: The Multinational Spread of U.S. Enterprises.* New York: Basic Books.

———. 1977. *Storm over the Multinationals: The Real Issues.* Cambridge: Harvard University Press.

Vietor, Richard H. K. 1984. *Energy Policy in America since 1945: A Study of Business-Government Relations.* Cambridge: Cambridge University Press.

Vogel, David. 1986. *National Styles of Regulation: Environmental Policy in Great Britain and the United States.* Ithaca: Cornell University Press.

Wallerstein, Immanuel. 1976. *Modern World System: Capitalist Agriculture and the Origins of the European Economy in the Sixteenth Century.* New York: Academic.

——. 1980. *Modern World System II: Mercantilism and the Consolidation of the European World Economy, 1600–1750.* New York: Academic.

——. 1982. "Crisis as Transition." In *Dynamics of Global Crisis,* pp. 11–54. New York: Monthly Review Press.

——. 1984. *The Politics of the World Economy: The States, the Movements, the Civilizations.* Cambridge: Cambridge University Press.

Wallerstein, Immanuel, and Terence K. Hopkins. 1982. *World Systems Analysis: Theory and Methodology.* Beverly Hills, Calif.: Sage.

Waltz, Kenneth N. 1959. *Man, the State, and War: A Theoretical Analysis.* New York: Columbia University Press.

——. 1970. "The Myth of National Interdependence." In *The International Corporation,* ed. Charles P. Kindleberger, pp. 205–226. Cambridge: MIT Press.

——. 1979. *Theory of International Politics.* New York: Random House.

Ward, Thomas E. 1965. *Negotiations for Oil Concessions in Bahrain, El Hasa (Saudi Arabia), the Neutral Zone, Qatar and Kuwait.* New York: Ardlee Service. Limited edition, copies of Harvard University and Library of Congress.

Weber, Alfred. 1971. *Theory of the Location of Industries.* Trans. Carl J. Friedrich. New York: Russell and Russell. Originally published 1929.

Weber, Max. 1927. *General Economic History.* Trans. Frank H. Knight. Glencoe, Ill.: Free Press.

——. 1968. *Economy and Society.* Ed. Guenther Roth and Claus Wittich. Berkeley and Los Angeles: University of California Press. Originally published 1922.

Wendt, Alexander. 1988. "The Agent-Structure Problem in International Relations Theory." *International Organization* 41 (summer): 335–370.

Whigam, H. J. 1903. *The Persian Problem.* New York: Charles Scribner's Sons.

White, Gerald T. 1962. *The Formative Years in the Far West: A History of Standard Oil Company of California and Predecessors through 1919.* New York: Appleton Century Crofts.

Williamson, Harold F., Ralph L. Andreano, R. Arnold Daum, and Gilbert C. Klose. 1963. *The American Petroleum Industry, 1899–1959.* Vol. 2, *The Age of Energy.* Evanston, Ill.: Northwestern University Press.

Womack, James P., Daniel T. Jones, and Daniel Roos. 1990. *The Machine That Changed the World.* New York: Macmillan.

Yergin, Daniel. 1991. *The Prize: The Epic Quest for Oil, Money, and Power.* New York: Simon and Schuster.

Zeldin, Theodore. 1973. *France 1848–1945: Politics and Anger.* Oxford: Oxford University Press.

Index

After page numbers, the letter "f" indicates a figure or text to a figure; the letter "t" indicates a table.

Achnacarry agreements. *See* "As is" agreements
Admiralty, British, 49, 54, 61, 71, 184, 271
Africa, 131
African and Eastern Concessions, Ltd., 67
After Hegemony . . . (Keohane), 27
Agriculture lobby, 250
Albania, 194
Alcohol fuel: gasohol, 250–51; methanol, 234, 235
Alcohol lobby, 47–48, 72, 115, 289
Algeria, 121, 125, 131, 264n
American oil interests, 185–86, 188–89, 218; French WWI oil policy and, 104–5, 115, 142; Texas oil, 58, 74, 170, 197, 202. *See also* Standard Oil; United States
American Petroleum Institute, 140
Anarchy, international. *See* International relations theory
Anglo-French alliance, 115–16, 122–23, 126–27, 129; disintegration of, 132–39, 141–44, 173
Anglo-Persian Oil Company, 53–56, 78, 98, 142, 273, 283; as Anglo-Iranian Oil Company, 271, 274; Burmah Oil and, 56, 61, 69; competition with Deutsche Bank, 68, 70, 92; Mesopotamian oil and, 184, 188–89; Royal Dutch–Shell and, 54, 121, 122–24, 130, 193–94
Arendt, Hannah, 29
Armenia, 83f, 118f, 127, 174, 175
Asian oil market, 52, 83f, 193
Asiatic Petroleum, 52, 53, 56–58
"As is" agreements, 191–202, 204, 209, 213, 216, 240, 249; durable effects of, 219–20, 221–22
Automobile industry, 226, 289

Baghdad, 98
Baghdad railway plan, 64–65, 67, 69, 78, 92, 120
Baku-Batum pipeline, 51
Banks and capital, 29, 51–52, 130, 152, 183, 205, 227, 270; banking interests in Pétrofina, 153, 157, 202, 219; financing of technological transitions, 297–98, 299; Nobel family oil interests, 49, 50, 77, 138, 150; Rothschild oil interests, 45, 48, 67, 73, 101, 134. *See also* Deutsche Bank; Pétrofina; Rothschild banking family; Soviet debt and oil
Banque de l'Union Parisienne, 153–54, 173, 230, 231
Banque de Paris et des Pays Bas, 83f, 120, 135, 140, 173, 179, 206, 270; Royal Dutch–Shell competition with, 121, 135, 218, 254; Standard Oil's alliance with, 83f, 135, 173, 179. *See also* Compagnie Française des Pétroles; Omnium International des Pétroles
Baron, Charles, 168, 212, 214, 280
Baruch, Bernard, 136
Beard, Charles A., 12, 20, 291
Bedford, A. C., 101, 104, 135–36, 165, 266

Bedford, Henry, 155, 201, 202
Belgian-French coalition, 153–58, 300
Benzol production, 227–28, 250–51, 277
Bérenger, Henry, 98n, 100–102, 141, 165, 174, 255; role during oil "crisis" of 1917, 105, 109, 110, 113–14, 117, 125–26; role during oil crisis of 1919–1920, 135–37, 144, 173
Bérenger-Mellon talks, 165
Bergius, Friedrich, 228
Bergius hydrogenation process, 72, 75, 228–29, 234, 242–43
Blum, Léon, 214, 252–53, 275
Bnito company, 57, 59, 76
Bone, Homer T., 241, 280
Borneo, 91, 106, 113–14
Bosch, Carl, 243
Briand, Aristide, 120
Britain: Anglo-Persian and, 55, 56, 61, 68, 71, 77–78, 93–94; British Admiralty, 49, 54, 184, 271; guaranteeing "freedom of the seas," 258, 260–61, 264, 271, 278; Mesopotamian oil interests of, 53–55, 69, 122, 131; oil policy of, 54–56, 74, 257, 283; Royal Dutch–Shell and, 92–93, 106, 107. *See also* Anglo-French alliance
British Petroleum, 60f, 93, 219
Burmah Oil Company, 49, 61, 69, 115; British oil policy and, 56, 106
Business interest groups, 10–11, 13n19, 285–86, 298. *See also* Banks and capital

Cadman, John, 202, 273
"Capitulations" treaty, 70
Cartel accords, 6, 50, 92; Asian market agreement of 1905, 52, 60f; Epu–Standard Oil accord, 59–61; French agreement of 1925, 159, 160f. *See also* World oil cartel
Cartel des Dix, 47, 73, 95
Cartel des Gauches, 161, 180
Catholic Church, 63, 203
Caucasus oil development, 60, 73, 97, 139, 150–52, 155
Cayrol, Robert, 167, 205, 274
Chemical industry: excluded from hydro-patent agreements, 239, 277, 281; synthetic fuel production and, 228, 229, 230f, 232, 236, 239, 240
Churchill, Winston, 55, 92, 97
Cilicia, 128f, 129
Clemenceau, Georges, 80, 81, 108, 124, 148
Clémentel, Etienne, 88, 105, 110, 111, 112–14, 124, 136
Coal industry, 1, 46, 115, 162, 223, 225, 230f, 283; benzol marketing and, 250–51; decline of, 224–27, 277–79; excluded from synthetic fuel patents, 232–33, 235–40, 242; oil imports a threat to, 231, 233, 234; oil industry's competition with, 242, 246–49, 277–78, 283–84, 297; oil involvements of, 229–31; synthetic fuel development strategy of, 72, 231–32. *See also* Synthetic fuel; World hydrocarbon cartel
Cohen, Robert Waley, 91, 92, 107
Colbert, Jean-Baptiste, 37n, 289
Cole, Charles Woolsey, 290, 291
Comité Général des Pétroles, 101, 102, 110, 112
Commissariat General aux Essences et Combustibles, 114–15, 141, 145
Compagnie de la Nacelle de Saint Pierre Fleurdelysée proposal, 291
Compagnie Française de Raffinage, 206, 207t
Compagnie Française des Pétroles, 135, 148, 154, 159, 170–73, 277; financial and oil interests represented in, 173–85, 190, 200, 215; left-wing vs., 210–17; struggle among shareholders, 205–10, 217–19, 252–53; revised tax schedule of, 208–12, 218, 235
Comptoir des Essences Synthétiques, 248, 249t
Conseil Supérieur de la Défense National, 256–57, 261–62, 264
Consolidated Petroleum, 193
"Containment" policy on Soviet oil, 152, 156, 157
Crédit Mobilier, 153

DAPG, 63, 244f
D'Arcy, William Knox, 53–54
Dardanelles, 153; closure of, 80–81, 83f, 90, 146, 262; negotiations on, 151–52
Defense. *See* Military; National security
Dégouy, Contre-Amiral, 75
Deregulation, 5, 16, 43, 90, 145, 146, 299; transnational structuring and, 286, 289, 292–93, 299
Desmarais Frères, 47–48, 102n47, 159, 164, 207t, 249t; nationalist activities of, 263–64; opposing Mesopotamian oil interests, 167, 202; refining and distribution, 205
Deterding, Henry, 157, 192, 201, 271–72; heading Royal Dutch–Shell, 52, 57, 63, 67; role in international negotiations, 69–70, 98, 122, 123, 152
Deutsch de la Meurthe, 102n, 103, 158
Deutsche Bank, 45, 58, 60f, 61–62, 142,

184; competition with Anglo-Persian, 68, 70, 92; competition with Royal Dutch–Shell, 64–65, 67, 73, 145. *See also* Epu (European Petroleum Union); Steaua Romana
Direction des Carburants, 219
Direction des Essences et Pétroles, 152, 154, 161
Dutch East Indies, 60, 83f, 105, 139

Economic regulation. *See* International economic regulation; State economic intervention
Electrification and coal complex, 226–27
Energy self-sufficiency, 39–40, 223–24, 275–76
Epu (European Petroleum Union), 60f, 77, 81, 83f, 217; Deutsche Bank and, 63, 67, 71, 93; Epu–Standard Oil accord, 59–61; Royal Dutch–Shell/Rothschild interests in, 56–59, 67, 71, 74
Etatism: French tradition of, 4–5, 37–39, 144–45, 148, 170, 245, 276, 300–301; theoretical critique of, 285–86, 288–90, 293. *See also* States
Excess profits tax, 86–88
Exxon, 219

Federal Trade Commission report, 1, 236, 272–73, 295
Ferrier, R. W., 55, 56, 70, 78
Financial interests. *See* Banks and capital
Fischer-Tropsch process, 242–43, 244f, 246
Flandin, Pierre-Etienne, 243, 245
France, etatist tradition of, 4–5, 37–39, 144–45, 148, 170, 245, 276, 300–301; military of, 74, 97, 162, 163, 209, 219, 262, 267, 275–76, 278; synthetic fuel program of, 243, 245–46, 248–50; war debt of, 165. *See also* Anglo-French alliance; Franco-Belgian coalition; French national oil firms; French oil supply
France, oil policy of, 129, 171, 174, 209, 219–22; absence of, before WWI, 45, 48, 71–76, 79, 115; favoritism to Royal Dutch–Shell, 62, 71–76, 82, 97–98, 103, 124–26, 137; import quotas and law of 1928, 203–4, 209, 212, 233, 253, 254–56, 266, 268, 277; internal dissension over, 102–3, 124–25, 140, 145, 181, 201–2; during interwar period, 260–61; Mesopotamian oil and, 98–99, 123, 129, 173, 182–84, 191, 203; military report on, 256–57, 261–62, 264; reserve stock requirement, 74, 103, 108–11, 252, 254–60; Soviet oil faction and, 156, 158–65, 216–17, 219; Standard Oil and, 108–14, 132–41, 174, 257, 260–61, 264, 266–67; world hydrocarbon cartel and, 251–52, 264; world oil cartel and, 200–202, 216–22, 275, 277–78; during WWI, 83f, 94–96, 109, 111–16, 117, 120–24, 130. *See also* State economic intervention
France and Her Capacity to Pay (Bérenger), 165
Franco-Belgian coalition, 153–58, 161–63, 167, 300
French left wing, 180, 210–17
French national oil firms, 97, 102n47, 164, 177f, 178f; cartel arrangements of, 47–48, 73, 95, 160f; efforts to break patent pool by, 243, 245, 246–47, 249; multinationals and, 158–61, 181, 209, 218, 278; refining industry, 45–46, 264–68, 276. *See also* Desmarais Frères
French oil supply, 81–82, 91t, 109, 110, 159, 161, 163, 268–69; oil crisis of 1919–1920, 135–41, 144, 253, 257; reserve stock requirement, 74, 103, 252, 254–59. *See also* Oil "crisis" of 1917
Fuel oil market, 54, 158, 162–63, 226

Gabon, 263
Galicia, 48, 74. *See also* Poland
Gallipoli expedition, 97, 146
Gasohol legislation, 250–51
Gasoline markets, 54, 158, 223, 228, 233
General Motors, 244f
Genoa negotiations, 151, 155
Germany, 63, 124, 283; hydrogenation process development in, 228, 233, 242; Nazi regime, 220–21, 241–42, 271–72; pre-WWI oil monopoly plan of, 45, 61–65, 67, 71, 79; WWI submarine campaign of, 96, 103, 144, 146. *See also* Deutsche Bank; I. G. Farben
Gilpin, Robert, 33
Government policy. *See* State economic intervention
Greece, 151–52, 220
Greenway, Charles, 54, 94, 98, 122
Gulbenkian, Calouste S., 65–67, 69–72, 98–101, 158; role in Mesopotamian negotiations, 116, 121, 182, 186, 188–90, 213
Gulf Oil, 58, 64, 165, 185n, 197
Gwinner, Arthur von, 73

Haggard, Stephan, 26
Hegemonic power, 27
Herriot, Edouard, 161, 216
Hexner, Ervin, 2n, 228n, 236n, 294

Hilferding, Rudolf, 30, 298
Hirschman, Albert O., 22
Hobson, J. A., 15, 30, 288
Huntington, Samuel, 24
Hydrogenation process, 228–29, 233, 234, 235–40, 242–43. *See also* Synthetic fuel
Hydro-patents: control and suppression of, 228, 231, 236–40, 249t, 267–68, 275, 281; international agreements on, 232–33, 246–49
Hydro Patents Corporation, 238, 239

I. G. Farben, 233, 234, 237–43, 244f; Standard Oil arrangements with, 235–38, 240, 266, 299
Import monopoly plans, 62–63, 72–73, 75–76, 83f, 96, 124, 136, 140, 145, 166, 175, 212
Import quotas of 1928, 203–4, 209, 212, 233, 253, 254–56, 266, 268, 277; evasion of, 259–60
Independents, 64, 138, 170, 197, 200, 232, 259. *See also* French national oil firms
India. *See* Burmah Oil Company; Dutch East Indies
Indonesian oil development, 190
Interallied Petroleum Conference, 110, 111–13, 115
Interest groups, 5, 13, 17, 37, 82, 86, 89, 146–47, 161, 183, 197, 201, 203, 204, 211, 213, 217, 222, 234, 245, 250, 276, 278–79
International Bergin Company, 228, 237
International cartel accords. *See* World oil cartel
International Cartels (Hexner), 236n, 294
International economic regulation, 17–20, 28–30; convergence based on technical optimality, 23, 29; convergence of state policies, 197–200, 201, 276–79, 288; state-centered theories of, 20–23. *See also* International relations theory
International Hydro Patents, 239, 240, 243, 246
International patent pool, 239, 240, 243, 244f, 246, 249; coal industry excluded from, 232–33, 235–40, 242; hydro-patent agreements, 232–33, 246–49; patent accords of 1928, 1, 10, 37, 222, 225, 288. *See also* World hydrocarbon cartel; Fischer-Tropsch process; Bergius hydrogenation process
International relations theory, 27, 30–31, 280, 287; concept of international anarchy, 6, 7, 24, 32, 146, 282–85; critique of etatism, 285–86, 288–90, 292; critique of realism-mercantilism, 280–86, 292; regime theory, 12, 21, 25–28; "sovereignty at bay" model, 12, 32–34; transnational structuring and, 27, 222, 278, 280, 282–88. *See also* Etatism; International economic regulation; Realist-mercantilist analysis
Iranian oil development, 190, 199
Iraq, 131, 146, 188, 190, 191, 271, 272–75. *See also* Mesopotamian oil
Iraq Petroleum Company, 66, 191, 202, 206, 207, 213, 252, 254, 263. *See also* Turkish Petroleum Company
Italy, 194n, 200, 246, 257–58, 261

Japan, 200, 220, 221, 242, 246

Kautsky, Karl, 30
Kemal, Mustafa, 83f, 129, 271
Keohane, Robert O., 26, 27
Kerosene market, 61, 63. *See also* Lamp fuel market
Key, V. O., 15, 291
Kirkuk oil fields, 270
Krasner, Stephen D., 21n, 22, 28, 37, 39–40
Kuisel, Richard F., 40, 292

Labarrière, Guillaume de, 40, 165
Lamp fuel market, 46–47, 54, 62, 94–95, 107, 158, 225, 226
Lane, Frederick, 67, 97
Lausanne negotiations, 155
Left wing, French, 180, 209–17. *See also* Soviet oil faction
Lenin, V. I., 30, 288
Lobbies. *See* Interest groups
Logic of Collective Action . . . (Olson), 294
Long, Walter, 113
Long-Bérenger accord, 126–27, 129
Loucheur, Louis, 174, 175, 180, 235, 253, 275
Loucheur amendment, 202
Luxemburg, Rosa, 30

Maginot Line, 276
Margaine, Alfred, 155–56, 165, 166–68, 191, 211
Markets, 6, 32, 52, 60f, 82–83f, 193, 225–27, 244f; deregulation of, 16, 43, 289, 292, 299; unregulated, 5, 90, 146. *See also* World oil market
Marxist analysis, 11, 13n19, 30–32, 288
Mellon, Andrew, 58, 165
Mercantile states. *See* States
Mercier, Ernest, 134, 166, 171, 175, 176,

179, 180, 259, 263, 273, 292; Compagnie Française, 292 des Pétroles and, 201, 202–3, 207t, 208; policy struggle with Pineau, 181–82, 213–14; Standard Oil connections of, 134, 218, 267. *See also* Omnium International des Pétroles
Mesopotamian oil, 45, 50–51, 65, 67, 81, 188; agreement of 1914, 54, 70–71, 76, 78, 122, 126, 216, 217; agreement of 1928, 188–91, 202; Anglo-Persian and, 184, 188–89; British oil policy and, 54–55, 69, 122, 131; Deutsche Bank interest in, 62, 67; effort to limit supply of, 231, 233, 234, 243, 263, 272–75; French oil policy and, 98–99, 102, 123, 128f, 129, 173, 182–84, 191, 206–7; Royal Dutch- Shell and, 62, 68, 72–76; vs. Russian oil, 150, 161, 162, 166–68, 180–88, 191, 203, 220; transnational interests in, 62, 68, 72–76, 116, 121, 300. *See also* Iraq Petroleum Company; Middle East; Mosul negotiations; Turkish Petroleum Company
Methanol fuel, 234, 235
Mexico, 74, 83f, 190, 261, 262
Middle East, 119f, 122, 128f, 129, 270–75; Zionist interests, 271–72
Military: British Admiralty, 49, 54, 184, 271; of France, 74, 97, 162, 163, 209, 219, 262, 267, 275–76, 278; Italian admiralty, 257–58; preference for synfuels, 278. *See also* National security
Millerand, Alexandre, 141
Ministère de la Guerre, 97
Monopolies, 54, 76. *See also* State-sanctioned monopolies; World oil cartel
More, Thomas, 10
Morgenthau, Hans, 28, 222
Mosul negotiations, 116–17, 120–21; Sykes-Picot accord, 118f, 122, 127
Muffelmann, Leo, 62, 63, 75–76
Multinational companies, 2, 4, 32–34, 185–88; British oil policy and, 53, 54–55; Compagnie Française des Pétroles and, 173–85; competition among, 17, 49–50, 58–59, 60, 64–65, 81, 90, 130; "containment" policy toward Russian oil, 152, 156, 157; national governments and, 7, 33, 56–57, 115, 124–26, 196–200, 294; national oil firms and, 158–61, 181, 209, 218, 243, 278. *See also* Anglo-Persian Oil Company; Cartel accords; Royal Dutch–Shell; Standard Oil
Muskogee-Goldshell incident, 103–4, 107, 111
National Bank of Turkey, 65–70
National interest, 5, 11, 23, 42, 86, 144–45; interacting with transnational interests, 56–57, 253; as rhetoric or mask, 20, 35, 43, 55, 80, 104, 113–14, 148, 149, 172–73, 175, 182, 216, 282, 287. *See also* Realist-mercantilist analysis; States
National security, 12, 13, 34, 130; energy self-sufficiency and, 223–24; oil policy and, 40–42, 55–57, 97, 275; strategic commodities and, 23, 40, 254–55, 258–59; wartime state intervention, 80–89. *See also* Military
Nazi regime, 220–21, 241–42, 271–72
Nobel family, 49, 50–51, 60, 77, 138, 150, 226

Office National des Combustibles Liquides, 148, 161–62, 170, 203–4, 219, 250, 255, 269–70, 276; agency name history, 101, 141, 154, 219; multinational oil interests and, 234–35, 245–46
Oil "crisis" of 1917, 94, 102–105, 108–10, 149, 253; Clemenceau's telegram to Wilson, 81, 108, 124, 148; *Muskogee-Goldshell* incident, 103–4, 107, 111
Oil crisis of 1919–1920, 81–82, 135–41, 149, 253, 257
Oil depletion allowance, 291
Oil exploration, 83f, 127–29, 173, 192, 262–63, 264n97, 273–74
Oil hearings of 1927–28, 166–70, 233
Oil market. *See* World oil market
Oil-producing regions, 48, 60, 82f, 128f, 150; emergence of, 49, 50–53; regional production conflicts, 77, 150, 183–85, 197, 211; WWI impact on, 83f, 90–92
Oil transport. *See* Pipeline routes; Railroads; Tankers
Olson, Mancur, 294
Omnium International des Pétroles, 133–34, 145, 173–76, 177t, 180
Open door plan, 185–86
Ottoman Empire, division of, 65, 117, 118f, 142, 188, 190

Patents. *See* Hydro-patents
Pechelbronn company, 230–31
Pennsylvanian oil fields, 46
Persian Gulf, plan for pipeline link to Caucasus, 50–54, 139, 142
Persian oil concessions, 53, 126, 142–44
Pétrofina, 154, 158, 176, 180, 195, 200, 230–31, 257; banking interests in, 153, 157, 202, 219; French national firms' alliance with, 161, 162, 167, 204–5, 212;

Pétrofina (*cont.*)
Royal Dutch–Shell split with, 157–58, 168–69, 193. *See also* Soviet debt and oil
Petroleum Committee, 101, 113
Peyerhimoff, Henri de, 229, 230, 233, 250
Peyerhimoff-Schneider coal syndicate, 134, 237
Phillips Petroleum, 170
Phoenix Oil and Transport, 271
Pineau, Louis, 154, 155–56, 162, 163–64, 202, 245, 275, 276; multinational constituency of, 245–46, 277; policy struggle with Mercier, 181–82, 213–14
Pipeline routes, 270–75; Persian Gulf vs. Black Sea, 50–54, 139, 142; through Syria, 131, 278
Poincaré, Raymond, 149, 161, 171, 172, 180, 204, 208, 209, 218, 253, 262
Poland (Galicia), 48, 74, 183, 192, 261
Polanyi, Karl, 29–30
Political Order in Changing Societies (Huntington), 24
Posner, Richard A., 11
Price controls, 86–88
Price fixing, basing point system, 198–99, 221–22
Pricing wars, 52, 58–59, 158–59, 162, 196

Quota system. *See* Import quotas of 1928

Railroads, 119f, 226; Baghdad railway, 64–65, 67, 68, 69, 71n, 78, 92
Realist-mercantilist analysis: critique of, 280–86, 292; focusing on states as primary actors, 8, 15, 20–23, 24–25, 34–36, 42, 280–82, 284–85; national interest concept and, 12–13, 14–15, 21–25, 35, 36, 39–43; of national security, 55–56, 80, 82, 254; of oil policy during WWI, 80–81, 146, 148–49, 254; of oil policy in interwar period, 149, 254, 261, 268, 279; of pre-WWI oil policy, 55–56, 76–78
Red Line agreement, 78, 188–91
Refining industry, 46, 47, 48, 98, 99, 134, 158, 195, 220, 260, 264–68, 276
Régie, definition, 164n
Regime theory, 12, 21, 25–28, 32, 222
Regulatory policy. *See* State economic intervention
Reichstag project, 45, 61–65, 68, 71, 77
Rent-seeking theory, 11, 12, 15–17, 34, 35, 36, 41, 290, 301
Riedemann, Heinrich von, 63–64, 68, 246, 247
Rockefeller, John D., Jr., 195–96
Romania: oil production in, 48, 60, 67, 73, 74, 76, 127, 130, 131, 133, 138, 139, 141, 142, 153, 176, 192, 197, 199, 202, 211, 260; Steaua Romana, 58, 59, 60f, 61; during WWI, 83f, 90, 97, 113
Rondot, Jean, 170–71, 181
Rothschild banking family: Asiatic Petroleum and, 52–53, 57; Bnito company of, 57, 59, 76; influencing French oil policy, 47; oil interests of, 45, 48, 67, 73, 101, 134; Royal Dutch–Shell connections of, 62, 72, 73, 74, 115, 120–21, 134, 176, 271; Russian oil holdings of, 50, 150. *See also* Bnito company
Royal Dutch, 52–53
Royal Dutch–Shell, 45, 58, 77–79, 83f, 141, 174–75, 283; Anglo-Persian and, 54–55, 121, 122–24, 130, 193–94; British Admiralty and, 92–93, 106, 107; competition with Deutsche Bank, 64–65, 67, 73; competition with Standard Oil, 17, 73, 83f, 90, 110–11, 114, 130, 136–39; Epu and, 60f, 71, 74; French national firms and, 158–59, 200; hydrogenation licensing agreements and, 237–38; loss of dominance in France, 132, 135, 139–41, 173; Mesopotamian oil and, 62, 68, 69, 72–76, 121, 184; regional interests of, 101, 106–7, 115; Rothschild connections to, 62, 72, 73, 74, 115, 120–21, 134, 176; Soviet oil interests and, 156, 157–58, 168–69, 192–93; during WWI, 90–92, 93, 94, 113, 116, 141, 145. *See also* Anglo-French alliance; Deterding, Henry; Multinational oil companies; World oil cartel
Ruhrchemie, 242, 244f
Russia, 51, 116, 127, 131, 203; Caucasus region, 60, 73, 97, 126–27, 130, 139, 141, 150–52, 155, 175, 257, 262; high cost of importing oil from, 231; Nobel family holdings in, 49, 50, 77, 138, 150; oil development in, 50–52, 83f, 92, 113, 194–95; politics affecting world oil supply, 60–61, 83f, 92, 95, 141, 202; Rothschild family holdings in, 47, 50, 150; Russian vs. Mesopotamian oil, 150, 161, 162, 166–68, 180–88, 191–92, 203, 216, 253–54, 261; Soviet debt and oil, 147, 148, 149–50, 157, 163, 184; Soviet oil policy, 151–52, 163, 169–70, 196–97, 220–21; withdrawal from world oil market, 194, 210t, 220–21, 300. *See also* Witte, Sergei; Pétrofina

Saint Gobain group, 212–17, 221, 249t, 254, 266, 277
Samuel, Marcus, 56–58, 75, 92, 271

San Remo Treaty, 129–32, 138, 141, 142, 144, 145, 173
Schattschneider, R. E., 34, 291, 301
Schneider industrial group, 134, 229–30, 234, 235, 243, 245
Schvartz commission report, 42, 209, 218, 301
Self-denying clause, 186–87, 190
Senate Patent Hearings, 1–2, 236, 268
Shale oil industry, 46
Shell Trading and Transport, 52, 53, 56–58. *See also* Royal Dutch–Shell; Samuel, Marcus
Simmons, Beth, 26
Smith, Adam, 21n, 290
Smith, G. H. E., 12, 20, 291
Social contract theory, 25–26
Société pour l'Exploitation des Pétroles (Royal Dutch–Shell), 134–35, 173, 174–75
Socony, 105, 106, 193, 195–96
"Sovereignty at bay" model, 12, 32–34
Soviet debt and oil, 147, 148, 149–50, 157, 163, 184
Soviet oil faction, 210–17, 234, 253, 254, 262; French oil policy and, 156, 158–65, 216–17, 219. *See also* Pétrofina
Soviet Union. *See* Russia
Spain, 169n, 221
Stalin, Josef, 221
Standard Oil, 76, 93, 101; Banque de Paris alliance with, 83f, 132–35, 140, 173, 179; campaign against Anglo-French alliance, 132–40, 142, 144, 145, 174, 176, 184, 253, 257, 264; defeats German oil monopoly plan, 62–64, 68, 79; Epu accord with, 59–61; French oil policy and, 257, 259–61, 264, 266–67; French treatment of, during WWI, 80, 82, 93, 95–96, 103–5, 109, 112; I. G. Farben arrangements with, 235–38, 240, 266, 299; Mesopotamian oil and, 68, 184; vs. Pétrofina and French nationals, 162–63, 165, 180, 182–85; pre-WWI market domination challenged, 48–50, 52, 57–58, 71, 73, 76; WWI competition with Royal Dutch–Shell, 83f, 90, 98, 110–11, 114, 130, 136–39. *See also* Multinational oil companies; World oil cartel
Standard Oil-Epu accord, 59–61
State-centrism. *See* Etatism
State economic intervention, 5, 11, 21, 24–25, 43, 51, 235, 242, 291, 300; assigning oil market shares, 73, 77–78; excess profits tax during WWI, 86–88; import quotas of 1928, 203–4, 209, 212, 233, 253, 254–56, 266, 268, 277; politics of, 6, 34–38, 43, 51–52, 102, 114–16, 216; price controls, 86–88; realist-mercantilist analysis of, 76–77, 78, 80–81, 281, 292; transnational interests influencing, 33, 165–66, 216–17, 221, 286–88; in wartime, 80–89. *See also* France, oil policy of
States: multinational oil interests and, 81, 115, 267–68, 300; "sovereignty at bay" model of, 12, 32–34; strong state–weak state concept, 37, 40–41, 222, 254, 261, 279, 300; viewed as unitary actors, 7, 8, 15, 20–23, 24–25, 34–36, 42, 281, 286, 292; world hydrocarbon cartel and, 240, 243, 246–47, 251–52, 276–77. *See also* Etatism; National interest; National security
State-sanctioned monopolies: import monopoly plans during WWI, 83f, 88, 96, 124, 136, 140, 145, 175, 255; interwar proposals for, 163–70, 210, 211, 214; pre-WWI plans for, 45, 61–65, 68, 71, 73–76
Steaua Romana, 58, 59, 60f, 61, 176
Strange, Susan, 18
Suez Canal, 50, 52
Sykes-Picot accord, 118f, 122, 127
Synthetic fuel: Bergius hydrogenation process, 72, 75, 228–29, 234, 242–43; coal-based, 223–24, 227, 228–29, 230f, 244f; Fischer-Tropsch process, 241–42; lack of state support for, 235, 277–78; world hydrocarbon cartel blocks production of, 242, 246–49, 260, 283–84, 297. *See also* Coal industry
Synthetic rubber, 242, 244f, 281–82, 284, 285, 287
Syria, 118f, 128f, 174, 175, 251, 262–63; oil pipeline through, 129, 131, 278

Tankers, 96, 97–98, 101, 139, 226; French WWII fleet, 268–69, 278; tanker conflicts during WWI, 103–10, 111, 113–14
Tardieu, André, 104–5, 107, 109, 113, 208, 212, 276, 277
Tariffs, 46–47, 48, 115, 297. *See also* Import quotas of 1928
Taxes, 235, 291, 297; excess profits tax during WWI, 86–88; revised tax schedule of 1928, 208–12, 218, 235
Teagle, Walter, 101, 142, 166, 195, 218, 266, 267
Technical optimality, 23, 29
Technology, 33, 44; transitions in, 223, 276, 277, 267, 285–86, 297–301

Texas oil development, 58, 74, 165, 170, 190, 197, 199, 200, 202, 221, 288
Toluol production, 106–7, 113, 230f
Transcaucasus, 60, 73, 97, 126–27, 130, 139, 141, 150–52, 155, 175, 257, 262
Transcaucasus–Persian Gulf pipeline plan, 50–54, 139, 142
Transnational ideology, 240–41
Transnational structuring, 36–39, 280, 291–301; analytic theory of, 3, 4–11, 15, 31–32, 43–44, 76, 289–91; convergence of state policies and, 197–200, 201, 276–79, 288; difficulty in documenting, 5–7, 35–37, 149, 290–92, 295–97; durable effects of, 219–22, 241, 278–79, 282; games analogy of, 282–85, 295–96; international relations theory and, 282–88; during interwar period, 149, 198–200, 216, 219–22; before WWI, 45, 76–79, 155; during WWI, 80–81, 82, 144–47. *See also* Banks and capital; Multinational oil companies; State economic intervention; World oil cartel
Transport of oil. *See* Pipeline routes; Railroads; Tankers
Tronchère, Louis, 207t, 263
Truck industry, 226, 276
Turkey, 65, 67, 90, 97, 151, 152, 174, 181; consortium plan re, 68, 188; division of the Ottoman Empire, 65, 116, 117, 118f, 122, 127, 142, 151, 190. *See also* Dardanelles; Mesopotamian oil
Turkey, National Bank of, 65–72
Turkish Petroleum Company, 68, 69, 70, 76, 116, 121, 122, 144, 186–87; founding of, 68, 69, 70, 76; during interwar period, 171, 186–89, 271, 273; as Iraq Petroleum Company, 66, 186, 191, 202, 206, 207, 213, 252, 254, 263, 272; during WWI period, 116, 121, 144. See also Mesopotamian oil

United States, 76, 83f, 101, 107, 110, 136, 142, 155, 165, 182, 185–86, 190, 238, 239, 261, 277, 281–82, 285, 291; Federal Trade Commission report, 1, 236, 272–73, 295; Senate Patent Hearings, 236, 268. *See also* American oil interests
Unregulated markets. *See* Markets

Vegetable oil industry, 46–47
Venezuela, 187, 190, 197, 200, 274–75
Vernon, Raymond, 32–33

Wallerstein, Immanuel, 31, 288
Waltz, Kenneth N., 22, 281–82, 284–85
Wartime, state economic intervention in, 80–89
Wealth of Nations (Smith), 290
Weber, Max, 30, 301n
Wenger, Léon, 161, 167, 257
Wilson, Woodrow, 80, 81, 105, 108, 148
Witte, Sergei, 50–53
Wood gasification, 276, 278
World hydrocarbon cartel, 1, 27–28, 222, 223, 225, 240–41, 244f, 282; blocking synfuel production from coal, 242, 246–49, 260, 283–84, 297; distinguished from world oil cartel, 225; French state policy and, 37, 247–49, 274, 277; patent accords of 1928, 1, 10, 37, 222, 225, 235–38, 288; world patent pool, 235–38, 239, 240, 242, 243, 244f, 245–250
World oil cartel, 191–200; "as is" agreements of 1928, 191–200, 213, 240, 249; the first (pre-WWI), 60f, 59–61; French oil policy and, 200–202, 216–22, 249, 274, 275, 277–78, 296–97; Mesopotamian oil agreement of 1928, 188–91, 202; Red Line agreement of 1928, 188–91
World oil market, 5, 48, 78, 163, 272; events in Russia affecting, 60–61, 83f, 92, 141; fuel oil, 54, 158, 162–63, 226; pre-WWI domination by Standard Oil, 48–50, 52, 57–58, 71, 73, 76; as a regime, 26–27; withdrawal of Soviets from, 220–21; WWI impact on, 76, 79, 83f, 90–92, 108, 113. *See also* French oil supply; Markets; Monopolies; Multinational oil companies
World patent pool. *See* World hydrocarbon cartel
World systems theories, 30–31, 287, 288
World War I, 80, 117, 118f; German submarine campaign of, 96, 103, 144, 146; impact on oil market, 76, 79, 83f, 90–92, 108, 113; oil tanker wars during, 103–7, 111, 113–14, 139
World War II, 268–69, 270, 278; interwar period (1922–1939), 66, 98–99, 129, 146, 149

Zaharoff, Basil, 125n, 152n
Zaku, Ahmed, 194
Zédet, Emile, 276

Cornell Studies in Political Economy

Edited by Peter J. Katzenstein

Collapse of an Industry: Nuclear Power and the Contradictions of U.S. Policy, by John L. Campbell

Power, Purpose, and Collective Choice: Economic Strategy in Socialist States, edited by Ellen Comisso and Laura D'Andrea Tyson

The Political Economy of the New Asian Industrialism, edited by Frederic C. Deyo

Dislodging Multinationals: India's Strategy in Comparative Perspective, by Dennis J. Encarnation

Rivals beyond Trade: America versus Japan in Global Competition, by Dennis J. Encarnation

Democracy and Markets: The Politics of Mixed Economies, by John R. Freeman

The Misunderstood Miracle: Industrial Development and Political Change in Japan, by David Friedman

Patchwork Protectionism: Textile Trade Policy in the United States, Japan, and West Germany, by H. Richard Friman

Ideas, Interests, and American Trade Policy, by Judith Goldstein

Ideas and Foreign Policy: Beliefs, Institutions, and Political Change, edited by Judith Goldstein and Robert O. Keohane

Monetary Sovereignty: The Politics of Central Banking in Western Europe, by John B. Goodman

Politics in Hard Times: Comparative Responses to International Economic Crises, by Peter Gourevitch

Closing the Gold Window: Domestic Politics and the End of Bretton Woods, by Joanne Gowa

Cooperation among Nations: Europe, America, and Non-tariff Barriers to Trade, by Joseph M. Grieco

Pathways from the Periphery: The Politics of Growth in the Newly Industrializing Countries, by Stephan Haggard

The Politics of Finance in Developing Countries, edited by Stephan Haggard, Chung H. Lee, and Sylvia Maxfield

Rival Capitalists: International Competitiveness in the United States, Japan, and Western Europe, by Jeffrey A. Hart

The Philippine State and the Marcos Regime: The Politics of Export, by Gary Hawes

Reasons of State: Oil Politics and the Capacities of American Government, by G. John Ikenberry

The State and American Foreign Economic Policy, edited by G. John Ikenberry, David A. Lake, and Michael Mastanduno

The Paradox of Continental Production: National Investment Policies in North America, by Barbara Jenkins

Pipeline Politics: The Complex Political Economy of East-West Energy Trade, by Bruce W. Jentleson

The Politics of International Debt, edited by Miles Kahler

Corporatism and Change: Austria, Switzerland, and the Politics of Industry, by Peter J. Katzenstein

Industry and Politics in West Germany: Toward the Third Republic, edited by Peter J. Katzenstein

Small States in World Markets: Industrial Policy in Europe, by Peter J. Katzenstein

The Sovereign Entrepreneur: Oil Policies in Advanced and Less Developed Capitalist Countries, by Merrie Gilbert Klapp

International Regimes, edited by Stephen D. Krasner

Business and Banking: Political Change and Economic Integration in Western Europe, by Paulette Kurzer

Power, Protection, and Free Trade: International Sources of U.S. Commercial Strategy, 1887–1939, by David A. Lake

State Capitalism: Public Enterprise in Canada, by Jeanne Kirk Laux and Maureen Appel Molot

France after Hegemony: International Change and Financial Reform, by Michael Loriaux

Economic Containment: CoCom and the Politics of East-West Trade, by Michael Mastanduno

Mercantile States and the World Oil Cartel, 1900–1939, by Gregory P. Nowell

Opening Financial Markets: Banking Politics on the Pacific Rim, by Louis W. Pauly

The Limits of Social Democracy: Investment Politics in Sweden, by Jonas Pontusson

The Fruits of Fascism: Postwar Prosperity in Historical Perspective, by Simon Reich

The Business of the Japanese State: Energy Markets in Comparative and Historical Perspective, by Richard J. Samuels

Crisis and Choice in European Social Democracy, by Fritz W. Scharpf, translated by Ruth Crowley

In the Dominions of Debt: Historical Perspectives on Dependent Development, by Herman M. Schwartz

Europe and the New Technologies, edited by Margaret Sharp

Europe's Industries: Public and Private Strategies for Change, edited by Geoffrey Shepherd, François Duchêne, and Christopher Saunders

Ideas and Institutions: Developmentalism in Brazil and Argentina, by Kathryn Sikkink

The Cooperative Edge: The Internal Politics of International Cartels, by Debora L. Spar

Fair Shares: Unions, Pay, and Politics in Sweden and West Germany, by Peter Swenson

Union of Parts: Labor Politics in Postwar Germany, by Kathleen A. Thelen

Democracy at Work: Changing World Markets and the Future of Labor Unions, by Lowell Turner

National Styles of Regulation: Environmental Policy in Great Britain and the United States, by David Vogel

International Cooperation: Building Regimes for Natural Resources and the Environment, by Oran R. Young

Polar Politics: Creating International Environmental Regimes, edited by Oran R. Young and Gail Osherenko

Governments, Markets, and Growth: Financial Systems and the Politics of Industrial Change, by John Zysman

American Industry in International Competition: Government Policies and Corporate Strategies, edited by John Zysman and Laura Tyson